Antimicrobial Activity of Different Plant Extracts, Plant-Derived Compounds and Synthetic Derivatives of Natural Compounds on Pathogenic Microorganisms, 2nd Edition

Antimicrobial Activity of Different Plant Extracts, Plant-Derived Compounds and Synthetic Derivatives of Natural Compounds on Pathogenic Microorganisms, 2nd Edition

Guest Editors

Anna Duda-Madej
Joanna Kozłowska
Katarzyna Wińska

MDPI

Basel • Beijing • Wuhan • Barcelona • Belgrade • Novi Sad • Cluj • Manchester

Guest Editors
Anna Duda-Madej
Department of Microbiology
Faculty of Medicine
Wroclaw Medical University
Wroclaw
Poland

Joanna Kozłowska
Department of Food
Chemistry and Biocatalysis
Wroclaw University of
Environmental and Life
Sciences
Wroclaw
Poland

Katarzyna Wińska
Department of Food
Chemistry and Biocatalysis
Wroclaw University of
Environmental and Life
Sciences
Wroclaw
Poland

Editorial Office
MDPI AG
Grosspeteranlage 5
4052 Basel, Switzerland

This is a reprint of the Special Issue, published open access by the journal *Antibiotics* (ISSN 2079-6382), freely accessible at: https://www.mdpi.com/journal/antibiotics/special_issues/IO32O463Z8.

For citation purposes, cite each article independently as indicated on the article page online and as indicated below:

Lastname, A.A.; Lastname, B.B. Article Title. *Journal Name* **Year**, *Volume Number*, Page Range.

ISBN 978-3-7258-3633-8 (Hbk)
ISBN 978-3-7258-3634-5 (PDF)
https://doi.org/10.3390/books978-3-7258-3634-5

Contents

antibiotics

Review

Meropenem/Vaborbactam: β-Lactam/β-Lactamase Inhibitor Combination, the Future in Eradicating Multidrug Resistance

Anna Duda-Madej [1,*,†], Szymon Viscardi [2,†] and Ewa Topola [2]

1 Department of Microbiology, Faculty of Medicine, Wroclaw Medical University, Chałubińskiego 4, 50-368 Wrocław, Poland
2 Faculty of Medicine, Wroclaw Medical University, Ludwika Pasteura 1, 50-367 Wrocław, Poland; szymon.viscardi@student.umw.edu.pl (S.V.); ewa.topola@student.umw.edu.pl (E.T.)
* Correspondence: anna.duda-madej@umed.wroc.pl
† These authors contributed equally to this work.

Abstract: Due to the fact that there is a steadily increasing trend in the area of antimicrobial resistance in microorganisms, there is a need to look for new treatment alternatives. One of them is the search for new β-lactamase inhibitors and combining them with β-lactam antibiotics, with the aim of increasing the low-dose efficacy, as well as lowering the resistance potential of bacterial strains. This review presents the positive effect of meropenem in combination with a vaborbactam (MER-VAB). This latest antibiotic-inhibitor combination has found particular use in the treatment of infections with the etiology of carbapenem-resistant *Enterobacterales* (CRE), Gram-negative bacteria, with a high degree of resistance to available antimicrobial drugs.

Keywords: β-lactam/β-lactamase inhibitor; carbapenem-resistant *Enterobacterales* (CRE); *Klebsiella pneumoniae* carbapenemase (KPC); meropenem; multidrug resistance; vaborbactam

Citation: Duda-Madej, A.; Viscardi, S.; Topola, E. Meropenem/Vaborbactam: β-Lactam/β-Lactamase Inhibitor Combination, the Future in Eradicating Multidrug Resistance. *Antibiotics* **2023**, *12*, 1612. https://doi.org/10.3390/antibiotics12111612

Academic Editor: Marc Maresca

Received: 12 October 2023
Revised: 4 November 2023
Accepted: 8 November 2023
Published: 10 November 2023

1. Introduction

Antimicrobial resistance (AMR) is an ever-growing problem in medical care. It reduces the effectiveness of the therapy, increases the risk of death, and accelerates the spread of infections with the etiology of multidrug-resistant (MDR) bacteria. In addition, it creates the need for very powerful alternative therapies, resulting in negative side effects on a patient's health and quality of life. This phenomenon is driven by the excessive use of antibiotics (1) in clinically unjustified cases (e.g., in viral contamination) and (2) outside the "ACCESS" list: according to the WHO (World Health Organization) AWaRe (Access, Watch, Reserve) classification, effective against a wide range of infections, and induces low resistance levels [1]. In response to such irresponsible behaviors, bacteria evolve and naturally develop resistance to drugs to which they had previously shown sensitivity.

In recent years, the antibiotic consumption has increased dramatically; hence, The Global Antimicrobial Resistance and Use Surveillance System (GLASS) has decided to introduce analyses for AMR indicators in the context of antimicrobial drugs consumption [2]. According to the collected data, the transition to the 21st century led to a 65% increase in the defined daily dose (DDD) of antimicrobial substances. In contrast, antibiotic usage increased by 39% (from 11.3 to15.7 DDD per 1000 inhabitants per day) [3]. Moreover, the consumption of DDDs in pediatric patients raises concerns. A study by Browne et al. found that the DDDs increased by 46% in 2018 (37.2–43.7 DDD per 1000 inhabitants per day) compared to 2000 (9.2–10.5 DDD per 1000 inhabitants per day) [4]. However, although these figures have been reported as the pre-COVID-19 (coronavirus disease 2019) pandemic conditions, they have already raised major concern. The post-COVID-19 era seems to break down under all the rules. The analyses conducted by Sulayyim et al. showed that the antibiotic resistance during the COVID-19-related pandemic increased among both Gram-negative (i.e., *Acinetobacter baumannii*, *Klebsiella pneumoniae*, *Escherichia coli*, and *Pseudomonas aeruginosa*) and Gram-positive (i.e., *Staphylococcus aureus* and *Enterococcus faecium*) bacteria. The observed resistance was mainly against antibiotics used in the empirical therapy of in-hospital (colistin and ciprofloxacin) and also out-of-hospital (ampicillin and erythromycin) pneumonia [5]. However, it is disturbing to note that resistance to the so-called "last chance" antibiotic (colistin) has enhanced in the absence of effectiveness of other antimicrobial agents. The results of other studies also confirmed this alarming phenomenon, even though they showed increased resistance to other antibiotics, i.e., imipenem and ceftazidime, used in the treatment of hospital-acquired pneumonia (including ventilator-treated pneumonia) with an etiology of Gram-negative bacteria of the *Enterobacteriaceae* family [6].

The resistance problem is a huge obstacle to appropriate therapy because only a few clinically available antibiotics can be used for the treatment. Moreover, AMR is proving to be one of the most common causes of death worldwide. According to data presented by Murray et al., more than 1.2 million people worldwide died in 2019 from a direct infection with an etiology of MDR bacteria. In turn, such contamination was an indirect cause of nearly 5 million deaths. The resistance problems particularly affect patients with pneumonia, blood infections, and appendicitis [7]. Hospital-acquired pneumonia (HAP) is one of the most common infections occurring in the hospital environment with a high potential for mortality [8–10]. However, although the era of antibiotics may still be in its golden age, this view is at a high risk of change. This situation, which had a protective effect for many years, could become a killer of the population. It may come into a complete circle and lead to situations in which the infections become deadly again. This could have catastrophic consequences all over the world.

Gram-negative bacteria show higher resistance than Gram-positive bacteria. They have various abilities to learn new ways to affect the failure of therapy and easily transfer their genetic material (including antibiotic resistance genes) to other bacteria. The World Health Organization has created a list of antibiotic-resistant "priority pathogens", the critical group including MDR Gram-negative, carbapenem-resistant, and ESBLs-producing bacteria (extended-spectrum β-lactamases), i.e., *A. baumannii*, *P. aeruginosa*, *Klebsiella* spp., *E. coli*, *Serratia* spp., and *Proteus* spp. [11]. These bacterial strains are associated with serious, often fatal, health conditions, including bloodstream infection and pneumonia occurring in patients (1) residing in hospitals [12,13] and (2) long-term-care facilities [14], (3) requiring artificial ventilation [15], and (4) with an intravenous catheter in situ [16]. A recent study of this group of bacteria was more alarming because it showed that in the post-COVID-19 era, the resistance of the most critical group of bacteria increased dramatically, for example, levels for *E. coli* and *K. pneumoniae* on ampicillin raised to 89.6% and 98%, respectively; for *P. aeruginosa* on imipenem to 91.8%; and for *A. baumannii* to ceftazidime to 94.6%. In addition, an increment of about 30% has been shown for *K. pneumoniae* on ceftriaxone, a representative of the third-generation cephalosporins, which is the most effective antibiotic against multidrug-resistant strains [6]. These frightening data are thought to be underestimated by the inadequacy of some countries' bacterial resistance monitoring systems. Thus, the obtained results and, therefore, the international response are not comparable to the current risk situation [17].

The cornerstone of antibiotic therapy is the largest and safest group referring to β-lactam antibiotics. They inactivate transpeptidases, carboxypeptidases, and endopeptidases involved in the synthesis of the bacterial cell wall. Carbapenems, which belong to this group of antibiotics, represent a last resort in the therapy of infections caused by Gram-negative *Enterobacteriaceae* [18]. However, since the β-lactam ring structure is sensitive to the action of bacterial enzymes, β-lactamases, which degrade and inactivate the antibiotic, inhibitors of these enzymes are a key factors. Thus far, several of them have been recognized, i.e., clavulanic acid, sulbactam, tazobactam, and avibactam. However, in many cases, they have already lost their effectiveness, as bacteria have adapted to the new conditions by producing newer and newer enzymes (including carbapenemases). This situation is a clear indication of the need to find and develop new antibiotics that include multidrug-resistant

Gram-negative bacteria in their spectrum of action. To win this race against antibiotic resistance, it is important to use what is available by introducing modifications or enriching them with an additional substances. These new approaches aim to combat the intrinsic and acquired resistance of Gram-negative bacteria. A promising trend is natural bactericidal agents, i.e., bacteriophages, but also modified peptide antibiotics, peptide benzimidazoles, quorum sensing (QS) inhibitors, and metal-based antibacterial agents [19]. Moreover, one of such methods that gives hope for the future is to obtain the activity of these resistant bacteria by deactivating the mechanism responsible for the occurrence of resistance as a result of the action of antibiotic adjuvants that are β-lactamase inhibitors.

In this review, we compile the data on the combination of a β-lactam antibiotic with an inhibitor, meropenem with a vaborbactam. This latest interconnection, through its unique action, has found application in the therapy of critically ill patients, mainly in the ICU (intensive care unit). Furthermore, it has shown to be an effective therapy for treating very difficult infections with CRE (carbapenem-resistant *Enterobacterales*) etiology.

2. Characteristics

As the prevalence of multidrug-resistant strains increases, the effectiveness of older β-lactamase inhibitors (BLIs) used with β-lactams (BLs) decreases. Resistance to carbapenems is very important in bacteria such as *Pseudomonas* spp., *Acinetobacter* spp., and the *Enterobacterales* and is an indicator of selection for the appropriate antibiotic [20]. The increased demand for new methods of treating resistant microorganisms has led to the creation of a combination of vaborbactam, which is a non-β-lactam β-lactamase inhibitor, with meropenem—an antibiotic from the carbapenem group [21,22].

2.1. Chemical Structure of Vaborbactam

Vaborbactam (VAB) (formerly RPX7009), with the molecular formula: 2-[(3R,6S)-2-hydroxy-3-[(2-thiophen-2-ylacetyl)amino]oxaborinan-6-yl]acetic acid (Figure 1), is a cyclic boronic acid pharmacophore inhibitor with serine protease activity (PubChem). The activity of the β-lactamase inhibitor is increased by the 2-thienoacytel group of vaborbactam. The compound has the ability to block enzymes of KPC (*K. pneumoniae* carabapenemase) bacterial strains and improves the activity of carbapenem antibiotics, especially in groups of CRE strains that carry genes encoding β-lactamases [23].

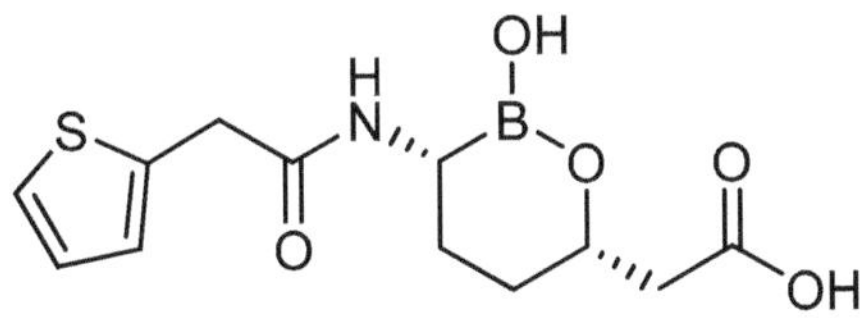

Figure 1. The 2D (two-dimensional) chemical structure of vaborbactam.

Vaborbactam is an inhibitor of type A and C β-lactamases [24]. The mechanism of its action is to enter the bacterial cell through porins OmpK35 and OmpK36. Therefore, the AcrAB-TolC multidrug efflux pump has a negligible effect on the action of vaborbactam [25].

2.2. Chemical Structure of Meropenem

Meropenem (MER), with the molecular formula (4R,5S,6S)-3-[[(3S,5S)-5-dimethylcarbamoylpyrrolidin-3-yl]thio]-6-[(1R)-1-hydroxyethyl]-4-acid methyl-7-oxo-1-azabicyclo[3,2,0]hept-2-ene-2-carboxylic acid (Figure 2), is a carbapenem that acts on the cell wall by inhibiting the cross-linking of its peptidoglycan, resulting in lysis and subsequent cell death. It is unstable in aqueous solution and has bactericidal activity against both Gram-positive and Gram-negative bacteria [26,27]. For maximum antibacterial potential, the %T > MIC (MIC—minimal inhibitory concentration; %T > MIC is the percentage of time that free-drug concentrations are higher than the MIC) of the free drug should be approximately 40%. In turn, for bacteriostatic activity, %T > MIC is approximately 20% [28,29]. The advantage of meropenem is that it is a weaker inducer of AmpC β-lactamases (cephalosporinases encoded on the chromosomes) produced by *Enterobacterales*, compared to other carbapenems, which prevents rapid development of resistance. It has high affinity for PBP 2, PBP3, and PBP4 (PBP; penicillin binding proteins). Moreover, it quickly penetrates the outer cell membrane of Gram-negative bacteria and has a highly stable hydrolytic effect [30]. Meropenem remains active against cephalosporinases and penicillinases, but loses stability against carbapenemases.

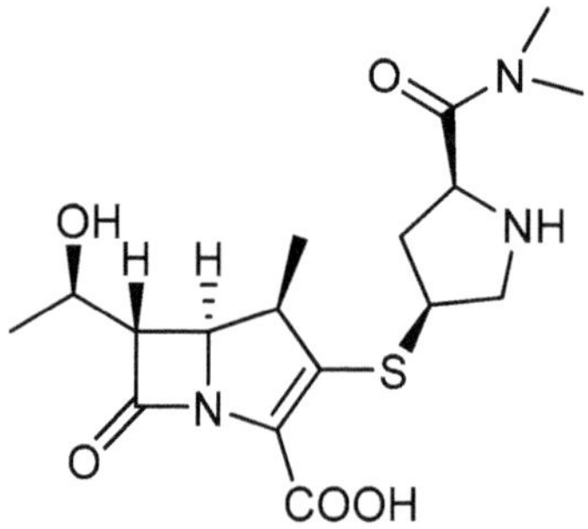

Figure 2. The 2D (two-dimensional) chemical structure of meropenem.

2.3. *Combination of Meropenem with Vaborbactam*

Meropenem/vaborbactam (MER-VAB) is a combination of two compounds under the trade name Vabomere® (Menarini International Operations Luxembourg S.A., Luxembourg, Luxembourg). Vaborbactam reduces the MIC of meropenem by blocking the production of β-lactamases. A vaborbactam concentration of 8 μg/mL restores the activity of meropenem with an MIC value of ≤2 μg/mL for the most mutated bacterial strains [25]. MER-VAB was approved by the FDA (Food and Drug Agency) in 2017 for the treatment of patients over 18 years of age. It is registered for the treatment of urinary tract infections including pyelonephritis. Due to increasing multidrug resistance, the drug should only be used against sensitive bacteria [31].

2.4. *Pharmacokinetic Properties*

According to the study by Sabet et al., the half-life for a therapeutic dose (2 g MER/2 g VAB), administered in a 3 h infusion every 8 h, is 1.36 ± 0.07 and 1.47 ± 0.14 for MER and VAB, respectively. The maximum concentration (Cmax; mg/L) reaches the value of 34.47 ± 3.95 and 24.1 ± 5.84 for MER and VAB, respectively, whereas AUC (area under the ROC curve) 0–8 h/0–24 h (mg/L) is 132.63 ± 15.23/397.88 ± 45.70 and 105.57 ± 4.42/316.7 ± 13.26 for MER vs. VAB [32]. Both compounds have a similar half-life in the final process of drug elimination from the plasma [33]. It is important that meropenem/vaborbactam has a low degree of binding to plasma proteins, and administration in multiple doses does not cause accumulation in the body [34]. About 1/4 of the β-lactam ring of meropenem is metabolized in the liver. Hydrolysis metabolites and MER-VAB are excreted unchanged by the kidneys [35]. MER-VAB monotherapy shows a high degree of cure for CRE infections, as well as reduced nephrotoxicity in relation to the use of BAT (best available treatment) [36]. The use of MER-VAB affects glomerular filtration rate (GFR) by increasing the renal clearance value of 8.6–17.6 L/h, respectively, for VAB and 6.9–13.5 L/h for MER, where the range in healthy people is 5.4–7.2 L/h. This indicates active glomerular secretion and renal filtration [33]. It is worth noting that the pleural penetration is sufficient when the MER-VAB is used [37]. Taking into account the total drug concentration in plasma and based on the AUC0-8 ELF (epithelial lining fluid) values for meropenem, the intrapulmonary penetration is as much as 63%, while for vaborbactam it is 53% (3 doses of 2 g MER/2 g VAB in a 3 h prolonged infusion). The persistent ELF concentration above the MIC limit of 1 μg/mL and good penetration suggest the possibility of using MER-VAB therapy to treat HAP and ventilator-associated pneumonia [38].

2.5. Spectrum of Action

Meropenem is highly effective against microorganisms with ESBLs resistance. It is safe and very well tolerated, which contributes to the use of this antibiotic in the treatment of infections caused by Gram-negative bacteria [39]. Vaborbactam, on the other hand, has no antimicrobial effect, but blocks class A bacterial β-lactamases (TEM—plasmid-encoded β-lactamase in Gram-negative bacteria, CTX-M—preferential hydrolytic activity against cefotaxime, SHV—sulf-hydryl variable penicilinase) and C (FOX—AmpC β-lactamase gene, MIR—novel plasmid mediated β-lactamase, P99—β-lactamase from *Enterobacter cloacae* P99), as well as class A carbapenemase [25]. Data indicate that the addition of vaborbactam at a dose of 8 mg/L results in a reduction in the MIC by as much as 8–64 times. This dose of vaborbactam also reduces the MIC of meropenem to $\leq$8 mg/mL for all strains. Furthermore, combined with MER-VAB (MIC $\leq$ 1 μg/mL), it is active against *E. coli*, including the genes of the latest *bla*$_{KPC}$ and mutations, obtaining significantly better results than in the case of meropenem alone (sensitivity 99.9%) with a decrease in MIC values. Compared to *K. pneumoniae*, the activity of MER-VAB increased from 88% to 98.8% at doses of 1 μg/mL MER and 4 μg/mL VAB. Additionally, in this case, MER-VAB lowers the MIC value and affects the KPC. A similar relationship has been observed in the species *Enterobacter cloacae* and *Enterobacter aerogenes*. However, for *A. baumannii*, *Stenotrophomonas maltophila*, and *P. aeruginosa*, the MER-VAB combination slightly changes the MIC value, which indicates the ineffectiveness of the therapy [40]. This was confirmed by in vitro studies, which proved the effect of MER-VAB on *Citrobacter freundii*, *Citrobacter koserii*, *E. cloacae*, *E. coli*, *E. aerogenes*, *K. pneumoniae*, *Klebsiella oxytoca*, *Morganella morganii*, *Proteus mirabilis*, *P. aeruginosa*, *Serratia marcescens*, and *Providencia* spp. Nevertheless, no effect has been shown on *A. baumannii* and *S. maltophilia* [31].

2.6. MER-VAB against Bacterial Resistance

Based on the available research, it can be concluded that MER-VAB is an alternative treatment for infections caused by bacteria belonging to the *Enterobacterales* producers of ESBLs, KPC, and AmpC, and, thus, effectively reduces the level of resistance. Moreover, it shows activity against MDR, XDR (extensively drug-resistant), and CRE microorganisms with MIC50 (MIC that inhibited 50% of tested microorganisms) and MIC90 (MIC that inhibited 90% of tested microorganisms) in comparison to other tested antibiotics [41].

MER-VAB is used in infections with MDR microorganisms that produce carbapenemases of types A (TEM-116, CTX-M, SHV, ESBLs) and C (FOX, MIR, P99), but is not used in the treatment of infections with strains producing carbapenemase types B and D. Therefore, it is not used in infections with bacteria showing MBL (metallo-β-lactamase)-type resistance (NDM—New Delhi metallo-β-lactamase, IMP—imipenemase, VIM—Verona-integron-imipenemase) and/or OXA-48 (oxacilinase-48)-type resistance, because the limited activity of this combination has been demonstrated against them [42,43]. However, the research of Biaga et al. clearly indicates that MER-VAB in combination with aztreonam and/or with ceftazidime and avibactam simultaneously (CAZ-AVI) may constitute an effective therapeutic option against aztreonam-resistant NDM strains and other serine-producing members of the *Enterobacterales* [44]. Moreover, the effect of MER-VAB on OXA-2 (oxacilinase-2), OXA-30 (oxacilinase-30), i.e., narrow-spectrum oxacillinases, as well as on carbapenemases of the KPC, FRI-1 (carbapenem-hydrolyzing Class A β-lactamase), and SME-2 (carbapenem-hydrolyzing β-lactamase Sme-2) type has been demonstrated [45]. A summary of the activity of the MER-VAB combination is shown in Table 1.

Table 1. Scheme of MER-VAB action on individual types of resistance in microorganisms.

Class According to Ambler	Effect of the Vaborbactam	β-Lactamases
Class A	Inhibition	Serine carbapenemases: KPC, NMC-A, SME-2
		ESBL: SHV-2, PER-1
		Narrow spectrum: TEM-1, TEM-2
Class B	Lack of inhibition	Metalo-β-lactamases: VIM, NDM-1
Class C	Inhibition	Cephalosporinases: AmpC, P99, ACT-1
Class D	Lack of inhibition	OXA 48

Legend: KPC—*K. pneumoniae* carabapenemase; NMC-A—chromosomal-encoded class A β-lactamases first isolated from *Enterobacter cloacae* (clinical strain NOR-1); SME-2—carbapenem-hydrolyzing β-lactamase Sme-2; ESBL—extended-spectrum β-lactamases; SHV-2—sulf-hydryl variable penicillinase; PER-1—Pseudomonas extended resistant β-lactamase; TEM-1—plasmid-encoded β-lactamase in Gram-negative bacteria; TEM-2—plasmid-encoded β-lactamase in Gram-negative bacteria; VIM—Verona-integron-imipenemase; NDM-1—New Delhi metallo-β-lactamase-1; AmpC—cephalosporinases encoded on the chromosomes of many of the *Enterobacteriaceae*; P99—β-lactamase from *Enterobacter cloacae* P99; ACT-1—plasmid-encoded AmpC β-lactamase; OXA-48—oxacilinase-48.

Interestingly, MER-VAB shows very high activity against carbapenem-resistant (CRE) *K. pneumoniae* strains with or without KPC-type resistance in the standard variant, or, additionally, in the CAZ-AVI combination. This may be an alternative to treatment for infections resistant to this combination, which is used as a first-line drug. The in vitro effects of MER-VAB against *Enterobacterales* KPC(+) are stronger than commonly used antibiotics, i.e., CAZ-AVI, tygecycline, and others [46]. According to research conducted by Schields et al., microorganisms belonging to the *Enterobacterales* (mainly *K. pneumoniae*) with CRE resistance isolated from critically ill patients (mainly with bacteremia and pneumonia) staying in the ICU were successfully eradicated in 65% of cases. This therapy prolonged survival in 90% of patients after 30 days of use [47].

The only critical point in using MER-VAB is that *K. pneumoniae* can withstand this combination due to mutations in the bla_{KPC} gene [48,49]. Since these genes are located in transposons, clonal expansion, i.e., the transmission of microorganisms with these genes, is the most dominant method of global KPC distribution.

Inactivation against bacteria with porin mutations related to excessive expression of efflux pumps has been demonstrated. Reduced activity occurs in microorganisms with inactivated OmpK35 and OmpK36, and overexpressing AcrA strains [49].

3. Clinical Application

3.1. Dosage

MER-VAB is administered at a dose of 4 g (2 g MER/2 g VAB) every 8 h as a 3 h prolonged infusion. The treatment period is up to 14 days [35]. Dosage for impaired renal function eGFR (mL/min/1.73 m^2) is as follows: for eGFR $\geq$ 50, 2 g MER, 2 g VAB every 8 h; for eGFR 30–49, dosage has to be decreased to 1 g MER, 1 g VAB every 8 h; for eGFR 15–29, 1 g MER and 1 g VAB time has to be extended to 12 h; for eGFR < 15, dosage has to be decreased to 0.5 g MER and 0.5 g VAB every 12 h [50].

Due to the absence of data and safety evaluation, the dose for children has not been determined. Safety dosing research is ongoing [51]. The literature describes a case of the use of MER-VAB in a 4-year-old male child with short intestine syndrome, colostomy and gastrojejunal tube, bronchopulmonary dysplasia, and central line. Multiple central lines caused blood infections, which were treated with MER-VAB at a dose of 40 mg/kg every 6 h for 3 h as an infusion. The treatment lasted 14 days and resulted in clinical success, with *K. pneumoniae* KPC(+) being eliminated from the bloodstream [52].

3.2. Indications

According to the guidelines of the European Medicines Agency (EMA), MER-VAB is used for (i) complicated abdominal infections; (ii) hospital acquired pneumonia, including ventilator-associated pneumonia; (iii) blood infections; (iv) treatment of infections caused by aerobic Gram-negative microorganisms in patients with limited therapeutic options. Furthermore, according to FDA's guidelines, MER-VAB is recommended for the therapy of complex urinary tract infections (cUTIs), including pyelonephritis kidney infection. Studies conducted by Shoulders et al. have demonstrated its efficacy in these types of infections [53]. The dose of meropenem and vaborbactam (2 g/2 g) every 8 h must be administered with uninterrupted renal function in each of the above cases, depending on the duration of the treatment [54].

3.3. Interactions

Administration of meropenem with valprionic acid reduces its concentration, which lowers the seizure threshold with the risk of an epileptic fit. Probenecid competes with meropenem for tubular secretion and increases the concentration of meropenem in the blood [35].

4. Antimicrobial Activity of Meropenem–Vaborbactam

MER-VAB's antimicrobial activity in clinical environments was evaluated for the first time in phase 3 of the TANGO I (Targeting Antibiotic Non-susceptible Gram-Negative Organisms) multicenter, randomized, double-blind, double-dummy, active-control trial. The study was conducted with adults who were afflicted with complicated urinary tract infections (cUTI) and acute pyelonephritis (AP). Patients received meropenem/vaborbactam (2 g/2 g via 3 h infusion, every 8 h) or piperacilin/tazobactam (PIP-TAZ) (4 g/0.5 g via 30 min infusion every 8 h). Results showed that MER-VAB was noninferior to PIP-TAZ. Moreover overall success in the end of intravenous treatment indicated superiority of MER-VAB over PIP-TAZ in the range of clinical cure [55]. Primary end-points of TANGO I trial are shown in Table 2.

Table 2. TANGO I trial primary end-points.

Primary End-Points	Meropenem/ Vaborbactam	Piperacilin/ Tazobactam
Overall success defined as clinical cure or improvement and microbial eradication composite * ($p < 0.001$)	98.4%	94.0%
Microbiological eradication in mMITT ($p < 0.001$)	66.7%	57.7%
Microbiological eradication in ME ($p < 0.001$)	66.3%	60.4%

* Clinical cure defined as complete subsidence or significant improvement of initial signs and symptoms of cUTIs or AP and microbial eradication defined as baseline pathogens reduced to $<10^4$ CFU/mL. Primary end-point by FDA. Trial was performed among population of patients, divided into two groups: research sample n = 274 (MER-VAB therapy) and control n = 276 (PIP-TAZ therapy); mMITT—microbiological modified-intent-to-treat population, which included all the patients who presented baseline pathogens of $\geq 10^5$ CFU/mL in urine culture or the same bacterial pathogen present in concurrent blood and urine cultures. Primary end-point by EMA (European Medicines Agency); ME—microbiological evaluable analysis.

The incidence of side effects in both groups was comparable: 39.0% and 35.5%, respectively, in MER-VAB and PIP-TAZ. The most commonly isolated uropathogens in mMITT (microbiological modified-intent-to-treat) populations were *E. coli* (65.1% vs. 64.3%) and *K. pneumoniae* (15.6% vs. 15.4%). Other isolated pathogens (*P. mirabilis*, *E. cloacae*, and *P. aeruginosa)* were less numerous. Clinical cure efficiency in mMITT populations was 98.4% vs. 94.0% for *E. coli*, 100.0% vs. 100.0% for *P. mirabilis*, *E. cloacae*, and *P. aeruginosa*, and 96.7% vs. 100.0% for *K. pneumoniae* (respectively, for MER-VAB and PIP–TAZ), whereas microbiological eradication ratio was 71.2% vs. 58.1% *E. coli*, 63.3% vs. 50.0% *K. pneumoniae*,

50% vs. 75% *P. mirabilis*, 90% vs. 60% *E. cloacae*, and 100% vs. 30% *P. aeruginosa* (respectively, for MER-VAB and PIP-TAZ) [55].

MER-VAB's clinical effectiveness was also evaluated in three phase randomized clinical trials of TANGO II. MER-VAB was compared to BAT (best available treatment—which was polytherapy of polymixin/aminoglycosides/carbapenems/tigecycline or monotherapy using CAZ-AVI) in the therapy of infections (cUTI, HABP/VABP—hospital-acquired-bacterial-pneumonia/ventilator-acquired-bacterial-pneumonia, AP, BSI—bloodstream infection, cIAI—complicated intra-abdominal infection) due to CRE [34]. The end-points of the TANGO II trial are shown in Table 3.

Table 3. End-points of the TANGO II trial.

Primary End-Points of TANGO II Trial	Meropenem/ Vaborbactam	BAT
Overall clinical success * in EOT ($p = 0.03$)	65.6%	33.3%
Microbiologic cure in EOT	65.6%	40.0%
Overall clinical success in TOC	59.4%	26.7%
Microbiologic cure in TOC	53.1%	33.3%
Day-28 mortality in mCRE-MITT population	15.6%	33.3%

* Overall clinical success was defined as a composite result of clinical and microbiological cure at EOT and TOC. Clinical cure (in accordance with FDA) was defined as a total resolution of baseline signs and symptoms of infection without need of further antimicrobial treatment or surgery in case of cIAI; microbiologic cure was defined as microbiological eradication or presumed eradication (clinical cure in absence of sample for repeat culture) at EOT and TOC; mCRE-MITT (microbiologic-CRE-modified-intent-to-treat population)—population of patients (with microbiologically confirmed CRE-pathogen related infection) who received $\geq$1 dose of study drug. MER-VAB: n = 32, BAT: n = 15; EOT—end of treatment; TOC—test of cure (7 $\pm$ 2 after EOT).

Patients were divided into two groups: those receiving MER-VAB therapy and those receiving BAT. Infections that were the most prevalent among patients in both groups were BSI, cUTI/AP, HABP/VABP, and cIAI. *K. pneumoniae* and *E. coli* were the most frequently isolated pathogens. CRE pathogens in both groups represented, respectively, 71.9% and 93.3% of the isolates. MER-VAB was more effective in overall clinical success in both EOT (end of treatment) and TOC (test of cure) in the mCRE-MITT population. Patients with confirmed cUTI/AP who received MER-VAB therapy scored a 75% result in overall clinical success vs. 50% outcome in BAT population. The clinical efficiency of administered therapy in TOC was, respectively, 33.3% vs. 50.0% (MER-VAB vs. BAT). Patients from the MER-VAB population also obtained a lower rate of day-28 mortality vs. BAT. The least mortality rate occurred among patients with HABP/VABP who received MER-VAB (22.2%) vs. BAT (44.4%). Treatment emergent adverse events (TEAE) were less frequent in the MER-VAB population (84%) vs. BAT (92%), as well as severe TEAE (14% vs. 28%) [36].

Shields et al. conducted a large-scale study on the clinical use of MER-VAB at the University of Pittsburgh Medical Center from 2017 to 2019. In this research, MER-VAB was used as a first-line treatment option against the CRE producing KPC. Clinical success was defined as a composite result with the following outcomes: (i) complete resolution of the symptoms of infection; (ii) absence of recurrence of infection; (iii) absence of microbiologic failure (isolating the same bacterial species for 7 days after MER-VAB treatment) within 30 days from the appearance of the infection. The study involved a cohort of 20 patients with several types of bacterial infections, i.e., BSI, bacterial pneumonia including VABP, tracheobronchitis, SSTI (skin and soft tissues infection), cIAI, and AP. The most commonly isolated pathogens were *K. pneumoniae* (n = 14), *Klebsiella oxytoca* (n = 2), and *E. coli* (n = 2). The monotherapy of MER-VAB was administered by i.v. infusion (2 g + 2 g every 8 h) in 16/20 cases. Only four patients additionally received another drug for >48 h (gentamicin inhalation (n = 2), gentamicin i.v. and inhalation (n = 1), and ciprofloxacin i.v. (n = 1)). The 30- and 90-day survival rates presented were, respectively, 90% and 80%. The overall clinical success was obtained in 13/20 patients (65%), whereas in the case of BSI and pneumonia, it amounted, respectively, to 63% and 67%. In the entire population, only one (1/20) case of severe TEAE was reported. Meanwhile, six patients developed the microbiologic failure (recurrent CRE infection (n = 3), persistent bacteremia, respiratory tract colonization, breakthrough infection during treatment). It is important to note that one of the recurrent bacterial isolates was resistant to MER-VAB. The overall clinical success and day-30 survival ratio achieved 65% and 90%, respectively, which are comparable to the TANGO II trial outcomes (59% and 84%) [46]. MER-VAB seems to be noninferior to CAZ-AVI, which obtained, respectively, 59% and 76% in the study of Shields et al. [56]. Finally, their study confirmed that TANGO II trial results were related to clinical efficiency of MER-VAB in the treatment of severe VABP. In the future, MER-VAB could be a possible alternative treatment for polymyxin in CRE pneumonia [47].

Tumbarello et al. described the clinical use of MER-VAB in a retrospective, observational study conducted in 12 Italian hospitals from 2020 to 2021. The study group consisted of 37 patients with confirmed *K. pneumoniae* producing KPC infection. Patients were hospitalized for BSI (23/37), LRTI (lower-respiratory-tract infection) (10/37), IAI (1/37), cUTI (2/37), and ABSSI (acute bacterial skin and skin structure infection) (1/37). MER-VAB (2 g + 2 g) was administered via 3 h i.v. infusion every 8 h for $\geq$72 h. Bacterial isolates were resistant to penicillins, extended spectrum cephalosporines, ciprofloxacin, and meropenem. In addition, 22 out of 37 patients' isolates were confirmed to be resistant to CAZ-AVI. Clinical cure was achieved in 28/37 cases, while microbiologic eradication occurred in 25/37 cases. In-hospital mortality was observed for 24.3% of the patients. Deaths appeared only in patients with LRTI (n = 3) and BSI (n = 6) and affected mainly sufferers above 60 years old. The recurrent in-hospital infection was observed in three cases but was successfully treated using MER-VAB with colistin or fosfomycin combined therapy. Tumbarello's study revealed that MER-VAB is very efficient in the treatment of CAZ-AVI—resistant, KPC(+) *K. pneumoniae* infections (CAZ-AVI—resistant pathogens represented at least 60% of all isolates). This is a significant condition in the era of increasing microbial resistance to CAZ-AVI [57].

Outcomes of Carvalhaes et al.'s analysis about in vitro susceptibility of bacterial isolates collected from patients cured in USA hospitals in 2014–2018 revealed 99.9% antimicrobial efficiency of MER-VAB against *Enterobacterales* (4790 isolates) and 85.9% against *P. aeruginosa* (3193 isolates). Isolates were collected from patients with confirmed HABP/VABP infections. MER-VAB's in vitro activity against CRE pathogens (resistant to many other antimicrobial agents) oscillated above 99%. MER-VAB's ability to treat CRE etiology HABP/VABP is noteworthy as it also has the potential to treat MDR/XDR *P. aeruginosa* isolates [58].

Castanheira et al., in a similar study involving CRE isolates (with a predominance of *K. pneumoniae* 77/152 and *E. cloacae* 27/152) collected in USA hospitals in 2016–2018, revealed that MER-VAB inhibited growth of 95.4% isolates, whereas β-lactam comparators only inhibited 6.6%. Comparable results to MER-VAB in the inhibition of growth of CRE were achieved only by tigecycline (96.7%). The study confirmed that MER-VAB is more efficient than combination therapy and has greater effectiveness than comparators commonly used in the treatment of CRE etiology infections [59].

Alosaimy et al. carried out a multicenter, real-world analysis of the clinical outcomes of MER-VAB therapy for GNB (Gram-negative bacteria). Patients who were qualified in the sample population (n = 40) reported serious bacterial infections. The most commonly confirmed types of infections were pneumonia (32.5%) and UTI (20%) caused particularly by *Enterobacterales* (*K. pneumoniae* 46.7%, *E. cloacae* 20.0%, and *E. coli* 13.3%). Clinical success was achieved in 70% of all cases, and in-hospital mortality rate was 7.5% (3/40). In all other cases, recurrent infection or presence of sign or symptoms of infection defined clinical failure. In post hoc analysis, day-60 and day-90 mortality rates reached, respectively, 15.0% and 22.5%. In terms of clinical outcomes and safety of MER-VAB, the analysis complements the results of the TANGO II trial [60].

Another multicenter, retrospective analysis of clinical efficiency of MER-VAB was conducted in USA in 2017–2020. Patients who qualified for the study (n = 126) were infected by MDR GNB, including CRE pathogens (mainly *K. pneumoniae, E. coli,* and *Enterobacter* spp.). The overall number of patients with confirmed CRE infection was 99. In addition, patients (n = 10) with confirmed non-CRE infections (*A. baumannii* and *P. aeruginosa*) were included in the study population and also received MER-VAB therapy. Infections in hospitals accounted for 58.7% of all cases. The most prevalent infections were respiratory tract (38.1%), abdominal cavity (19.0%), and urinary tract (13.5%). The mortality rate on day 30 was 18.3% (23/126), and 15 cases (11.9%) of repeated infection were reported. On the other hand, the mortality rate on day 90 eventually reached 31.7%. The positive clinical efficacy of MER-VAB against *P. aeruginosa* (day-90 mortality rate 1/8, recurrence of infection within 30 days to 2/8) is encouraging; however, the population of the study was not sufficiently large to draw some clear conclusions [61].

Comparisons of the clinical efficacy of CAZ-AVI and MER-VAB for CRE infections were carried out in a retrospective multicenter study between 2015 and 2018. The research included 131 patients in two cohorts receiving CAZ-AVI or MER-VAB (respectively, n = 105 and n = 26 patients) *K. pneumoniae* and *E. cloacae* were the most frequently isolated pathogens in both groups. The primary end-point of the study was overall clinical success (defined as a composite result of microbiologic eradication and resolution of the signs and symptoms of infection) with scores of, respectively, 62% vs. 69%. Meanwhile, the secondary end-points of trial (day-30, day-90 mortality ratio) revealed no crucial fluctuation, much like the issue of adverse events (36/105 vs. 6/26). Recurrent infection was confirmed in 14.3% of patients in the CAZ-AVI cohort and in 11.5% in the MER-VAB cohort. Post hoc analysis showed that bacterial drug resistance developed in patients with recurrent infections receiving CAZ-AVI monotherapy (n = 3). The treatment of MER-VAB and polytherapy of CAZ-AVI with other drugs did not cause this problem. In addition, CAZ-AVI monotherapy was associated with a greater need to switch to combined therapy (64/105 vs. 4/26). Moreover, the rate of drug resistance development among pathogens in CAZ-AVI monotherapy was higher than in MER-VAB treatment. Overall, MER-VAB clinical success was similar to the results of the TANGO II trial (respectively 69% vs. 64%) [62].

Interesting conclusions were provided by the study of the in vitro activity of vaborbactam combined with β-lactam antibiotics against clinical isolates of the MDR *Mycobacterium abscessus complex* (MABC). The antimicrobial activity of carbapenems and certain cephalosporines (particularly cefuroxime-II generation cephalosporine) against MABC was increased by vaborbactam. Vaborbactam, like relebactam and avibactam, is more efficient than classical inhibitors (e.g., sulbactam) against MB (*Mycobacterium* spp.) because it does not have a β-lactam ring in its structure. In the future, MER-VAB may be considered as a treatment strategy against MABC because of the restoring susceptibility to, for example, meropenem and cefuroxime among mycobacteria [63].

Gainey et al. described antimicrobial therapy implemented in the case of a pediatric patient suffering from cystic fibrosis (CF) complicated with MDR *Achromobacter* spp. etiology pneumonia who was successfully treated via a combination of MER-VAB, cefiderocol, and phage therapy. The therapy was well tolerated by the patient. An in vitro study by Caverly et al. revealed that MER-VAB is an effective method for managing CF complicated with *Achromobacter* spp., *Burkholderia gladioli,* and *Burkholderia cepacia* (including MDR/XDR strains) infection. MER-VAB had low antimicrobial activity (like other tested BL + BLI (β-lactam + β-lactamase inhibitor)) against *Pandoraea* spp. and *S. maltophilia* [64,65].

Antimicrobial activity of MER-VAB against clinical isolates of *Enterobacterales* and *P. aeruginosa* was tested in the study of Dee Shortridge et al. Pathogens were collected in 21 European countries in the years 2014–2019. The total amount of isolates was 6846 (*K. pneumoniae*, n = 1877; *E. coli*, n = 1646; *P. aeruginosa*, n = 3567) [66]. Bacterial strains were collected mainly from patients hospitalized for HABP/VABP. Total microbial susceptibility to MER-VAB reached 98% within *Enterobacterales* and 82.1% in *P. aeruginosa*. Authors reported that sensitivity to MER-VAB was higher in strains from Western Europe than Eastern

Europe. The study result could possibly be explained by the fact that OXA-48 resistance is more prevalent than KPC in Eastern Europe. MER-VAB inhibited the growth of 99.1% KPC(+) isolates vs. 40.5% OXA-48(+) isolates. Moreover MER-VAB obtained the highest efficiency (98%) among comparators used in the study against *Enterobacterales* isolates. Furthermore, MER-VAB was also the most active β-lactam antibiotic against *P. aeruginosa* (82.1%). The higher result reached only two comparators: colistin and amikacin [66].

In vitro antimicrobial activity of MER-VAB and other BL + BLI combinations (CAZ-AVI, I-R-C (imipenem-relebactam-cilastatine), C-TZB (ceftolozane/tazobactam)) against *P. aeruginosa* strains isolated from Western (WE) and Eastern Europe (EE) patients was evaluated in the study of Sader et al. Isolates were collected from patients with confirmed bacterial pneumonia. All of the BL+BLI revealed antimicrobial activity against >90% WE isolates, and MER-VAB showed the lowest efficiency 91% (MIC50 vs. 90 = 0.5 vs. 8 mg/L) versus comparators (CAZ-AVI = 97.2%; I-R-C = 94.5%; C-TZB = 94.3%). MER-VAB was also the least effective (83.5%) vs. comparators against EE strains. Moreover, the case revealed that MER-VAB had also the lowest antimicrobial efficiency (among BL + BLI comparators) against a *P. aeruginosa* strain resistant to PIP–TAZ, CAZ, MER, and IMI (imipenem) [67].

A clinical study by Belati et al. revealed that the combination of MER-VAB and aztreonam could be used as a treatment option for KPC(+), CAZ-AVI-resistant *K. pneumoniae*. All isolates were collected from the patients with confirmed severe BSI. Other authors described the case of MER-VAB as a salvage therapy against KPC(+), *K. pneumoniae* resistant to CAZ-AVI, and cefiderocol. Results showed that complete resolution and absence of recurrent infection were observed in a previously critically ill patient [68,69].

Sabet et al. described the potential clinical use of MER-VAB against *A. baumannii* and *P. aeruginosa*. In the study, authors proved in vivo efficiency of MER-VAB and meropenem therapy using a murine model of GNB infection complicated with neutropenia. MER-VAB achieved a higher eradication rate than meropenem alone against *P. aeruginosa* isolates. MER-VAB obtained excellent antimicrobial activity in *P. aeruginosa* and *A. baumannii* etiology infections. The results of this study correlate with similar articles, e.g., Sabet et al. and Weiss et al. (murine model of pyelonephritis) [70–72].

Bovo et al. conducted a study between 2018 and 2020 on the in vitro sensitivity of clinical bacterial isolates (collected from patients with bacteremia) to MER-VAB and other novel BL + BLI. In total, 133 bacterial strains of *K. pneumoniae* (within 87.2% strains were KPC(+)) were evaluated in this study. During the years 2018–2020, the in vitro antimicrobial activity of MER-VAB fluctuated from 85% to 97% [73].

Tan X et al. provided a comprehensive overview of the current knowledge on the treatment of cUTI with carbapenemes or their combinations with BLI. The authors revealed that MER-VAB, ertapenem, and biapenem are characterized by higher overall clinical success rate in this type of infection. MER-VAB and ertapenem also achieved a higher microbiologic eradication rate than other comparators. On the other hand, MER-VAB, similarly to imipenem/cilastatin, obtained a higher percentage of adverse events. The authors postulated that MER-VAB should be used as a first-choice drug in cUTIs of CRE etiology [74]. An overview of the antimicrobial activity of MER-VAB is summarized in Table 4.

Table 4. Antimicrobial activity of meropenem–vaborbactam.

Microbiological Characteristics	No. of Patients	Infection Type	No. of CRE Pathogens	Median Duration of Treatment	Clinical Cure	Therapy Scheme	In-Hospital Mortality	Severe TEAE	References
K. pneumoniae (14/20), *K. oxytoca* (2/20), *E. coli* (2/20), *E. cloacae* (1/20), *C. freundii* (1/20)	n = 20	- BSI (8/20), - Pneumonia (6/20 - incl. 5/6 HABP) - Tracheobronchitis - (2/20 incl. 1VATB) - SSTI (2/20) - AP (1/20) - cIAI (1/20)	18/20 (90%)	8 days	65% (13/20)	MER-VAB MT in 80% (16/20); PT > 48 h (4/20) (GM inh. (n = 2), GM i.v. and inh. (n = 1), CIP i.v. (n = 1))	2/20	1/20 eosinophilia	[46]
K. pneumoniae (100%)	n = 37	- BSI (23/37), - LRTI (10/37), - IAI (1/37), - cUTI (2/37), - SSTI (1/37)	100%	13.5 days	75.6% (28/37)	MER-VAB MT (14/37); PT ((+1 drug), n = 17: FOS (6), COL (6), TGC (3), GM (1), AKN (1); (+ ≥2 drugs) n = 6)	9/37	1/37 leukopenia, thrombocytopenia	[56]
Enterobacterales (86.7%) 39/45: *K. pneumoniae* (46.7%) *E. cloacae* (20%) *E. coli* (13.3%) *B. cepacia* (6.6%) *A. baumannii* (2.2%) *P. aeruginosa* (4.4%)	n = 40	- Nosocomial infections 18/40 - cUTI 8/40 - IAI 5/40 - SSTI 5/40 - BSI 2/40	33/39 (84.6%)	12 days	70% (28/40)	MER-VAB MT; PT ((+ 1 drug): MNO/LEV/AKN 37.5% (12/40); 8/40 TOB/COL; 4/40 changed after 72 h MER-VAB for MNO/CAZ-AVI)	3/40	1/40 SJS/TEN	[59]
K. pneumoniae (42.1%) *E. coli* (19.8%) *E. cloacae* (16.7%) *P. aeruginosa* (8.7%) other (12.7%)	n = 129	- Pneumonia (38.1% incl. VABP 19.8%) - IAI (19.0%) - UTI (13.5%) - SSTI (10.3%) - BSI (9.5%) - Other (9.6%)	78.6%	11.7 days	ND	MER-VAB MT; PT (34.1% + ≥1 drug for ≥48 h)	30/126	4/126 SJS/TEN; hepatotoxiticy, AKI	[60]
K. pneumoniae, *E. coli*, *Enterobacter* spp., *Citrobacter* spp., *Serratia* spp.	n = 131 (incl. n = 105 CAZ-AVI n = 26 MER-VAB)	- MER-VAB cohort: - Primary BSI 1/26 - Secondary bacteriemia 8/26 - Nosocomial infections 10/26 - IAI 5/26 - SSTI 2/26	ND	MER-VAB cohort: 12.3 days CAZ-AVI cohort: 10.8 days	MER-VAB cohort: 69.2% CAZ-AVI cohort: 61.9%	MER-VAB cohort: MER-VAB MT; PT (4/26) (+ ≥1 drug) PXB/COL/TGC	MER-VAB cohort: 3/26	6/26 (3/6 AKI; 2/6 leukopenia;1/6 SJS)	[61]

Antimicrobial agents: GM (gentamycin), AKN (amikacin), TOB (tobramycin), TGC (tygecycline), COL (colistin), PXB (polymixin B), MNO (minocyclin), LEV (levofloxacin), FOS (fosfomycin), CIP (ciprofloxacin). Drug administration: i.v.—intravenous; inh.—inhalation. Severe TEAE: SJS/TEN—Stevens–Johnson syndrome/toxic epidermal necrolysis; AKI—acute kidney injury; infection type: VATB—ventilatory acquired tracheobronchitis. Abbreviations: ND—no data; MT—monotherapy; PT—polytherapy; incl.—include.

5. Adverse Events

Adverse events (Aes) related to MER-VAB therapy were described in clinical trials of phase 3: TANGO I and TANGO II. In the first case (study population n = 272), AE occurred in 39% of patients. The most common AE was headache (8.8%), and other AEs were diarrhea (3.3%), nausea (1.8%), asymptomatic bacteriuria (1.5%), catheter site phlebitis (1.8%), infusion site phlebitis (2.2%), UTI (1.5%), hypokalemia (1.1%), vaginal infection (0.4%), alanine aminotransferase increased (1.8%), aspartate aminotransferase increased (1.5%), anemia (0.7%), and pyrexia (1.5%). MER-VAB was well tolerated by patients, and only 2.6% of them had to discontinue therapy because of severe TEAE [54]. In the TANGO second trial, TEAEs were reported in 84% (42/50) patients. The most commonly represented AEs were diarrhea (12%), anemia (10%), hypokalemia (10%), hypotension (8%), sepsis (4%), septic shock (2%), and AKI (2%). MER-VAB also showed a lower nephrotoxicity rate than BAT [36].

Similar amounts of AEs were described in a phase 1 study of the safety, tolerability, and pharmacokinetics of MER-VAB. TEAEs occurred in 62/80 patients (77.5%); 26 of them (study population n = 37) were connected with MER-VAB administration and 6 of them (study population n = 8) with administration vaborbactam alone. The most commonly observed AEs were infusion site phlebitis and infusion site pain [33].

Shields et al. indicated a single case of severe TEAE that occurred in the study population (n = 20). One patient developed severe eosinophilia after 19 days of MER-VAB i.v. therapy [47]. Tumbarello et al. recorded a single case of severe TEAE (1/37) in the form of leukopenia and thrombocytopenia [57].

Alosaimy et al. revealed that MER-VAB therapy involved low risk of nephrotoxicity (AKI) and diarrhea or *Clostridioides difficile* infectious complications. However, one of the patients developed a severe dermatological reaction—Stevens–Johnson syndrome/toxic epidermal necrolysis (SJS/TEN)—which was a fatal complication of the therapy [60]. Moreover, the authors proved that MER-VAB treatment had low risk of hepatotoxicity (0.8%). A total of 3.1% of patients experienced AEs, which is a lower number in comparison to the TANGO I and TANGO II trials [61].

Ackley and co-authors analyzed the safety of MER-VAB and CAZ-AVI treatment. In their study, MER-VAB showed lower rate of AEs than CAZ-AVI, respectively, 23.1% (6/26) vs. 34.3% (36/105). Nephrotoxicity (defined as AKI) was the most commonly observed form of AE (14.3%). Other AEs observed in the study were leukopenia (7.7%) and rash (3.9%) [62].

6. Conclusions

The emergence of multidrug-resistant strains represents one of the greatest dangers to health and life around the world. The antibiotic resistance rate is increasing rapidly, leading to the shortage of therapeutic options. Therefore, antibiotic use in medicine must be rationalized and the development of new compounds for future drug should be increased. A promising trend is the preparation of a method to deactivate the resistance mechanism due to the actions of antibiotic inhibitors. In this review, we took a closer look at the most recent combination of meropenem, a β-lactam antibiotic, with vaborbactam as an inhibitor (MER-VAB). In the time of antimicrobial-drug-resistance growth, novel combinations of BL+BLI may become important treatment options against Gram-negative bacteria. MER-VAB is predicted by many authors to play a greater role in the treatment of critically ill patients, particularly those with HABP and VABP. Many of them postulate using MER-VAB as a first-choice treatment option for CRE-related pneumonia (including hospital and ventilator-acquired pneumonia). The good tolerability and clinical efficiency confirmed in the TANGO I trial presents MER-VAB as an interesting antimicrobial therapy option in cUTIs caused by CRE pathogens. Moreover, there are suggestions for the use of MER-VAB in cUTIs as a treatment for CRE KPC(+), with AP included. Based on the data provided in the article, it is possible to predict the trends in the growing role of novel BL + BLI medicines. In the near future, it will be noted that combinations of drugs such as MER-VAB,

CAZ-AVI, and cefiderocol will become the first-line antimicrobials against the spread of CRE pathogens, especially in patients in ICUs.

Author Contributions: Conceptualization, S.V. and A.D.-M.; software, E.T.; investigation, A.D.-M., S.V. and E.T.; resources, A.D.-M., S.V. and E.T.; data curation, A.D.-M., S.V. and E.T.; writing—original draft preparation, A.D.-M., S.V. and E.T.; writing—review and editing, A.D.-M., S.V. and E.T.; visualization, A.D.-M., S.V. and E.T.; supervision, A.D.-M. All authors have read and agreed to the published version of the manuscript.

Funding: This research received no external funding.

Institutional Review Board Statement: Not applicable.

Informed Consent Statement: Not applicable.

Data Availability Statement: Not applicable.

Conflicts of Interest: The authors declare no conflict of interest.

References

1. Zanichelli, V.; Sharland, M.; Cappello, B.; Moja, L.; Getahun, H.; Pessoa-Silva, C.; Sati, H.; van Weezenbeek, C.; Balkhy, H.; Simão, M.; et al. The WHO AWaRe (Access, Watch, Reserve) Antibiotic Book and Prevention of Antimicrobial Resistance. *Bull. World Health Organ.* **2023**, *101*, 290–296. [CrossRef]
2. *WHO Report on Surveillance of Antibiotic Consumption: 2016–2018 Early Implementation*; World Health Organization: Geneva, Switzerland, 2018.
3. Klein, E.Y.; Van Boeckel, T.P.; Martinez, E.M.; Pant, S.; Gandra, S.; Levin, S.A.; Goossens, H.; Laxminarayan, R. Global Increase and Geographic Convergence in Antibiotic Consumption between 2000 and 2015. *Proc. Natl. Acad. Sci. USA* **2018**, *115*, E3463–E3470. [CrossRef] [PubMed]
4. Browne, A.J.; Chipeta, M.G.; Haines-Woodhouse, G.; Kumaran, E.P.A.; Hamadani, B.H.K.; Zaraa, S.; Henry, N.J.; Deshpande, A.; Reiner, R.C.; Day, N.P.J.; et al. Global Antibiotic Consumption and Usage in Humans, 2000–18: A Spatial Modelling Study. *Lancet Planet. Health* **2021**, *5*, e893–e904. [CrossRef] [PubMed]
5. Al Sulayyim, H.J.; Ismail, R.; Al Hamid, A.; Ghafar, N.A. Antibiotic Resistance during COVID-19: A Systematic Review. *Int. J. Environ. Res. Public Health* **2022**, *19*, 11931. [CrossRef] [PubMed]
6. Khoshbakht, R.; Kabiri, M.; Neshani, A.; Khaksari, M.N.; Sadrzadeh, S.M.; Mousavi, S.M.; Ghazvini, K.; Ghavidel, M. Assessment of Antibiotic Resistance Changes during the COVID-19 Pandemic in Northeast of Iran during 2020–2022: An Epidemiological Study. *Antimicrob. Resist. Infect. Control* **2022**, *11*, 121. [CrossRef] [PubMed]
7. Giuliano, K.K.; Baker, D.; Thakkar-Samtani, M.; Glick, M.; Restrepo, M.I.; Scannapieco, F.A.; Heaton, L.J.; Frantsve-Hawley, J. Incidence, Mortality, and Cost Trends in Nonventilator Hospital-Acquired Pneumonia in Medicaid Beneficiaries, 2015–2019. *Am. J. Infect. Control* **2023**, *51*, 227–230. [CrossRef]
8. Murray, C.J.; Ikuta, K.S.; Sharara, F.; Swetschinski, L.; Robles Aguilar, G.; Gray, A.; Han, C.; Bisignano, C.; Rao, P.; Wool, E.; et al. Global Burden of Bacterial Antimicrobial Resistance in 2019: A Systematic Analysis. *Lancet* **2022**, *399*, 629–655. [CrossRef]
9. Kollef, M.H. Prevention of Hospital-Associated Pneumonia and Ventilator-Associated Pneumonia. *Crit. Care Med.* **2004**, *32*, 1396–1405. [CrossRef]
10. Poovieng, J.; Sakboonyarat, B.; Nasomsong, W. Bacterial Etiology and Mortality Rate in Community-Acquired Pneumonia, Healthcare-Associated Pneumonia and Hospital-Acquired Pneumonia in Thai University Hospital. *Sci. Rep.* **2022**, *12*, 9004. [CrossRef]
11. WHO. Media Centre. News Release. WHO Published List of Bacteria for Which New an-Tibiotics Are Urgently Needed. 2017. Available online: http://www.who.int/Mediacentre/News/Releases/2017/Bacteria-Antibiotics-Needed/En/ (accessed on 27 February 2017).
12. Chen, C.Y.; Yang, K.Y.; Peng, C.K.; Sheu, C.C.; Chan, M.C.; Feng, J.Y.; Wang, S.H.; Chen, C.M.; Zheng, Z.R.; Liang, S.J.; et al. Clinical Outcome of Nosocomial Pneumonia Caused by Carbapenem-Resistant Gram-Negative Bacteria in Critically Ill Patients: A Multicenter Retrospective Observational Study. *Sci. Rep.* **2022**, *12*, 7501. [CrossRef]
13. Peleg, A.Y.; Hooper, D.C. Hospital-Acquired Infections Due to Gram-Negative Bacteria. *N. Engl. J. Med.* **2010**, *362*, 1804–1813. [CrossRef]
14. Santoro, A.; Franceschini, E.; Meschiari, M.; Menozzi, M.; Zona, S.; Venturelli, C.; Digaetano, M.; Rogati, C.; Guaraldi, G.; Paul, M.; et al. Epidemiology and Risk Factors Associated with Mortality in Consecutive Patients with Bacterial Bloodstream Infection: Impact of MDR and XDR Bacteria. *Open Forum Infect. Dis.* **2020**, *7*, ofaa461. [CrossRef] [PubMed]
15. Lakbar, I.; Medam, S.; Ronflé, R.; Cassir, N.; Delamarre, L.; Hammad, E.; Lopez, A.; Lepape, A.; Machut, A.; Boucekine, M.; et al. Association between Mortality and Highly Antimicrobial-Resistant Bacteria in Intensive Care Unit-Acquired Pneumonia. *Sci. Rep.* **2021**, *11*, 16497. [CrossRef] [PubMed]

16. Calo, F.; Retamar, P.; Perez-Crespo, P.M.M.; Lanz-Garcia, J.; Sousa, A.; Goikoetxea, J.; Reguera-Iglesias, J.M.; Leon, E.; Arminanzas, C.; Mantecon, M.A.; et al. Catheter-Related Bloodstream Infections: Predictive Factors for Gram-Negative Bacteria Aetiology and 30 Day Mortality in a Multicentre Prospective Cohort. *J. Antimicrob. Chemother.* **2020**, *75*, 3056–3061. [CrossRef] [PubMed]
17. Patel, J.; Harant, A.; Fernandes, G.; Mwamelo, A.J.; Hein, W.; Dekker, D.; Sridhar, D. Measuring the Global Response to Antimicrobial Resistance, 2020–21: A Systematic Governance Analysis of 114 Countries. *Lancet Infect. Dis.* **2023**, *23*, 706–718. [CrossRef]
18. Nordmann, P. Carbapenemase-producing Enterobacteriaceae: Overview of a major public health challenge. *Med. Mal. Infect.* **2014**, *44*, 51–56. [CrossRef] [PubMed]
19. Breijyeh, Z.; Jubeh, B.; Karaman, R. Resistance of Gram-Negative Bacteria to Current Antibacterial Agents and Approaches to Resolve It. *Molecules* **2020**, *25*, 1340. [CrossRef] [PubMed]
20. Theuretzbacher, U. Global Antimicrobial Resistance in Gram-Negative Pathogens and Clinical Need. *Curr. Opin. Microbiol.* **2017**, *39*, 106–112. [CrossRef]
21. Freire-Moran, L.; Aronsson, B.; Manz, C.; Gyssens, I.C.; So, A.D.; Monnet, D.L.; Cars, O. Critical Shortage of New Antibiotics in Development against Multidrug-Resistant Bacteria—Time to React Is Now. *Drug Resist. Updat.* **2011**, *14*, 118–124. [CrossRef]
22. Logan, L.K.; Weinstein, R.A. The Epidemiology of Carbapenem-Resistant Enterobacteriaceae: The Impact and Evolution of a Global Menace. *J. Infect. Dis.* **2017**, *215*, S28–S36. [CrossRef]
23. Hecker, S.J.; Reddy, K.R.; Totrov, M.; Hirst, G.C.; Lomovskaya, O.; Griffith, D.C.; King, P.; Tsivkovski, R.; Sun, D.; Sabet, M.; et al. Discovery of a Cyclic Boronic Acid β-Lactamase Inhibitor (RPX7009) with Utility vs Class A Serine Carbapenemases. *J. Med. Chem.* **2015**, *58*, 3682–3692. [CrossRef] [PubMed]
24. Sun, D.; Rubio-Aparicio, D.; Nelson, K.; Dudley, M.N.; Lomovskaya, O. Meropenem-Vaborbactam Resistance Selection, Resistance Prevention, and Molecular Mechanisms in Mutants of KPC-Producing *Klebsiella pneumoniae*. *Antimicrob. Agents Chemother.* **2017**, *61*, e01694-17. [CrossRef] [PubMed]
25. Lomovskaya, O.; Sun, D.; Rubio-Aparicio, D.; Nelson, K.; Tsivkovski, R.; Griffith, D.C.; Dudley, M.N. Vaborbactam: Spectrum of Beta-Lactamase Inhibition and Impact of Resistance Mechanisms on Activity in Enterobacteriaceae. *Antimicrob. Agents Chemother.* **2017**, *61*, e01443-17. [CrossRef] [PubMed]
26. Mendez, A.; Chagastelles, P.; Palma, E.; Nardi, N.; Schapoval, E. Thermal and Alkaline Stability of Meropenem: Degradation Products and Cytotoxicity. *Int. J. Pharm.* **2008**, *350*, 95–102. [CrossRef]
27. Morrill, H.J.; Pogue, J.M.; Kaye, K.S.; LaPlante, K.L. Treatment Options for Carbapenem-Resistant Enterobacteriaceae Infections. *Open Forum Infect. Dis.* **2015**, *2*, OFV050. [CrossRef]
28. Nicolau, D.P. Pharmacokinetic and Pharmacodynamic Properties of Meropenem. *Clin. Infect. Dis.* **2008**, *47*, S32–S40. [CrossRef]
29. Herald, F.; Burgos, R.M. Clinical Evaluation of Meropenem-Vaborbactam Combination for the Treatment of Urinary Tract Infection: Evidence to Date. *Infect. Drug Resist.* **2023**, *16*, 555–568. [CrossRef]
30. Franceschini, N. Meropenem Stability to Beta-Lactamase Hydrolysis and Comparative In Vitro Activity against Several Beta-Lactamase-Producing Gram-Negative Strains. *J. Antimicrob. Chemother.* **2002**, *49*, 395–398. [CrossRef]
31. Vabomere (Meropenem and Vaborbactam) for Iniection; Approval by FDA's Center for Drug Evaluation and Research. Available online: https://www.accessdata.fda.gov/Drugsatfda_docs/Label/2017/209776lbl.pdf (accessed on 6 October 2023).
32. Sabet, M.; Tarazi, Z.; Rubio-Aparicio, D.; Nolan, T.; Parkinson, J.; Lomovskaya, O.; Dudley, M.; Griffith, D. Activity of simulated human dosage regimens of meropenem and vaborbactam against carbapenem-resistant enterobacteriaceae in an in vitro hollow-fiber model. *Antimicrob. Agents Chemother.* **2018**, *62*, e01969-17. [CrossRef]
33. Rubino, C.M.; Bhavnani, S.M.; Loutit, J.S.; Morgan, E.E.; White, D.; Dudley, M.N.; Griffith, D.C. Phase 1 Study of the Safety, Tolerability, and Pharmacokinetics of Vaborbactam and Meropenem Alone and in Combination Following Single and Multiple Doses in Healthy Adult Subjects. *Antimicrob. Agents Chemother.* **2018**, *62*, e02228-17. [CrossRef]
34. Griffith, D.C.; Loutit, J.S.; Morgan, E.E.; Durso, S.; Dudley, M.N. Phase 1 Study of the Safety, Tolerability, and Pharmacokinetics of the β-Lactamase Inhibitor Vaborbactam (RPX7009) in Healthy Adult Subjects. *Antimicrob. Agents Chemother.* **2016**, *60*, 6326–6332. [CrossRef] [PubMed]
35. Lee, Y.; Kim, J.; Trinh, S. Meropenem–Vaborbactam (Vabomere™): Another Option for Carbapenem-Resistant Enterobacteriaceae. *Pharm. Ther.* **2019**, *44*, 110–113.
36. Wunderink, R.G.; Giamarellos-Bourboulis, E.J.; Rahav, G.; Mathers, A.J.; Bassetti, M.; Vazquez, J.; Cornely, O.A.; Solomkin, J.; Bhowmick, T.; Bishara, J.; et al. Effect and Safety of Meropenem–Vaborbactam versus Best-Available Therapy in Patients with Carbapenem-Resistant Enterobacteriaceae Infections: The TANGO II Randomized Clinical Trial. *Infect. Dis. Ther.* **2018**, *7*, 439–455. [CrossRef]
37. Novelli, A.; Del Giacomo, P.; Rossolini, G.M.; Tumbarello, M. Meropenem/Vaborbactam: A next Generation β-Lactam β-Lactamase Inhibitor Combination. *Expert Rev. Anti-Infect. Ther.* **2020**, *18*, 643–655. [CrossRef] [PubMed]
38. Wenzler, E.; Gotfried, M.H.; Loutit, J.S.; Durso, S.; Griffith, D.C.; Dudley, M.N.; Rodvold, K.A. Meropenem-RPX7009 Concentrations in Plasma, Epithelial Lining Fluid, and Alveolar Macrophages of Healthy Adult Subjects. *Antimicrob. Agents Chemother.* **2015**, *59*, 7232–7239. [CrossRef] [PubMed]
39. Venuti, F.; Romani, L.; De Luca, M.; Tripiciano, C.; Palma, P.; Chiriaco, M.; Finocchi, A.; Lancella, L. Novel Beta Lactam Antibiotics for the Treatment of Multidrug-Resistant Gram-Negative Infections in Children: A Narrative Review. *Microorganisms* **2023**, *11*, 1798. [CrossRef] [PubMed]

40. Lapuebla, A.; Abdallah, M.; Olafisoye, O.; Cortes, C.; Urban, C.; Quale, J.; Landman, D. Activity of Meropenem Combined with RPX7009, a Novel β-Lactamase Inhibitor, against Gram-Negative Clinical Isolates in New York City. *Antimicrob. Agents Chemother.* **2015**, *59*, 4856–4860. [CrossRef]
41. Castanheira, M.; Huband, M.D.; Mendes, R.E.; Flamm, R.K. Meropenem-Vaborbactam Tested against Contemporary Gram-Negative Isolates Collected Worldwide during 2014, Including Carbapenem-Resistant, KPC-Producing, Multidrug-Resistant, and Extensively Drug-Resistant Enterobacteriaceae. *Antimicrob. Agents Chemother.* **2017**, *61*, e00567-17. [CrossRef]
42. Gaibani, P.; Giani, T.; Bovo, F.; Lombardo, D.; Amadesi, S.; Lazzarotto, T.; Coppi, M.; Rossolini, G.M.; Ambretti, S. Resistance to Ceftazidime/Avibactam, Meropenem/Vaborbactam and Imipenem/Relebactam in Gram-Negative MDR Bacilli: Molecular Mechanisms and Susceptibility Testing. *Antibiotics* **2022**, *11*, 628. [CrossRef]
43. Tooke, C.L.; Hinchliffe, P.; Bragginton, E.C.; Colenso, C.K.; Hirvonen, V.H.A.; Takebayashi, Y.; Spencer, J. β-Lactamases and β-Lactamase Inhibitors in the 21st Century. *J. Mol. Biol.* **2019**, *431*, 3472–3500. [CrossRef]
44. Biagi, M.; Wu, T.; Lee, M.; Patel, S.; Butler, D.; Wenzler, E. Exploring Aztreonam in Combination with Ceftazidime-Avibactam or Meropenem-Vaborbactam as Potential Treatments for Metallo- and Serine-β-Lactamase-Producing Enterobacteriaceae. *Antimicrob. Agents Chemother.* **2019**, *63*, AAC.01426-19. [CrossRef] [PubMed]
45. Langley, G.W.; Cain, R.; Tyrrell, J.M.; Hinchliffe, P.; Calvopiña, K.; Tooke, C.L.; Widlake, E.; Dowson, C.G.; Spencer, J.; Walsh, T.R.; et al. Profiling Interactions of Vaborbactam with Metallo-β-Lactamases. *Bioorg. Med. Chem. Lett.* **2019**, *29*, 1981–1984. [CrossRef] [PubMed]
46. Hackel, M.A.; Lomovskaya, O.; Dudley, M.N.; Karlowsky, J.A.; Sahm, D.F. In Vitro Activity of Meropenem-Vaborbactam against Clinical Isolates of KPC-Positive Enterobacteriaceae. *Antimicrob. Agents Chemother.* **2018**, *62*, e01904-17. [CrossRef] [PubMed]
47. Shields, R.K.; McCreary, E.K.; Marini, R.V.; Kline, E.G.; Jones, C.E.; Hao, B.; Chen, L.; Kreiswirth, B.N.; Doi, Y.; Clancy, C.J.; et al. Early Experience With Meropenem-Vaborbactam for Treatment of Carbapenem-Resistant Enterobacteriaceae Infections. *Clin. Infect. Dis.* **2020**, *71*, 667–671. [CrossRef] [PubMed]
48. Findlay, J.; Poirel, L.; Nordmann, P. In Vitro-Obtained Meropenem-Vaborbactam Resistance Mechanisms among Clinical *Klebsiella pneumoniae* Carbapenemase-Producing *K. pneumoniae* Isolates. *J. Glob. Antimicrob. Resist.* **2023**, *32*, 66–71. [CrossRef]
49. Yahav, D.; Giske, C.G.; Gramatniece, A.; Abodakpi, H.; Tam, V.H.; Leibovici, L. New β-Lactam–β-Lactamase Inhibitor Combinations. *Clin. Microbiol. Rev.* **2020**, *34*, e00115-20. [CrossRef]
50. Jorgensen, S.C.J.; Rybak, M.J. Meropenem and Vaborbactam: Stepping up the Battle against Carbapenem-Resistant Enterobacteriaceae. *Pharmacother. J. Hum. Pharmacol. Drug Ther.* **2018**, *38*, 444–461. [CrossRef]
51. Chiotos, K.; Hayes, M.; Gerber, J.S.; Tamma, P.D. Treatment of Carbapenem-Resistant Enterobacteriaceae Infections in Children. *J. Pediatr. Infect. Dis. Soc.* **2020**, *9*, 56–66. [CrossRef]
52. Hanretty, A.M.; Kaur, I.; Evangelista, A.T.; Moore, W.S.; Enache, A.; Chopra, A.; Cies, J.J. Pharmacokinetics of the Meropenem Component of Meropenem-Vaborbactam in the Treatment of ^{KPC}-Producing *Klebsiella pneumoniae* Bloodstream Infection in a Pediatric Patient. *Pharmacother. J. Hum. Pharmacol. Drug Ther.* **2018**, *38*, e87–e91. [CrossRef]
53. Shoulders, B.R.; Casapao, A.M.; Venugopalan, V. An Update on Existing and Emerging Data for Meropenem-Vaborbactam. *Clin. Ther.* **2020**, *42*, 692–702. [CrossRef]
54. Andrei, S.; Valeanu, L.; Chirvasuta, R.; Stefan, M.-G. New FDA Approved Antibacterial Drugs: 2015–2017. *Discoveries* **2018**, *6*, e81. [CrossRef] [PubMed]
55. Kaye, K.S.; Bhowmick, T.; Metallidis, S.; Bleasdale, S.C.; Sagan, O.S.; Stus, V.; Vazquez, J.; Zaitsev, V.; Bidair, M.; Chorvat, E.; et al. Effect of Meropenem-Vaborbactam vs Piperacillin-Tazobactam on Clinical Cure or Improvement and Microbial Eradication in Complicated Urinary Tract Infection: The TANGO I Randomized Clinical Trial. *JAMA* **2018**, *319*, 788–799. [CrossRef] [PubMed]
56. Shields, R.K.; Potoski, B.A.; Haidar, G.; Hao, B.; Doi, Y.; Chen, L.; Press, E.G.; Kreiswirth, B.N.; Clancy, C.J.; Nguyen, M.H. Clinical Outcomes, Drug Toxicity, and Emergence of Ceftazidime-Avibactam Resistance among Patients Treated for Carbapenem-Resistant Enterobacteriaceae Infections. *Clin. Infect. Dis.* **2016**, *63*, ciw636. [CrossRef]
57. Tumbarello, M.; Raffaelli, F.; Cascio, A.; Falcone, M.; Signorini, L.; Mussini, C.; De Rosa, F.G.; Losito, A.R.; De Pascale, G.; Pascale, R.; et al. Compassionate Use of Meropenem/Vaborbactam for Infections Caused by KPC-Producing *Klebsiella pneumoniae*: A Multicentre Study. *JAC-Antimicrob. Resist.* **2022**, *4*, dlac022. [CrossRef]
58. Carvalhaes, C.G.; Shortridge, D.; Sader, H.S.; Castanheira, M. Activity of Meropenem-Vaborbactam against Bacterial Isolates Causing Pneumonia in Patients in U.S. Hospitals during 2014 to 2018. *Antimicrob. Agents Chemother.* **2020**, *64*, e02177-19. [CrossRef] [PubMed]
59. Castanheira, M.; Doyle, T.B.; Kantro, V.; Mendes, R.E.; Shortridge, D. Meropenem-Vaborbactam Activity against Carbapenem-Resistant *Enterobacterales* Isolates Collected in U.S. Hospitals during 2016 to 2018. *Antimicrob. Agents Chemother.* **2020**, *64*, e01951-19. [CrossRef]
60. Alosaimy, S.; Jorgensen, S.C.J.; Lagnf, A.M.; Melvin, S.; Mynatt, R.P.; Carlson, T.J.; Garey, K.W.; Allen, D.; Venugopalan, V.; Veve, M.; et al. Real-World Multicenter Analysis of Clinical Outcomes and Safety of Meropenem-Vaborbactam in Patients Treated for Serious Gram-Negative Bacterial Infections. *Open Forum Infect. Dis.* **2020**, *7*, ofaa051. [CrossRef]
61. Alosaimy, S.; Lagnf, A.M.; Morrisette, T.; Scipione, M.R.; Zhao, J.J.; Jorgensen, S.C.J.; Mynatt, R.; Carlson, T.J.; Jo, J.; Garey, K.W.; et al. Real-World, Multicenter Experience with Meropenem-Vaborbactam for Gram-Negative Bacterial Infections Including Carbapenem-Resistant *Enterobacterales* and *Pseudomonas aeruginosa*. *Open Forum Infect. Dis.* **2021**, *8*, ofab371. [CrossRef]

62. Ackley, R.; Roshdy, D.; Meredith, J.; Minor, S.; Anderson, W.E.; Capraro, G.A.; Polk, C. Meropenem-Vaborbactam versus Ceftazidime-Avibactam for Treatment of Carbapenem-Resistant Enterobacteriaceae Infections. *Antimicrob. Agents Chemother.* **2020**, *64*, e02313-19. [CrossRef]
63. Kaushik, A.; Ammerman, N.C.; Lee, J.; Martins, O.; Kreiswirth, B.N.; Lamichhane, G.; Parrish, N.M.; Nuermberger, E.L. In Vitro Activity of the New β-Lactamase Inhibitors Relebactam and Vaborbactam in Combination with β-Lactams against *Mycobacterium abscessus* Complex Clinical Isolates. *Antimicrob. Agents Chemother.* **2019**, *63*, e02623-18. [CrossRef]
64. Gainey, A.B.; Burch, A.K.; Brownstein, M.J.; Brown, D.E.; Fackler, J.; Horne, B.; Biswas, B.; Bivens, B.N.; Malagon, F.; Daniels, R. Combining Bacteriophages with Cefiderocol and Meropenem/Vaborbactam to Treat a Pan-Drug Resistant Achromobacter Species Infection in a Pediatric Cystic Fibrosis Patient. *Pediatr. Pulmonol.* **2020**, *55*, 2990–2994. [CrossRef] [PubMed]
65. Caverly, L.J.; Spilker, T.; Kalikin, L.M.; Stillwell, T.; Young, C.; Huang, D.B.; LiPuma, J.J. In Vitro Activities of β-Lactam–β-Lactamase Inhibitor Antimicrobial Agents against Cystic Fibrosis Respiratory Pathogens. *Antimicrob. Agents Chemother.* **2020**, *64*, e01595-19. [CrossRef] [PubMed]
66. Shortridge, D.; Carvalhaes, C.; Deshpande, L.; Castanheira, M. Activity of Meropenem/Vaborbactam and Comparators against Gram-Negative Isolates from Eastern and Western European Patients Hospitalized with Pneumonia Including Ventilator-Associated Pneumonia (2014-19). *J. Antimicrob. Chemother.* **2021**, *76*, 2600–2605. [CrossRef]
67. Sader, H.S.; Carvalhaes, C.G.; Shortridge, D.; Castanheira, M. Comparative Activity of Newer β-Lactam/β-Lactamase Inhibitor Combinations against *Pseudomonas aeruginosa* from Patients Hospitalized with Pneumonia in European Medical Centers in 2020. *Eur. J. Clin. Microbiol. Infect. Dis.* **2022**, *41*, 319–324. [CrossRef] [PubMed]
68. Belati, A.; Bavaro, D.F.; Saracino, A.; Diella, L.; De Gennaro, N.; Di Gennaro, F. Meropenem/Vaborbactam Plus Aztreonam as a Possible Treatment Strategy for Bloodstream Infections Caused by Ceftazidime/Avibactam-Resistant *Klebsiella pneumoniae*: A Retrospective Case Series and Literature Review. *Antibiotics* **2022**, *11*, 373. [CrossRef]
69. Tiseo, G.; Falcone, M.; Leonildi, A.; Giordano, C.; Barnini, S.; Arcari, G.; Carattoli, A.; Menichetti, F. Meropenem-Vaborbactam as Salvage Therapy for Ceftazidime-Avibactam-, Cefiderocol-Resistant ST-512 *Klebsiella pneumoniae*–producing KPC-31, a D179Y Variant of KPC-3. *Open Forum Infect. Dis.* **2021**, *8*, ofab141. [CrossRef]
70. Sabet, M.; Tarazi, Z.; Griffith, D.C. Activity of Meropenem-Vaborbactam against *Pseudomonas aeruginosa* and *Acinetobacter baumannii* in a Neutropenic Mouse Thigh Infection Model. *Antimicrob. Agents Chemother.* **2019**, *63*, e01665-18. [CrossRef]
71. Sabet, M.; Tarazi, Z.; Nolan, T.; Parkinson, J.; Rubio-Aparicio, D.; Lomovskaya, O.; Dudley, M.N.; Griffith, D.C. Activity of Meropenem-Vaborbactam in Mouse Models of Infection Due to KPC-Producing Carbapenem-Resistant Enterobacteriaceae. *Antimicrob. Agents Chemother.* **2018**, *62*, e01446-17. [CrossRef]
72. Weiss, W.J.; Pulse, M.E.; Nguyen, P.; Peterson, K.; Silva, J.; Simecka, J.W.; Valtierra, D.; Sabet, M.; Griffith, D.C. Activity of Meropenem-Vaborbactam against Carbapenem-Resistant Enterobacteriaceae in a Murine Model of Pyelonephritis. *Antimicrob. Agents Chemother.* **2018**, *62*, e01439-17. [CrossRef]
73. Bovo, F.; Lombardo, D.; Lazzarotto, T.; Ambretti, S.; Gaibani, P. Epidemiology and In Vitro Activity of Ceftazidime/Avibactam, Meropenem/Vaborbactam and Imipenem/Relebactam against KPC-Producing *K. pneumoniae* Collected from Bacteremic Patients, 2018 to 2020. *Antibiotics* **2022**, *11*, 1621. [CrossRef]
74. Tan, X.; Pan, Q.; Mo, C.; Li, X.; Liang, X.; Li, Y.; Lan, Y.; Chen, L. Carbapenems vs Alternative Antibiotics for the Treatment of Complicated Urinary Tract Infection: A Systematic Review and Network Meta-Analysis. *Medicine* **2020**, *99*, e18769. [CrossRef] [PubMed]

antibiotics

MDPI

Article

Efficacy of Willow Herb (*Epilobium angustifolium* L. and *E. parviflorum* Schreb.) Crude and Purified Extracts and Oenothein B Against Prostatic Pathogens

Alexia Barbarossa [1], Antonio Rosato [1], Alessia Carocci [1], Sabrina Arpini [2], Stefania Bosisio [2], Luca Pagni [2], Diletta Piatti [3], Eleonora Spinozzi [3], Simone Angeloni [3], Gianni Sagratini [3], Gokhan Zengin [4], Marco Cespi [3], Filippo Maggi [3,*] and Giovanni Caprioli [3]

1 Department of Pharmacy-Pharmaceutical Sciences, University of Bari "Aldo Moro", Via E. Orabona, 4, 70125 Bari, Italy; alexia.barbarossa@uniba.it (A.B.); antonio.rosato@uniba.it (A.R.); alessia.carocci@uniba.it (A.C.)

2 Indena S.p.A., Via Don Minzoni 6, 20090 Settala, Italy; sabrina.arpini@indena.com (S.A.); stefania.bosisio@indena.com (S.B.); luca.pagni@indena.com (L.P.)

3 Chemistry Interdisciplinary Project (ChIP) Research Center, School of Pharmacy, University of Camerino, Via Madonna delle carceri, 62032 Camerino, Italy; diletta.piatti@unicam.it (D.P.); eleonora.spinozzi@unicam.it (E.S.); simone.angeloni@unicam.it (S.A.); gianni.sagratini@unicam.it (G.S.)

4 Department of Biology, Science Faculty, Selcuk University, Konya 42130, Turkey; gokhanzengin@selcuk.edu.tr

* Correspondence: filippo.maggi@unicam.it

Citation: Barbarossa, A.; Rosato, A.; Carocci, A.; Arpini, S.; Bosisio, S.; Pagni, L.; Piatti, D.; Spinozzi, E.; Angeloni, S.; Sagratini, G.; et al. Efficacy of Willow Herb (*Epilobium angustifolium* L. and *E. parviflorum* Schreb.) Crude and Purified Extracts and Oenothein B Against Prostatic Pathogens. *Antibiotics* **2025**, *14*, 117. https://doi.org/10.3390/antibiotics14020117

Academic Editors: Katarzyna Wińska, Joanna Kozłowska and Anna Duda-Madej

Received: 6 November 2024
Revised: 6 January 2025
Accepted: 18 January 2025
Published: 23 January 2025

Citation: Barbarossa, A.; Rosato, A.; Carocci, A.; Arpini, S.; Bosisio, S.; Pagni, L.; Piatti, D.; Spinozzi, E.; Angeloni, S.; Sagratini, G.; et al. Efficacy of Willow Herb (*Epilobium angustifolium* L. and *E. parviflorum* Schreb.) Crude and Purified Extracts and Oenothein B Against Prostatic Pathogens. *Antibiotics* **2025**, *14*, 117. https://doi.org/10.3390/antibiotics14020117

Abstract: Background/Objectives: Plants species of the *Epilobium* genus are traditionally used to treat prostatitis and other urinary tract disorders and are particularly rich in ellagitannins and flavonol 3-*O*-glycosides. The aim of this work was to evaluate the inhibitory activity of different extracts from *E. angustifolium* L. and *E. parviflorum* Schreb. and their major bioactive compound, oenothein B, against a panel of Gram-positive (*Enterococcus faecalis* ATCC 29212, *Enterococcus faecalis* BS, *Staphylococcus aureus* ATCC 25923, *Staphylococcus aureus* ATCC 29213, and *Staphylococcus aureus* ATCC 43300) and Gram-negative (*Escherichia coli* ATCC 25922, *Escherichia coli* ATCC 35218, *Escherichia coli* BS, *Klebsiella pneumoniae* ATCC 13883, *Klebsiella pneumoniae* ATCC 70063, *Klebsiella pneumoniae* BS, *Proteus mirabilis* BS, and *Pseudomonas aeruginosa* ATCC 27853) bacteria responsible for prostatitis. **Methods**: Aqueous and ethanolic raw extracts were prepared, and the latter were further purified using the resin Amberlite^{TM} XAD7HP. Then, an HPLC-MS/MS method was developed for the quantification of the marker bioactives and their levels were correlated with the antimicrobial activity. **Results**: Purified extracts were richer in polyphenols (330.80 and 367.66 mg/g of dry extract for *E. angustifolium* and *E. parvifolium*, respectively) than the raw extracts. Oenothein B was the predominant compound in all the extracts (119.98 to 327.57 mg/g of dry extract). The minimum inhibitory concentrations (MICs) in the range of μg/mL indicated significant antibacterial activity, which was higher for the purified extracts and oenothein B (MIC values from 4 to 16 and 8 to 1024 μg/mL on Gram-positive and Gram-negative strains, respectively). **Conclusions**: These results outline the outstanding potential of *E. angustifolium* and *E. parviflorum* extracts and oenothein B as therapeutic alternatives or complementary agents to conventional antibiotic treatments of prostatitis and other urinary tract infections.

Keywords: *Epilobium angustifolium*; *Epilobium parviflorum*; prostatic pathogens; HPLC-MS/MS; oenothein B; gram-positive

1. Introduction

The genus *Epilobium*, belonging to the Onagraceae family, encompasses 200 species, mostly erect perennial herbs, distributed around the world, with twenty-six species occurring in Europe [1]. One of the most important traditional uses of these plants, mainly in Europe and North America, consists in the treatment of skin and mucosa infections and disorders of the genito-urinary tract such as prostatitis, with the latter being characterized by inflammatory process and/or benign prostatic hyperplasia (BPH) [2,3]. The marker compounds of the *Epilobium* genus responsible for the therapeutic properties (e.g., anti-inflammatory and antioxidant) are phenolics, mostly ellagitannins (oenothein B), and flavonol 3-*O*-glycosides (kaempferol, quercetin, and myricetin derivatives) [2,4–6].

BPH is a common human disease affecting men over 50 years [7], and its timely treatment is of pivotal importance to avoid recrudescence of the pathology leading to tumor formation [8]. Therapeutic preparations used in the treatment of prostatic disorders based on *Epilobium* species are mostly characterized by *E. angustifolium* L. and *E. parviflorum* Schreb. extracts [1,9]. Their use in such preparations is supported by studies that have investigated the safety profile and toxicity of these extracts, highlighting their low toxicity and safety at moderate doses [10,11]. *E. angustifolium*, also known as fireweed, is a perennial herb native to Northern and Eastern Europe, where it has been traditionally used as a wound-healing and anti-infective agent and for the treatment of prostatic disorders [12]. In 2019, the European Medicines Agency (EMA) claimed the traditional use of preparations made up of aerial parts as a reliever for urinary tract disorders with a particular efficacy towards BPH [13]. *E. parviflorum*, also known as small-flower willowherb, is distributed in Europe, Western Asia, North Africa, and North America. In traditional Chinese Medicine, *E. parviflorum* has been used for the treatment of inflammations and menstrual disorders [14]. In Europe, preparations made up of its aerial parts, which are approved by the European Medicinal Agency (EMA), are used for the treatment of prostatic and urinary tract disorders [11].

The efficacy of *E. angustifolium* and *E. parviflorum* extracts on prostatic ailments relies on modulation of androgens, inhibition of PSA (prostatic specific antigen) and metalloproteinases, and antiproliferative and pro-apoptotic activities [1].

Acute bacterial prostatitis is caused by opportunistic bacteria colonizing the urinary tract following an infection or obstruction such as *Escherichia coli* and others [15]. Acute bacterial prostatitis is an acute infection of the prostate gland that causes pelvic pain and urinary tract symptoms, such as dysuria, urinary frequency, and urinary retention, and may lead to systemic symptoms, such as fevers, chills, nausea, emesis, and malaise. While the exact prevalence is unclear, it is estimated that acute bacterial prostatitis accounts for about 10% of all prostatitis cases [16]. The infection typically arises from bacteria that ascend through the urethra, the backward flow of contaminated urine into the prostatic ducts, direct bacterial introduction during procedures like transrectal biopsy or urethral instrumentation, or through hematogenous spread [17]. Most cases (65–80%) are caused by *E. coli*, while other pathogens such as *Pseudomonas aeruginosa*, *Proteus mirabilis*, *Klebsiella* spp., and *Enterobacter* spp. account for the rest [18,19]. In individuals with compromised immunity, additional pathogens, including fungi like *Cryptococcus* spp. or *Histoplasma* spp., may also play a role in the infection [20,21]. The cure for bacterial prostatitis is hindered by the lack of a transport mechanism of antibiotics as well as by their scarce penetration into the prostatic tissue [22]. Only selected antibiotics are effective for treating bacterial infections of the prostate. While most antibiotics can penetrate the acutely inflamed prostate, this is not the case for a chronically inflamed gland. The capillaries in the prostate are nonporous and lack a mechanism for antibiotic transport. Fluoroquinolones possess the best pharmacological properties for treating bacterial prostatitis, achieving concentrations in the prostate that are 10 to 50% of those found in the serum [23]. Clindamycin, azithromycin, trimethoprim-sulfamethoxazole, and doxycycline are antibiotics encompassed with good penetration capacity into the tissues, as well as carbapenems, piperacillin, cephalosporins, and some of the aminoglycosides [24]. However, one of the major threats in managing bacterial prostatitis is the growing resistance of microorganisms, especially to fluoroquinolones. Indeed, since fluoroquinolones have a broad spectrum of activity, their use has increased, leading to resistance [25]. Hence, given the increasing prevalence of drug-resistant strains to conventional therapies, the study of new, effective antibacterial agents to combat bacterial prostatitis is an urgent priority. In this context, natural extracts, with their vast array of phytochemicals, offer a valuable resource for discovering novel compounds that could overcome these therapeutic challenges and provide sustainable solutions to this concern.

To the best of our knowledge, the activity of *E. angustifolium* and *E. parviflorum* extracts against opportunistic bacteria occurring during prostatic disorders has not been fully explored. So far, the inhibitory properties of *E. angustifolium* extracts against reference strains from collection associated with skin infections such as *Streptococcus pneumoniae*, *E. coli*, *Enterococcus faecalis*, *E. faecium*, *Sarcina lutea*, and *Bacillus pseudomycoides* have been evaluated [26].

Therefore, in an attempt to support the traditional uses of these species as anti-infective agents useful in treating urinary tract-related disorders, and to deepen knowledge of their antibacterial spectrum, in this work, we evaluated the inhibitory properties of *E. angustifolium* and *E. parviflorum* polar extracts against a panel of bacterial strains both derived from clinical isolation and ATCC collection, which are correlated with opportunistic infections of the urinary tract. To enhance the antibacterial potency, plant extracts were subjected to a purification procedure. An HPLC-MS/MS method was developed and used for the quantitative determination of fifteen bioactive compounds in *Epilobium* extracts, among which oenothein B was used as a quality marker and to correlate their levels with the antibacterial activity observed.

2. Results

2.1. HPLC-MS/MS Analysis

2.1.1. Optimization of the HPLC-MS/MS Analytical Conditions

An HPLC-MS/MS analysis was performed in dynamic-MRM mode, permitting us to simultaneously monitor the transitions of all the analytes according to their retention times. The MS/MS transitions of each analyte were optimized in flow injection analysis (FIA) using MassHunter Optimizer Triple Quad ver. B.04.01 software (Agilent, Santa Clara, CA, USA) through the injection of an individual standard solution (10 μg/mL) for each analyte. The precursor ions of all analytes except for oenothein B were the deprotonated molecules $[M - H]^-$ in negative polarity. Instead, the precursor ion of oenothein B was the molecule minus two protons, $[M - 2H]^{2-}$, as reported in Table 1 and in Figures S1–S4 of the Supplementary Materials.

Table 1. High-performance liquid chromatography–tandem mass spectrometry (HPLC-MS/MS) acquisition parameters, working in dynamic "multiple reaction monitoring" mode, including retention time (Rt) and delta retention time (ΔRt) for each optimized transition.

No.	Compound	Molecular Weight (g/mol)	Precursor Ion (*m*/*z*)	Product Ion (*m*/*z*)	Fragmentor (V)	Collision Energy (V)	Retention Time (Rt) (min)	Delta Retention Time (ΔRt)	Polarity
1	gallic acid	170	169	125	97	12	3.4	3	negative
2	chlorogenic acid	354	353	191 [1], 179	82	12, 12	5.6	3	negative
3	oenothein B	1569	783	783 [1], 765	40	0, 15	6.5	4	negative
4	myricetin-3-*O*-galactoside	480	479	316 [1], 271	170	24, 44	8.6	3	negative
5	myricitrin	464	463	316 [1], 271	155	24, 44	10.5	3	negative
6	quercetin-3-*O*-glucuronide	478	477	301 [1], 175	136	16, 16	10.7	3	negative
7	hyperoside	464	463	300 [1], 271	170	24, 48	10.7	3	negative
8	ellagic acid	302	301	301	170	0	11.8	4	negative
9	kaempferol-3-*O*-glucoside	448	447	284 [1], 255	170	24, 40	12.5	3	negative
10	kaempferol-3-*O*-glucuronide	462	461	285	126	16	12.7	3	negative
11	quercitrin	448	447	300 [1], 271	160	24, 44	12.9	3	negative
12	myricetin	318	317	179 [1], 193	150	16, 16	15.7	4	negative
13	kaempferol-3-*O*-rhamnoside	432	431	285 [1], 255	150	16, 30	15.3	3	negative
14	quercetin	302	301	151 [1], 179	145	16, 12	20.4	4	negative
15	kaempferol	286	285	285	160	0	25.1	4	negative

[1] These product ions were used for quantification; the others to confirm the analytes.

Figure 1 reports the overlay of all chromatograms of the standard mixture obtained by the extraction of the quantitative transition of each analyte.

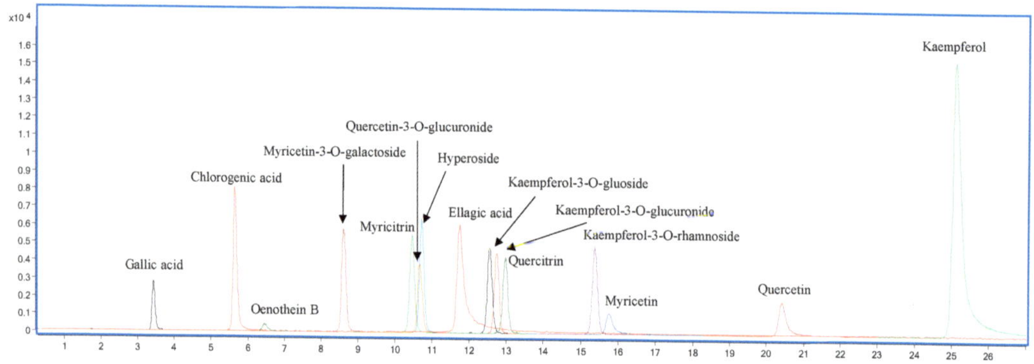

Figure 1. Overlay of all the chromatograms obtained using the extraction of the quantitative transition (dynamic-MRM) of each analyte of the standard mixture.

Several attempts have been carried out to optimize the analyte separation. In detail, two analytical columns, i.e., Kinetex PFP (100 × 2.1 mm, particle size 2.6 µm) and Sinergy Polar-RP 80A (150 mm × 4.6 mm, 4 µm), both from Phenomenex (Castel Maggiore, Bologna, Italy), were evaluated. With the Sinergy Polar column, a better resolution and analyte separation were achieved. Moreover, the shape and the width of the oenothein B peak were preferable to that resulting from the Kinetex PFP column. In addition, different mobile phases (water/methanol, water/acetonitrile both with/without formic acid) and several gradients were evaluated to optimize the separation of the analytes. Table S1 (Supplementary Materials) summarises the best gradients obtained, all with water/acetonitrile and formic acid. The best-performing one in terms of resolution and analyte separation was Gradient 4. This one was selected for the present study.

2.1.2. Method Validation

The analytical method validation was assessed in terms of linearity, limit of detection (LOD), limit of quantification (LOQ), and repeatability. Linearity was evaluated by constructing calibration curves with the respective determination coefficients (R^2) after the injection of different concentrations of the analytes. All the R^2 ranged from 0.998 to 1.000, displaying good linearity. LOD was calculated as the lowest standard concentration with a signal-to-noise ratio (SNR) = 3, whereas LOQ with SNR = 10 (Table 2).

Table 2. HPLC-MS/MS method validation parameters: regression equation, coefficient of determination (R^2), limits of detection (LODs), limits of quantification (LOQs), and reproducibility for the fifteen monitored compounds.

					Repeatability (% RSD)	
Compound	**Regression Equation**	**R^2**	**LOD (µg/L)**	**LOQ (µg/L)**	**Intraday**	**Inter-Day**
gallic acid	$y = 11{,}732x + 1871.5$	0.999	3	10	0.97	6.28
chlorogenic acid	$y = 19{,}982x - 175.02$	1	2	5	1.30	1.26
oenothein B	$y = 2006.7x - 680.97$	0.999	33	100	0.17	0.81
myricetin-3-O-galactoside	$y = 19{,}522x - 960.3$	0.999	2	5	0.76	12.73
myricitrin	$y = 21{,}108x - 787.66$	0.999	2	5	0.02	10.05

Table 2. *Cont.*

Compound	Regression Equation	R^2	LOD (μg/L)	LOQ (μg/L)	Repeatability (% RSD)	
					Intraday	Inter-Day
quercetin-3-*O*-glucuronide	$y = 22{,}667x - 2599.7$	0.999	3	10	0.35	11.85
hyperoside	$y = 14{,}458x + 75.328$	1	2	5	0.79	10.64
ellagic acid	$y = 28{,}444x - 4135.5$	0.999	17	50	4.67	3.09
kaempferol-3-*O*-glucoside	$y = 16{,}461x - 538.22$	1	2	5	0.42	12.25
kaempferol-3-*O*-glucuronide	$y = 15{,}941x - 1100.1$	0.999	3	10	0.45	7.89
quercitrin	$y = 18{,}629x - 202.71$	1	2	5	0.06	2.70
myricetin	$y = 6212.1x - 699.29$	0.999	3	10	2.92	1.28
kaempferol-3-*O*-rhamnoside	$y = 18{,}286x - 431.18$	1	2	5	2.94	7.56
quercetin	$y = 10{,}745x - 39.87$	1	3	10	5.29	4.82
kaempferol	$y = 94{,}006x - 33{,}517$	0.998	3	10	8.26	7.04

RSD% was employed to express the intraday repeatability or run-to-run precision and inter-day repeatability or day-to-day precision. Run-to-run precision ranged from 0.02 to 8.26%, whereas day-to-day precision was from 0.81 to 12.73%. High specificity was achieved using HPLC-MS/MS in dynamic-MRM mode. Method specificity was assessed by measuring retention time stability and setting up multiple precursor/product pairs.

2.1.3. Quantification of Phenolic Compounds

The validated HPLC-MS/MS method was applied for the quantification of fifteen phenolic compounds in the extracts of *E. angustifolium* and *E. parviflorum*. In detail, three diverse extracts were analyzed for each *Epilobium* species, namely, $RE1_{ang}/RE1_{parv}$, $RE2_{ang}/RE2_{parv}$, and PE_{ang}/PE_{parv}. The results obtained for the quantification of the marker polyphenols are reported in Table 3.

Table 3. Quantification of the 15 bioactive compounds (expressed as mg/g of dry weight extract) in the two *Epilobium* species extracts using HPLC-MS/MS analysis.

No.	Compounds	$RE1_{ang}$ [1]	$RE2_{ang}$	PE_{ang} [2]	$RE1_{par}$	$RE2_{par}$	PE_{par}
1	gallic acid	13.03	1.76	0.71	12.08	4.23	1.58
2	chlorogenic acid	2.23	2.13	6.34	0.82	1.29	2.57
3	oenothein B	119.98	127.78	255.86	153.02	168.25	327.57
4	myricetin-3-*O*-galactoside	0.36	0.57	1.88	3.52	5.58	10.16
5	myricitrin	0.07	0.11	0.31	2.97	5.01	9.56
6	quercetin-3-*O*-glucuronide	15.22	12.57	42.08	0.03	0.05	0.10
7	hyperoside	2.01	2.98	6.09	0.57	0.93	1.90
8	ellagic acid	4.52	1.15	4.16	6.14	5.42	11.54
9	kaempferol-3-*O*-glucoside	0.27	0.32	0.83	0.04	0.07	0.12
10	kaempferol-3-*O*-glucuronide	1.63	1.85	5.42	n.d.	n.d.	n.d.
11	quercitrin	0.86	2.24	3.81	0.32	0.66	1.32
12	myricetin	n.d.	n.d.	0.06	0.21	0.55	0.79
13	kaempferol-3-*O*-rhamnoside	0.71	1.12	2.73	0.07	0.14	0.23
14	quercetin	0.09	0.09	0.44	0.07	0.22	0.15
15	kaempferol	n.d. [3]	n.q. [4]	0.02	0.01	0.04	n.q.
	total	161.05	154.74	330.80	179.94	192.50	367.66

[1,2] RE, raw extract; PE, purified extract; *ang*, *Epilobium angustifolium*; *parv*, *Epilobium parviflorum*; [3] n.d., not detectable; [4] n.q. not quantifiable. Each sample was analyzed in triplicate (n = 3) and relative standard deviation (RSD) values were from 0.10% to 12.73%.

Generally, all REs displayed a comparable polyphenol content, while the PEs resulted in particularly enriched targeted compounds. Indeed, the total amount of polyphenols was 154.74 and 192.50 mg/g of dry extract for $RE2_{ang}$ and $RE2_{parv}$, respectively. Conversely, the PEs were enriched in polyphenols, with a content of 330.80 and 367.66 mg/g of dry extract for *E. angustifolium* and *E. parviflorum*, respectively. This result demonstrates the crucial importance of the purification process for the removal of undesired macromolecules and sugars and to produce enriched bioactive compounds. Both *E. angustifolium* and *E. parviflorum* PEs were dominated by high levels of oenothein B accounting for 255.86 and 327.57 mg/g of dry extract, respectively.

Regarding the other polyphenols, some differences between the two species were detected. In detail, *E. angustifolium* PE was characterized by higher contents of quercetin-3-*O*-glucuronide (42.08 mg/g of dry extract), chlorogenic acid (6.34 mg/g of dry extract), hyperoside (6.09 34 mg/g of dry extract), and kaempferol-3-*O*-glucuronide (5.42 mg/g of dry extract). On the other hand, *E. parviflorum* PE displayed a higher content of ellagic acid (11.54 mg/g of dry extract), myricetin-3-*O*-galactoside (10.16 mg/g of dry extract), myricitrin (9.56 mg/g of dry extract), and chlorogenic acid (2.57 mg/g of dry extract).

2.2. Antibacterial Activity

The presented study focused on the evaluation of the antibacterial activity of the different extracts against the strains that are the main culprits of bacterial prostatitis with the aim of finding safe therapeutic alternatives to overcome the problem of antibiotic resistance. The antibacterial activity of *E. parviflorum* and *E. angustifolium* extracts, along with their predominant compound oenothein B, was investigated following the CLSI guidelines to determine the Minimum Inhibitory Concentration (MIC) and Minimum Bactericidal Concentration (MBC) against various Gram-positive and Gram-negative bacterial strains (M07-A9, 2012). These strains comprised both those from the American Type Culture Collection (ATCC) and clinical isolates. The results demonstrated remarkable antibacterial activity of both botanical species against the tested strains (Table 4). Notably, MIC values were observed in the range of µg/mL, indicating significant potency. In some instances, the MIC values were comparable to those of the reference antibiotic used in the assay, highlighting the potential efficacy of these extracts. Moreover, no remarkable differences were observed between extracts from the two species. However, PEs of both species exhibited the highest antibacterial activity, with MICs of 4 µg/mL against *Staphylococcus aureus* ATCC 25923 and *S. aureus* ATCC 29213, and 8 µg/mL against the MRSA strain *S. aureus* ATCC 43300. Additionally, *E. parviflorum* PE showed a particularly interesting effect against *E. faecalis* ATCC 29212 and clinical isolate *E. faecalis* BS, with MICs that were sometimes comparable to those of levofloxacin. The MIC values for all extracts against Gram-negative bacteria ranged from 8 to 1024 µg/mL, thus indicating a wide variability depending on the strain considered. Remarkably, PEs of both species displayed a noteworthy activity against the pathogen *P. aeruginosa*, which was identified as the most sensitive Gram-negative strain, with a determined MIC of 8 µg/mL (Table 4). This result is extremely interesting as *P. aeruginosa* is a complex nosocomial pathogen, highly resistant to conventional therapies, and difficult to eradicate. It is worthy of note that the MBC values differed by only one dilution from the MICs, demonstrating the bactericidal action of these natural extracts. Overall, it is possible to observe that there was variability in activity depending on the extraction method and bacterial strain. In fact, the PEs resulted in much more active than REs towards the strains under study, highlighting the importance of purification methods in enhancing the content of bioactive compounds, especially oenothein B, and thus increasing the antibacterial properties of the plants. To support this evidence, the antibacterial effects of oenothein B alone have been investigated. Oenothein

B demonstrated potent activity against Gram-positive strains, with MIC values consistently lower than or comparable to those of the extracts. For example, against *E. faecalis* ATCC 29212, oenothein B exhibited a MIC of 8 µg/mL, while the MIC values for $RE1_{ang}$ and $RE2_{ang}$ were 16 µg/mL. Similarly, for *S. aureus* ATCC 29213, the MIC of oenothein B (4 µg/mL) was lower than the MIC values of both raw and purified extracts (8–16 µg/mL); thus, the high content of oenothein B in the purified extracts appears to be fundamental to their antimicrobial activity, particularly against Gram-positive bacteria. In contrast, as pointed out for the extracts, Gram-negative strains exhibited lower susceptibility to oenothein B, with MIC values ranging from 64 µg/mL (*K. pneumoniae* BS) to 1024 µg/mL (*E. coli* BS). These values were like or slightly lower than those observed for the crude extracts. For instance, against *E. coli* ATCC 25922, the MIC of oenothein B (256 µg/mL) was identical to that of $RE1_{ang}$ and $RE2_{ang}$. Indeed, it has been found a negative very strong ($r > 0.9$) or strong Pearson correlation ($0.7 < r < 0.89$) between oenothein B concentration and the MIC values for some specific strains such as *S. aureus* ATCC 25923, *S. aureus* ATCC 29213, *E. coli* BS, and *K. pneumoniae* BS (Table 5). While the comparable MIC values of oenothein B and the crude extracts against Gram-negative strains suggest a contribution from other phytochemical components or synergistic effects, the results underline the pivotal role of oenothein B as the primary active compound. However, the results achieved for Gram-negative strains suggest a potential synergistic effect of oenothein B with other minor components present in the extracts.

Table 4. Minimum inhibitory concentration (MIC, μg/mL) and minimum bactericidal concentration (MBC, μg/mL) of *Epilobium angustifolium* and *Epilobium parviflorum* extracts.

	$RE1_{ang}$		$RE2_{ang}$		PE_{ang}		$RE1_{par}$		$RE2_{par}$		PE_{par}		Oenothein B		Levofloxacin
Gram-Positive Strains	MIC	MBC	MIC	MBC	MIC	MBC	MIC	MBC	MIC	MBC	MIC	MBC	MIC	MBC	MIC
E. faecalis ATCC 29212	16	32	32	64	16	32	8	16	32	64	8	16	8	16	2
E. faecalis BS	16	32	64	128	16	32	16	32	32	64	8	16	8	16	8
S. aureus ATCC 25923	8	16	8	16	4	8	8	16	8	16	4	8	4	8	0.5
S. aureus ATCC 29213	8	16	8	16	4	8	8	16	8	16	4	8	4	8	0.5
S. aureus ATCC 43300	16	32	16	32	8	16	16	32	8	16	16	32	8	16	1
Gram-negative strains															
E. coli ATCC 25922	256	512	512	1024	256	512	256	512	512	1024	256	512	1024	2048	0.12
E. coil ATCC 35218	512	512	512	1024	512	1024	512	512	512	512	512	1024	1024	2048	0.12
E. coli BS	1024	2048	1024	2048	512	1024	1024	2048	1024	2048	512	1024	1024	2048	1
K. pneumoniae ATCC 13883	128	256	64	128	32	64	64	128	64	128	32	64	256	512	8
K. pneumoniae ATCC 70063	128	256	64	128	32	64	64	128	64	128	32	64	128	256	8
K. pneumoniae BS	128	256	128	256	32	64	64	128	128	256	32	64	64	128	32
P. mirabilis BS	128	256	64	128	32	64	128	256	128	256	32	64	128	256	8
P. aeruginosa ATCC 27853	64	128	16	32	8	16	32	64	32	64	8	16	64	128	4

$RE1_{ang}$, *Epilobium angustifolium* raw dry extract (RE) (H_2O); $RE2_{ang}$, *Epilobium angustifolium* raw dry extract (EtOH); PE_{ang}, *Epilobium angustifolium* purified dry extract (EtOH); $RE1_{par}$ *Epilobium parviflorum* raw dry extract (H_2O); $RE2_{par}$, *Epilobium parviflorum* raw dry extract (EtOH); PE_{par}, *Epilobium parviflorum* purified dry extract (EtOH).

Table 5. Pearson correlation analysis on the contribution of the target phenolic compounds to the antibacterial activity.

	E. faecalis ATCC 29212	*E. faecalis* BS	*S. aureus* ATCC 25923	*S. aureus* ATCC 29213	*S. aureus* ATCC 43300	*E. coli* ATCC 25922	*E. coil* ATCC 35218	*E. coli* BS	*K. pneumoniae* ATCC 13883	*K. pneumoniae* ATCC 70063	*K. pneumoniae* BS	*Proteus mirabilis* BS	*P. aeruginosa* ATCC 27853
gallic acid	NS [1]	NS	NS	NS	NS	NS	NS	NS	NS	NS	NS	**−0.817** *,2	**−0.868** *
chlorogenic acid	NS	NS	NS	NS	NS	NS	NS	NS	NS	NS	NS	NS	NS
oenothein B	NS	NS	**−0.938** **	**−0.938** **	NS	NS	NS	**−0.938** **	NS	NS	**−0.802** *	NS	NS
myricetin-3-*O*-galactoside	NS	NS	NS	NS	NS	NS	NS	NS	NS	NS	NS	NS	NS
myricitrin	NS	NS	NS	NS	NS	NS	NS	NS	NS	NS	NS	NS	NS
quercetin-3-*O*-glucuronide	NS	NS	NS	NS	NS	NS	NS	NS	NS	NS	NS	NS	NS
hyperoside	NS	NS	NS	NS	NS	NS	NS	NS	NS	NS	NS	NS	NS

Table 5. *Cont.*

	E. faecalis ATCC 29212	*E. faecalis* BS	*S. aureus* ATCC 25923	*S. aureus* ATCC 29213	*S. aureus* ATCC 43300	*E. coli* ATCC 25922	*E. coil* ATCC 35218	*E. coli* BS	*K. pneumoniae* ATCC 13883	*K. pneumoniae* ATCC 70063	*K. pneumoniae* BS	*Proteus mirabilis* BS	*P. aeruginosa* ATCC 27853
ellagic acid	NS	NS	NS	NS	NS	NS	NS	NS	NS	NS	NS	NS	NS
kaempferol-3-*O*-glucoside	NS	NS	NS	NS	NS	NS	NS	NS	NS	NS	NS	NS	NS
kaempferol-3-*O*-glucuronide	NS	NS	NS	NS	NS	NS	NS	NS	NS	NS	NS	NS	NS
quercitrin	NS	NS	NS	NS	NS	NS	NS	NS	NS	NS	NS	NS	NS
myricetin	NS	NS	NS	NS	NS	NS	NS	NS	NS	NS	NS	NS	NS
kaempferol-3-*O*-rhamnoside	NS	NS	NS	NS	NS	NS	NS	NS	NS	NS	NS	NS	NS
quercetin	NS	NS	NS	NS	**−0.847** *	NS	NS	NS	NS	NS	NS	NS	NS
kaempferol	NS	NS	NS	NS	NS	NS	NS	NS	NS	NS	NS	NS	NS
total	NS	NS	**−0.982** ***	**−0.982** ***	NS	NS	NS	**−0.982** ***	NS	NS	**−0.802** *	NS	NS

[1] NS, refer to not significant correlation; [2] the asterisk describes the test significance as follows: * for $0.05 < p < 0.01$; ** $0.01 < p < 0.001$ and *** for $p < 0.001$.

3. Discussion

The high levels of oenothein B detected for the two species, corresponding to 25 and 32% of dry extracts, respectively, were significantly higher than those reported in the literature for *Epilobium* extracts, which vary from 2 to 14% [2]. Thus, the purification procedure applied resulted in an added value for the enhancement of the biological activity of these extracts. The data obtained in this work are in accordance with those previously published. Indeed, oenothein B has been reported as the main polyphenol found in *E. angustifolium* and *E. parviflorum*, even if its levels vary according to different factors, such as flowering stage or geographic origin [5,27–30]. For instance, Kiss et al. [5] reported content of the compound of 225.8 and 326.7 mg/g for *E. angustifolium* and *E. parviflorum* dry aqueous extracts, respectively. These concentrations are slightly higher than those reported in the aqueous extracts analyzed in this work. Conversely, in the work of Baert et al. [29], the level of oenothein B in *E. angustifolium* methanolic extract ranged from 24.90 to 50.26 mg/g of dry extract, quite lower than that reported in this work. Concerning the other polyphenols, for *E. angustifolium*, the main polyphenols reported were also quercetin-3-*O*-glucuronide and gallic acid together with kaempferol-3-*O*-glucuronide [27]. Indeed, quercetin-3-*O*-glucuronide and kaempferol-3-*O*-glucuronide have been identified as chemotaxonomic markers of *E. angustifolium* [6,11]. On the contrary, besides oenothein B, ellagic acid has been reported as one of the main polyphenols of *E. parviflorum* [31], as myricetin-3-*O*-galactoside [30].

Bacterial prostatitis is an inflammatory condition of the prostate gland primarily caused by bacterial infections. The main strains responsible for this condition include *E. coli*, which is the most common pathogen associated with acute bacterial prostatitis, as well as other Gram-negative bacteria such as *K. pneumoniae* and *P. mirabilis* [32]. Additionally, Gram-positive bacteria, including *E. faecalis* and *S. aureus*, may also contribute to chronic forms of the diseases [33]. The antibacterial properties of various natural products used alone or in synergy with conventional antibiotic or non-antibiotic drugs [34] have garnered significant attention in recent years, particularly considering the increasing prevalence of antibiotic-resistant pathogens. Indeed, the understanding of the antibacterial properties of various natural extracts could provide alternative therapeutic options for managing bacterial infections, among which is prostatitis. Among the plant extracts, the genus *Epilobium* has emerged as a promising source of antimicrobial compounds [35]. Specifically, *E. angustifolium* and *E. parviflorum* have been the subjects of numerous studies, revealing their potent antibacterial activities against several strains of clinical interest [36,37]. This paper explored the antibacterial activities of crude and purified extracts from both *E. angustifolium* and *E. parviflorum*, contributing to the growing body of evidence supporting their medicinal potential. However, several studies have previously explored the antibacterial effects of *Epilobium* extracts, considering different Gram-positive and Gram-negative strains. Studies on *E. parviflorum* demonstrated its effectiveness against *P. mirabilis* (ATCC 21721), with methanolic and ethyl acetate extracts exhibiting MICs of 484 and 623 μg/mL, respectively, and the aqueous extract displaying moderate activity with an MIC of 1834 μg/mL [38]. These findings are consistent with our observations on the clinical isolate *P. mirabilis* BS, which exhibited sensitivity to both extracts. However, extracts obtained using ethanol as a solvent demonstrated superior activity against this strain (MIC values amounting to 32 μg/mL). Furthermore, *E. angustifolium*, on the other hand, has demonstrated a broader antibacterial activity. Indeed, significant differences in the MICs between strains were detected for *E. coli* (from 0.625 μg/mL to 16.2 mg/mL), *S. aureus* (from 0.625 μg/mL to 7.6 mg/mL), and *P. aeruginosa* (from 1.25 μg/mL to 9.1 mg/mL) [39–42]. These differences are mainly due to the method of extraction, the part of the plant used, and, consequently, the variability of the components contained in the extracts. Ellagitannins are compounds with documented antimicrobial properties [42–45]. Some of these compounds, including oenothein B, also displayed a synergistic effect with currently used antibiotics against drug-resistant bacteria, such as *S. aureus*. For instance, the combination of oenothein B and tellimagrandin I with oxacillin resulted in a significant lowering of the MICs of the 1/250 or 1/500 antibiotics [46]. Moreover, concerning the mechanism of action, these compounds, including oenothein B, are thought to disrupt bacterial membranes, leading to increased permeability and leakage of cellular contents [47,48]. In addition, several studies corroborated their ability to bind to proteins through non-covalent bonds, which may neutralize bacteria by interfering with their membrane proteins [49]. The other mechanisms proposed for antibacterial activity include the inhibition of extracellular microbial enzymes, oxidative phosphorylation, and disruption of cellular membrane permeability [50,51]. Moreover, some of the compounds present in the extracts may act as efflux pump inhibitors, enhancing the efficacy of antibiotics by preventing bacterial cells from expelling these drugs [38]. Thus, the high content of oenothein B in all tested extracts, particularly in PEs, could justify the promising antibacterial activity detected in this study.

The results presented with this study underscore the importance of exploring natural products for their therapeutic potential. Indeed, based on our findings, it is uncommon to encounter natural substances with a high level of effectiveness against resistant pathogens, especially those demonstrating an activity comparable to that of widely used antibiotics. This aspect outlines the potential value of *E. angustifolium* and *E. parviflorum* as therapeutic alternatives or complementary agents to conventional antibiotic treatments. Therefore, future studies could be directed at investigating a possible synergistic effect of these extracts with classical antibiotics to enhance their effect, decrease toxicity, and overcome antibiotic resistance.

4. Materials and Methods

4.1. Chemicals and Reagents

Gallic acid (CAS Number 149-91-7), chlorogenic acid (CAS Number 327-97-9), oenothein B (CAS Number 104987-36-2), myricitrin (CAS Number 17912-87-7), quercetin-3-*O*-glucuronide (CAS Number 22688-79-5), hyperoside (CAS Number 482-36-0), ellagic acid (CAS Number 476-66-4), kaempferol-3-*O*-glucoside (CAS Number 480-10-4), quercitrin (CAS Number 522-12-3), myricetin (CAS Number 529-44-2), kaempferol-3-*O*-rhamnoside (CAS Number 482-39-3), quercetin (CAS Number 117-39-5), kaempferol (CAS Number 520-18-3), and oenothein B were purchased from Merck (Milan, Italy). Myricetin-3-*O*-galactoside (CAS Number 15648-86-9) and kaempferol-3-*O*-glucuronide (CAS Number 22688-78-4) were purchased from Extrasynthese (Genay, France). Individual stock solutions of each analyte, at a concentration of 1000 μg/mL, were prepared by dissolving pure standard compounds in HPLC-grade methanol:water (80:20) and storing them in glass-stoppered bottles at 4 °C. Regarding ellagic acid, the stock solution was prepared by dissolving 5 mg in 25 mL of an HPLC-grade acetone:acetonitrile:ethanol mixture (1:1:1). HPLC-grade methanol was purchased by Carlo Erba (Milan, Italy), while HPLC-grade formic acid (99%) was acquired from Merck (Darmstadt, Germany), as well as the resin Amberlite™ XAD7HP. A Milli-Q SP Reagent Water System (Millipore, Bedford, MA, USA) was employed to produce ultrapure water. All the solutions were filtered with Phenex™ RC 4 mm 0.2 μM syringeless filters (Phenomenex, Castel Maggiore, Italy) prior to the analysis.

4.2. Plant Collection and Extract Preparation

E. angustifolium and *E. parviflorum* aerial parts were harvested from cultivated plants near Piacenza (Italy), the former and from wild growing plants in northeastern Hungary the latter, during the pre-flowering and flowering phase, respectively (May–July 2022) Then, the plant material, deprived of the damaged and foreign parts, was air dried and subsequently milled and poured through an electric grid to obtain a homogeneous material. The botanical identification of the two species was carried out at Indena's Botanical Laboratory, according to an internal ID method. Herbarium specimens are kept by the same lab, with protocol n. 5780 (*E. angustifolium*) and n. 5781 (*E. parviflorum*).

For the study, different extracts for each species of *Epilobium* were prepared, namely, two raw extracts ($RE1_{ang}/RE1_{parv}$ and $RE2_{ang}/RE2_{parv}$ for *E. angustifolium* and *E. parviflorum*, respectively) and one purified extract (PE_{ang}/PE_{parv} for *E. angustifolium* and *E. parviflorum*, respectively). In detail, the REs were prepared from the aerial parts (500 g) of both *Epilobium* species performing an extraction with a percolator with water at 70 °C (RE1) or ethanol 60% v/v at 25 °C (RE2) [52] covering with 3 L of solvent keeping in static conditions for at least 3 h. After discharging, additional extractions of about 1.5 L of solvent were repeated in the same conditions until complete exhaustion of the biomass (10 L of solvent employed in both cases). The resulting solutions were filtered and concentrated for dryness under vacuum using a rotary evaporator (to obtain 132 g of RE1*ang*, 134.5 g of RE1*par* and RE2*ang*, and, finally, 112 g of RE2*par*). The extraction yield (% w/w) was calculated on a dry weight basis. The extraction yields are reported in Table 6.

Table 6. Yields (% w/w) obtained for the extraction and purification processes.

	Yield (% w/w)
$RE1_{ang}$	26.4
$RE1_{par}$	26.9
$RE2_{ang}$	26.9
$RE2_{par}$	22.4
PE_{ang}	10.0
PE_{par}	9.0

RE, raw extract; PE, purified extract; ang, *Epilobium angustifolium*; parv, *Epilobium parviflorum*.

The purification was conducted only on ethanolic extracts since losses in oenothein B content have been observed for aqueous extracts after the clarification process. Moreover, the purification process of the aqueous extracts also led to precipitation issues in the purification column. Regarding PE_{ang} and PE_{par}, they were obtained via the purification of the $RE2_{ang}$ and $RE2_{parv}$ through an adsorption resin, namely, Amberlite™ XAD7-HP loading the extract in an aqueous solution to adsorb the nonpolar molecules and eliminating the more polar fraction of the total extract. The resin was selected after a screening of ten adsorbent resins (aliphatic and aromatic) with different characteristics (porosity and surfaces) due to its good capacity and extract profile. In detail, the aerial parts of the two *Epilobium* species (700 g) were extracted with ethanol 60% at 25 °C, obtaining the hydroalcoholic solutions containing the REs (about 190 g), which were concentrated until aqueous solution (about 1200–1300 g), clarified using centrifugation to separate a precipitate (with a dry residue of about 20 g containing only small amount of oenothein B) to be then loaded at a flow rate of 1.5 BV/h into a column filled with 1500 mL of Amberlite™ XAD7-HP sorbent resin previously conditioned in water. The initial aqueous eluate was discarded, and then the column was washed with 3000 mL of water. The aqueous eluates (total dry residue of about 100 g) were discarded. Then, the column was eluted with 4500 mL of 70% ethanol to desorb the PEs. The hydroalcoholic eluate was collected, concentrated to dryness under vacuum using a rotary evaporator, and dried at 50 °C under vacuum to yield (% w/w) the Pes (70 g PE_{ang} and 63 g PE_{par}). Table 6 reports the yield obtained for the preparation of the extracts.

4.3. HPLC-MS/MS Analysis

4.3.1. Sample Preparation

The dry extracts $RE2_{ang}$/$RE2_{parv}$ and PE_{ang}/PE_{parv} were dissolved in methanol:water (80:20) (1000 µg/mL) and sonicated for 10 min at room temperature. Conversely, the dry extracts $RE1_{ang}$/$RE1_{parv}$ were dissolved in methanol:water (50:50) (1000 µg/mL). Then, the solutions were filtered through a 0.2 µm filter before HPLC-MS/MS analysis. All samples were stored at −15 °C prior to the analysis. Each sample was analyzed in duplicate.

4.3.2. Analytical Conditions

The quantification of bioactive compounds was carried out using a 1290 Infinity series liquid chromatograph (Agilent, Santa Clara, CA, USA) coupled with a 6420 Triple Quadrupole (Agilent). The mass spectrometer was furnished with an electrospray ionization (ESI) source operating in negative ionization mode. The analytes were separated using a Synergi Polar-RP 80 Å (150 mm × 4.6 mm i.d., 4 µm) column (Phenomenex, Torrance, CA, USA). The separation was performed in gradient mode using water (A) and acetonitrile (B), both with formic acid 0.1%, as mobile phase. The flow was 0.8 mL min^{-1} and the injection volume 2 µL. The column was maintained at 25 °C, while the drying gas temperature in the ionization source was 350 °C. The gas flow was 12 L/min, the nebulizer pressure was 45 psi, and the capillary voltage was set at 4000 V. Detection was achieved in dynamic-MRM mode. The quantification was performed through the integration of the dynamic-MRM peak areas, employing the most abundant transitions of the analytes. The mass spectrometer parameters, including Rt and ΔRt, are reported in Table 1.

4.3.3. Method Validation

The analyses of the solutions were performed in duplicate (n = 2). The relative standard deviation percentage (RSD %) was calculated to assess data precision. The HPLC-MS/MS method was validated by the linearity, limit of detection (LODs), limit of quantification (LOQs), and precision. The linearity was assessed by injecting standard solutions at various concentrations of the compounds mentioned in Section 2.1. The calibration curves were obtained by plotting the analyte peak areas on the analyte concentrations, and they were employed for the quantification of the marker polyphenols. The method was also assessed in terms of repeatability injecting each standard solution 3 times in the same day (intraday), as well as 3 times in 3 consecutive days (inter-day). The repeatability was evaluated in terms of relative standard deviation (RSD %). In addition, LOD was evaluated, accepting a signal-to-noise ratio (S/N) of 3:1, while the LOQ was estimated, accepting a signal-to-noise ratio of 10:1.

4.4. Antimicrobial Assays

The in vitro minimum inhibitory concentrations (MICs, µg/mL) and the minimum bactericidal concentration (MBCs, µg/mL) were determined by using the broth microdilution method according to the Clinical and Laboratory Standards Institute (CLSI) guidelines [53]. The concentration of the six extracts and oenothein B under investigation was initially prepared at the highest value to formulate stock solutions. These stock solutions were subsequently diluted at a ratio of 1:10 using Cation-Adjusted Mueller Hinton Broth (Oxoid, Milan, Italy). This procedure resulted in a gradient of concentrations ranging from 2048 to 2 µg/mL within the wells, achieved by performing two-fold serial dilutions in the designated testing medium. The stock solutions were prepared using pure DMSO as the diluent. We used the following bacterial strains from the American Type Culture Collection (ATCC, Rockville, MD, USA) that were available as freeze-dried discs: *E. faecalis* (ATCC 29212), *S. aureus* (ATCC 25923, ATCC 29213, ATCC 43300), *E. coli* (ATCC 25922),

K. pneumoniae (ATCC 13883. ATCC 70063), *P. aeruginosa* (ATCC 27853), and some strains derived from clinical isolation: *E. faecalis* BS, *S. aureus* BS, *E. coli* BS (ESBL), *K. pneumoniae* BS, and *P. mirabilis* BS.

The clinical isolates were obtained from patients of the Department of Biomedical Science and Human Oncology intensive care unit, University of Bari, Italy. Established physiological and morphological methods (API systems: API 20S, API RapidStaph, API Rapid 20E) were employed for isolation and identification procedures, which were performed in the Hygiene Section of the Department. To maintain the integrity of microbial cultures and ensure their reproducibility, cryovials containing all bacterial strains were established in a suitable culture medium and subsequently stored at −80 °C. Pre-cultures for each bacterial strain were prepared using Mueller Hinton Broth (MHB) and incubated from 3 to 5 h at 37 °C. Following the CLSI M7-A9 guidelines, the turbidity of the bacterial cell suspension was adjusted to match the 0.5 McFarland Standard using a spectrophotometer (OD_{625} nm readings between 0.08 and 0.10). This standardized suspension was then diluted further (1:100) using MHB to achieve a concentration of 1–2 $\times$ 10^6 (colony-forming units) CFU/mL. Aliquots of 200 µL from the final inoculum were introduced into each well. Control wells containing only the inoculated broth were also prepared to monitor bacterial growth. The plates were incubated at 37 °C for 24 h, with the minimum inhibitory concentration (MIC) values determined as the lowest concentration of the compounds at which no visible bacterial growth was observed [53]. Each experiment was performed three times in duplicate. Levofloxacin (Sigma Aldrich, Milan, Italy) was used as the standard reference antibiotic for this study. Moreover, the antibacterial efficacy of DMSO alone (which served as a solvent for the studied extracts) was assessed; hence, solvent volumes below 5% concentration were utilized throughout the experiments and did not affect microbial growth [54]. To determine the MBC, bacterial cultures treated with the tested compounds at varying concentrations were incubated as described in the MIC assay. Following incubation, 10 µL of each well that showed no visible bacterial growth were plated onto fresh Mueller–Hinton agar plates. The plates were then incubated at 37 °C for 24 h. The MBC was defined as the lowest concentration of the compound at which no visible bacterial growth was observed on the agar plates, indicating that ≥99.9% of the initial bacterial population had been killed. All experiments were performed in triplicate to ensure reproducibility.

4.5. Statistical Analysis

A Pearson correlation analysis has been carried out to highlight possible relationships between compound concentrations and MIC values for the different bacterial strains. When correlation coefficients were found to be statistically significant, the interpretation of their absolute magnitude was reported according to the following cut-off (absolute value): $r > 0.9$ very strong correlation; $0.7 < r < 0.89$ strong correlation; $0.4 < r < 0.69$ moderate correlation; $0.1 < r < 0.39$ weak correlation, and $r < 0.09$ negligible correlation [55,56]. Pearson's correlation test was performed with the software Prism v. 6.01 (GraphPad Inc., Boston, MA, USA).

5. Conclusions

Bacterial prostatitis is caused by some Gram-positive and Gram-negative bacterial strains that have developed resistance to conventional antibacterial drugs. In this work, the antibacterial spectrum of different extracts from two *Epilobium* species, i.e., *E. angustifolium* and *E. parviflorum*, and their major bioactive compound, oenothein B, was investigated with outstanding results, thus confirming the traditional uses of these herbs in the treatment of infectious disorders of the genito-urinary tract and supporting their application in pharmaceutical and nutraceutical products. The results demonstrated significant antimicrobial activity for both the Gram-positive and Gram-negative species tested, with MIC values ranging from 4 to 512 μg/mL, often comparable to levofloxacin, the conventional antibacterial drug used for the assays as a reference. This work also demonstrated the crucial importance of the purification process to enhance the content of bioactive compounds, especially that of oenothein B, and thus to improve the antibacterial properties of the extracts. These results suggest the outstanding potential of both *Epilobium* species as therapeutic alternatives or complementary agents to conventional antibiotic treatments of prostatitis and other urinary tract infections. However, it is crucial to acknowledge that the current study has been conducted in vitro, and further in vivo studies are necessary to confirm the observed antimicrobial activity in relevant animal models of prostatitis. Further studies are also needed to assess the possible synergistic action with conventional antibacterial drugs and to develop formulations to boost the already promising antibacterial potential of the extracts.

Supplementary Materials: The following supporting information can be downloaded at: https://www.mdpi.com/article/10.3390/antibiotics14020117/s1, Table S1: elution gradients employed for HPLC-MS/MS method development. Figure S1. Total ion chromatogram and dynamic "multiple reaction monitoring" mode chromatograms of single compounds of the $RE2_{ang}$. Figure S2. Total ion chromatogram and dynamic "multiple reaction monitoring" mode chromatograms of single compounds of the $RE2_{parv}$. Figure S3. Total ion chromatogram and dynamic "multiple reaction monitoring" mode chromatograms of single compounds of the PE_{ang}. Figure S4. Total ion chromatogram and dynamic "multiple reaction monitoring" mode chromatograms of single compounds of the PE_{parv}.

Author Contributions: Conceptualization, A.R.; methodology, A.B., A.C., S.A. (Sabrina Arpini), S.B., E.S. and S.A. (Simone Angeloni); validation, L.P. and M.C.; investigation, A.B., A.C., A.R., D.P. and G.Z.; data curation, A.B. and A.R.; writing—original draft preparation, A.B., A.R., D.P. and E.S.; writing—review and editing, G.S., G.Z. and F.M.; visualization, G.C.; supervision, F.M. and G.C.; funding acquisition, G.S. and F.M. All authors have read and agreed to the published version of the manuscript.

Funding: This research received no external funding.

Institutional Review Board Statement: Not applicable.

Informed Consent Statement: Not applicable.

Data Availability Statement: All data are included in the manuscript.

Acknowledgments: Authors thank Indena S.p.A. for scientific collaboration and financial support.

Conflicts of Interest: Authors Sabrina Arpini, Stefania Bosisio and Luca Pagni were employed by the company Indena S.p.A. The remaining authors declare that the research was conducted in the absence of any commercial or financial relationships that could be construed as a potential conflict of interest.

References

1. Raven, P.H. Generic and sectional delimitation in *Onagraceae*, Tribe Epilobieae. In *Annals of the Missouri Botanical Garden*; Missouri Botanical Garden Press: St. Louis, MO, USA, 1976.
2. Granica, S.; Piwowarski, J.P.; Czerwińska, M.E.; Kiss, A.K. Phytochemistry, pharmacology and traditional uses of different *Epilobium* species (*Onagraceae*): A review. *J. Ethnopharmacol.* **2014**, *156*, 316–346. [CrossRef] [PubMed]
3. Stolarczyk, M.; Naruszewicz, M.; Kiss, A.K. Extracts from *Epilobium* sp. herbs induce apoptosis in human hormone-dependent prostate cancer cells by activating the mitochondrial pathway. *J. Pharm. Pharmacol.* **2013**, *65*, 1044–1054. [CrossRef] [PubMed]

4. Dacrema, M.; Sommella, E.; Santarcangelo, C.; Bruno, B.; Marano, M.G.; Insolia, V.; Saviano, N.; Campiglia, P.; Stornaiuolo, M.; Daglia, M. Metabolic profiling, in vitro bioaccessibility and in vivo bioavailability of a commercial bioactive *Epilobium angustifolium* L. extract. *Biomed. Pharmacother.* **2020**, *131*, 110670. [CrossRef] [PubMed]
5. Kiss, A.K.; Bazylko, A.; Filipek, A.; Granica, S.; Jaszewska, E.; Kiarszys, U.; Kosmider, A.; Piwowarski, J. Oenothein B's contribution to the anti-inflammatory and antioxidant activity of *Epilobium* sp. *Phytomedicine* **2011**, *18*, 557–560. [CrossRef]
6. Tóth, B.H.; Blazics, B.; Kéry, Á. Polyphenol composition and antioxidant capacity of *Epilobium* species. *J. Pharm. Biomed. Anal.* **2009**, *49*, 26–31. [CrossRef]
7. Lee, S.W.H.; Chan, E.M.C.; Lai, Y.K. The global burden of lower urinary tract symptoms suggestive of benign prostatic hyperplasia: A systematic review and meta-analysis. *Sci. Rep.* **2017**, *7*, 7984. [CrossRef]
8. Boyle, P.; Napalkov, P. Epidemiology of benign prostatic hyperplasia: Current perspectives. *Eur. Urol.* **1996**, *29*, 7–11. [CrossRef]
9. Al-Musawi, J.F. The Effect of *Epilobium parviflorum* in BPH, in Comparison to Tamsulosin and Prazosin. *J. Pharm. Negat. Result.* **2022**, *13*, 46–51. [CrossRef]
10. Izzo, A.A.; Hoon-Kim, S.; Radhakrishnan, R.; Williamson, E.M. A Critical Approach to Evaluating Clinical Efficacy, Adverse Events and Drug Interactions of Herbal Remedies. *Phytother. Res.* **2016**, *30*, 691–700. [CrossRef]
11. Vitalone, A.; Allkanjari, O. *Epilobium* spp: Pharmacology and Phytochemistry. *Phytother. Res.* **2018**, *32*, 1229–1240. [CrossRef]
12. Esposito, C.; Santarcangelo, C.; Masselli, R.; Buonomo, G.; Nicotra, G.; Insolia, V.; D'Avino, M.; Caruso, G.; Buonomo, A.R.; Sacchi, R.; et al. *Epilobium angustifolium* L. extract with high content in oenothein B on benign prostatic hyperplasia: A monocentric, randomized, double-blind, placebo-controlled clinical trial. *Biomed. Pharmacother.* **2021**, *138*, 111414. [CrossRef] [PubMed]
13. Karakaya, S.; Süntar, I.; Yakinci, O.F.; Sytar, O.; Ceribasi, S.; Dursunoglu, B.; Ozbek, H.; Guvenalp, Z. In vivo bioactivity assessment on *Epilobium* species: A particular focus on *Epilobium angustifolium* and its components on enzymes connected with the healing process. *J. Ethnopharmacol.* **2020**, *262*, 113207. [CrossRef] [PubMed]
14. EMA. Epilobii Herba. 2019. Available online: https://www.ema.europa.eu/en/medicines/herbal/epilobii-herba (accessed on 20 September 2019).
15. Ducrey, B.; Wolfender, J.L.; Marston, A.; Hostettmann, K. Analysis of flavonol glycosides of thirteen *Epilobium* species (*Onagraceae*) by LC-UV and thermospray LC-MS. *Phytochemistry* **1995**, *38*, 129–137. [CrossRef]
16. Steenkamp, V.; Gouws, M.C.; Gulumian, M.; Elgorashi, E.E.; Van Staden, J. Studies on antibacterial, anti-inflammatory and antioxidant activity of herbal remedies used in the treatment of benign prostatic hyperplasia and prostatitis. *J. Ethnopharmacol.* **2006**, *103*, 71–75. [CrossRef]
17. Xiong, S.; Liu, X.; Deng, W.; Zhou, Z.; Li, Y.; Tu, Y.; Chen, L.; Fu, B. Pharmacological interventions for bacterial prostatitis. *Front. Pharmacol.* **2020**, *11*, 504. [CrossRef]
18. Gill, B.C.; Shoskes, D.A. Bacterial prostatitis. *Curr. Opin. Infect. Dis.* **2016**, *29*, 86–91. [CrossRef]
19. Millán-Rodríguez, F.; Palou, J.; Bujons-Tur, A.; Musquera-Felip, M.; Sevilla-Cecilia, C.; Serrallach-Orejas, M.; Baez-Angles, C.; Villavicencio-Mavrich, H. Acute bacterial prostatitis: Two different sub-categories according to a previous manipulation of the lower urinary tract. *World J. Urol.* **2006**, *24*, 45–50. [CrossRef]
20. Schneider, H.; Ludwig, M.; Hossain, H.M.; Diemer, T.; Weidner, W. The 2001 Giessen Cohort Study on patients with prostatitis syndrome—An evaluation of inflammatory status and search for microorganisms 10 years after a first analysis. *Andrologia* **2003**, *35*, 258–262. [CrossRef]
21. Singh, S.; Singh, M.; Bains, L.; Sagar, T. *Candida* prostatitis: A rare entity. *Trop. Doct.* **2023**, *53*, 282–284. [CrossRef]
22. Wu, Y.; Jiang, H.; Tan, M.; Lu, X. Screening for chronic prostatitis pathogens using high-throughput next-generation sequencing. *The Prostate* **2020**, *80*, 577–587. [CrossRef]
23. Nickel, J.C.; Moon, T. Chronic bacterial prostatitis: An evolving clinical enigma. *Urology* **2005**, *66*, 2–8. [CrossRef] [PubMed]
24. Perletti, G.; Skerk, V.; Magri, V.; Markotic, A.; Mazzoli, S.; Parnham, M.J.; Wagenlehner, F.M.E.; Naber, K.G. Macrolides for the treatment of chronic bacterial prostatitis: An effective application of their unique pharmacokinetic and pharmacodynamic profile. *Mol. Med. Rep.* **2011**, *4*, 1035–1044. [CrossRef] [PubMed]
25. Marquez-Algaba, E.; Burgos, J.; Almirante, B. Pharmacotherapeutic interventions for the treatment of bacterial prostatitis. *Expert Opin. Pharmacother.* **2022**, *23*, 1091–1101. [CrossRef]
26. Matsumoto, M.; Hamasuna, R.; Wada, K.M.; Sadahira, T.; Shigemura, K.; Maeda, K.; Hiyama, Y.; Togo, Y.; Nagasawa, S.; Yamanaka, K.; et al. Fluoroquinolone resistance and clinical characteristics of acute bacterial prostatitis in Japan: A multicenter study by the Japanese research group for urinary tract infection. *J. Infect. Chemother.* 2024; *in press*. [CrossRef]
27. Nowak, A.; Zagórska-Dziok, M.; Perużyńska, M.; Cybulska, K.; Kucharska, E.; Ossowicz-Rupniewska, P.; Piotrowska, K.; Duchnik, W.; Kucharski, L.; Sulikowshi, T.; et al. Assessment of the anti-inflammatory, antibacterial and anti-aging properties and possible use on the skin of hydrogels containing *Epilobium angustifolium* L. Extracts. *Front. pharmacol.* **2022**, *13*, 896706. [CrossRef]
28. Agnieszka, G.; Mariola, D.; Anna, P.; Piotr, K.; Natalia, W.; Aneta, S.; Marcin, O.; Bogna, O.; Zdzislaw, L.; Aurelia, P.; et al. Qualitative and quantitative analyses of bioactive compounds from ex vitro *Chamaenerion angustifolium* (L.) (*Epilobium augustifolium*) herb in different harvest times. *Ind. Crops Prod.* **2018**, *123*, 208–220. [CrossRef]

29. Baert, N.; Kim, J.; Karonen, M.; Salminen, J.P. Inter-population and inter-organ distribution of the main polyphenolic compounds of *Epilobium angustifolium*. *Phytochemistry* **2017**, *134*, 54–63. [CrossRef]
30. Ivanauskas, L.; Uminska, K.; Gudžinskas, Z.; Heinrich, M.; Georgiyants, V.; Kozurak, A.; Mykhailenko, O. Phenological Variations in the Content of Polyphenols and Triterpenoids in *Epilobium angustifolium* Herb Originating from Ukraine. *Plants* **2023**, *13*, 120. [CrossRef]
31. Stolarczyk, M.; Piwowarski, J.P.; Granica, S.; Stefańska, J.; Naruszewicz, M.; Kiss, A.K. Extracts from *Epilobium* sp. herbs, their components and gut microbiota metabolites of *Epilobium* ellagitannins, urolithins, inhibit hormone-dependent prostate cancer cells-(LNCaP) proliferation and PSA secretion. *Phytother. Res.* **2013**, *27*, 1842–1848. [CrossRef]
32. Mohammadi Bazargani, M.; Falahati-Anbaran, M.; Rohloff, J. Comparative analyses of phytochemical variation within and between congeneric species of willow herb, *Epilobium hirsutum* and *E. parviflorum*: Contribution of environmental factors. *Front. Plant Sci.* **2021**, *11*, 595190. [CrossRef]
33. Coker, T.J.; Dierfeldt, D.M. Acute Bacterial Prostatitis: Diagnosis and Management. *Am. Fam. Physician* **2016**, *93*, 114–120. [PubMed]
34. Trinchieri, A.; Abdelrahman, K.M.; Bhatti, K.H.; Bello, J.O.; Das, K.; Gatsev, O.; Gergova, I.; Magri, V.; Mourmoras, N.; Mourmouris, P.; et al. Spectrum of Causative Pathogens and Resistance Rates to Antibacterial Agents in Bacterial Prostatitis. *Diagnostics* **2021**, *11*, 1333. [CrossRef] [PubMed]
35. Barbarossa, A.; Ceramella, J.; Carocci, A.; Iacopetta, D.; Rosato, A.; Limongelli, F.; Carrieri, A.; Bonofiglio, D.; Sinicropi, M.S. Benzothiazole-Phthalimide Hybrids as Anti-Breast Cancer and Antimicrobial Agents. *Antibiotics* **2023**, *12*, 1651. [CrossRef] [PubMed]
36. Vlase, A.M.; Toiu, A.; Tomuță, I.; Vlase, L.; Muntean, D.; Casian, T.; Fizeșan, I.; Nadăș, G.C.; Novac, C.S.; Tămaș, M.; et al. *Epilobium* Species: From Optimization of the Extraction Process to Evaluation of Biological Properties. *Antioxidants* **2023**, *12*, 91. [CrossRef]
37. Dreger, M.; Adamczak, A.; Foksowicz-Flaczyk, J. Antibacterial and Antimycotic Activity of *Epilobium angustifolium* L. Extracts: A Review. *Pharmaceuticals* **2023**, *16*, 1419. [CrossRef]
38. Batten, J.; Cock, I.E. Antibacterial Activity and Toxicity Profiles of *Epilobium parviflorum* (Schreb.) Schreb. Extracts and Conventional Antibiotics against Bacterial Triggers of Some Autoimmune Diseases. *Phcog. Commn.* **2023**, *13*, 34–44. [CrossRef]
39. Ferrante, C.; Chiavaroli, A.; Angelini, P.; Venanzoni, R.; Angeles Flores, G.; Brunetti, L.; Petrucci, M.; Politi, M.; Menghini, L.; Leone, S.; et al. Phenolic Content and Antimicrobial and Anti-Inflammatory Effects of *Solidago virga-aurea*, *Phyllanthus niruri*, *Epilobium angustifolium*, *Peumus boldus*, and *Ononis spinosa* Extracts. *Antibiotics* **2020**, *9*, 783. [CrossRef]
40. Kowalik, K.; Polak-Berecka, M.; Prendecka-Wróbel, M.; Pigoń-Zając, D.; Niedźwiedź, I.; Szwajgier, D.; Baranowska-Wójcik, E.; Waśko, A. Biological Activity of an *Epilobium angustifolium* L. (Fireweed) Infusion after In Vitro Digestion. *Molecules* **2022**, *27*, 1006. [CrossRef]
41. Neumann, N.; Honke, M.; Povydysh, M.; Guenther, S.; Schulze, C. Evaluating Tannins and Flavonoids from Traditionally Used Medicinal Plants with Biofilm Inhibitory Effects against MRGN *E. coli*. *Molecules* **2022**, *27*, 2284. [CrossRef]
42. Șachir, E.E.; Pușcașu, C.G.; Caraiane, A.; Raftu, G.; Badea, F.C.; Mociu, M.; Albu, C.M.; Sachelarie, L.; Hurjui, L.L.; Bartok-Nicolae, C. Studies Regarding the Antibacterial Effect of Plant Extracts Obtained from *Epilobium parviflorum* Schreb. *Appl. Sci.* **2022**, *12*, 2751. [CrossRef]
43. Berlinck, R.G.S.; Hatano, T.; Okuda, T.; Yoshida, T.; Okuda, T.; Yoshida, T.; Hatano, T. Hydrolyzable tannins and related polyphenols. In *Fortschritte der Chemie organischer Naturstoffe/Progress in the Chemistry of Organic Natural Products*; Wien-Springer: New York, NY, USA, 1995.
44. Kolodziej, H.; Kayser, O.; Latte, K.P.; Kiderlen, A.F. Enhancement of antimicrobial activity of tannins and related compounds by immune modulatory effects. In *Plant Polyphenols 2*; Gross, G.G., Hemingway, R.W., Yoshida, T., Eds.; Kluwer Academic/Plenum Publishers: New York, NY, USA, 1999; pp. 575–594. [CrossRef]
45. Okuda, T.; Yoshida, T.; Hatano, T. Pharmacologically active tannins isolated from medicinal plants. In *Plant Polyphenols*; Hemingway, R.W., Lacks, P.E., Eds.; Plenum Press: New York, NY, USA, 1992; pp. 530–569. [CrossRef]
46. Yoshida, T.; Hatano, T.; Ito, H.; Okuda, T. Highly Oxidized Ellagitannins and Their Biological Activity. In *Plant Polyphenols 2*; Gross, G.G., Hemingway, R.W., Yoshida, T., Eds.; Kluwer/Plenum: New York, NY, USA, 1999; pp. 127–144.
47. Farha, A.K.; Yang, Q.-Q.; Kim, G.; Li, H.-B.; Zhu, F.; Liu, H.-Y.; Gan, R.-Y.; Corke, H. Tannins as an alternative to antibiotics. *Food Biosci.* **2020**, *38*, 100751. [CrossRef]
48. Manso, T.; Lores, M.; de Miguel, T. Antimicrobial Activity of Polyphenols and Natural Polyphenolic Extracts on Clinical Isolates. *Antibiotics* **2022**, *11*, 46. [CrossRef] [PubMed]
49. Pizzi, A. Tannins medical/pharmacological and related applications: A critical review. Sustain. *Chem. Pharm.* **2021**, *22*, 100481. [CrossRef]
50. Aderogba, M.A.; Ndhlala, A.R.; Rengasamy, K.R.R.; Van Staden, J. Antimicrobial and Selected In Vitro Enzyme Inhibitory Effects of Leaf Extracts, Flavonols and Indole Alkaloids Isolated from Croton menyharthii. *Molecules* **2013**, *18*, 12633–12644. [CrossRef]

51. Hatano, T.; Tsugawa, M.; Ohyabu, T.; Kusuda, M.; Shiota, S.; Tsuchiya, T.; Yoshida, T. Effects of polyphenols in tea and foods on methicillin-resistant *Staphylococcus aureus* and the sustainability of the antibacterial effects in the presence of food additives. *Assoc. J. Jpn. Soc. Med. Use Funct. Foods* **2006**, *4*, 43–48. [CrossRef]
52. European Medicines Agency, Committee on Herbal Medicinal Products (HMPC). *Assessment Report on Epilobium angustifolium L., and/or Epilobium parviflorum Schreb., Herba*; EMA/HMPC/712510/2014; European Medicines Agency, Committee on Herbal Medicinal Products (HMPC): Amsterdam, The Netherlands, 10 March 2015.
53. Barbarossa, A.; Rosato, A.; Carrieri, A.; Fumarola, L.; Tardugno, R.; Corbo, F.; Fracchiolla, G.; Carocci, A. Exploring the Antibiofilm Effect of Sertraline in Synergy with *Cinnamomum verum* Essential Oil to Counteract *Candida* Species. *Pharmaceuticals* **2024**, *17*, 1109. [CrossRef]
54. Barbarossa, A.; Sblano, S.; Rosato, A.; Carrieri, A.; Corbo, F.; Clodoveo, M.L.; Fracchiolla, G.; Carocci, A. Synergistic Action of *Cinnamomum verum* Essential Oil with Sertraline. *Antibiotics* **2022**, *11*, 1617. [CrossRef]
55. Barbarossa, A.; Rosato, A.; Carrieri, A.; Tardugno, R.; Corbo, F.; Clodoveo, M.L.; Fracchiolla, G.; Carocci, A. Antifungal Biofilm Inhibitory Effects of Combinations of Diclofenac and Essential Oils. *Antibiotics* **2023**, *12*, 1673. [CrossRef]
56. Schober, P.; Boer, C.; Schwarte, L.A. Correlation coefficients: Appropriate use and interpretation. *Anesth. Analg.* **2018**, *126*, 1763–1768. [CrossRef]

antibiotics

Article

Restoring Multidrug-Resistant *Escherichia coli* Sensitivity to Ampicillin in Combination with (−)-Epigallocatechin Gallate

Anong Kiddee [1,2], Atchariya Yosboonruang [1,2], Achiraya Siriphap [1,2], Grissana Pook-In [1,2], Chittakun Suwancharoen [1,2], Acharaporn Duangjai [2,3], Ratsada Praphasawat [2,4], Masami Suganuma [5] and Anchalee Rawangkan [1,2,*]

1 Division of Microbiology and Parasitology, School of Medical Sciences, University of Phayao, Phayao 56000, Thailand; anong.ki@up.ac.th (A.K.); atchariya.yo@up.ac.th (A.Y.); achiraya.si@up.ac.th (A.S.); grissana.po@up.ac.th (G.P.-I.); chittakun.su@up.ac.th (C.S.)
2 Unit of Excellence on Research and Application of Natural Products for Health and Well-Being, University of Phayao, Phayao 56000, Thailand; acharaporn.du@up.ac.th (A.D.); ratsada.pr@up.ac.th (R.P.)
3 Division of Physiology, School of Medical Sciences, University of Phayao, Phayao 56000, Thailand
4 Department of Pathology, School of Medicine, University of Phayao, Phayao 56000, Thailand
5 Graduate School of Science and Engineering, Saitama University, Saitama 338-8570, Japan; masami_suganuma@icloud.com
* Correspondence: ewmedsci@gmail.com or anchalee.up.ac.th; Tel.: +66-54-466-666 (ext. 3824) or +66-86-926-2448; Fax:+66-54-466-690

Abstract: Multidrug-resistant (MDR) bacteria, especially *Escherichia coli*, are a major contributor to healthcare-associated infections globally, posing significant treatment challenges. This study explores the efficacy of (−)-epigallocatechin gallate (EGCG), a natural constituent of green tea, in combination with ampicillin (AMP) to restore the effectiveness of AMP against 40 isolated MDR *E. coli* strains. Antimicrobial activity assays were conducted to determine the minimum inhibitory concentrations (MIC) of EGCG using the standard microdilution technique. Checkerboard assays were employed to assess the potential synergistic effects of EGCG combined with AMP. The pharmacodynamic effects of the combination were evaluated through time-kill assays. Outer membrane disruption was analyzed by measuring DNA and protein leakage and with assessments using N-phenyl-1-naphthylamine (NPN) and rhodamine 123 (Rh123) fluorescence dyes. Biofilm eradication studies involved biofilm formation assays and preformed biofilm biomass and viability assays. Scanning electron microscopy (SEM) was used to examine changes in cellular morphology. The results indicated that EGCG demonstrated activity against all isolates, with MICs ranging from 0.5 to 2 mg/mL, while AMP exhibited MIC values between 1.25 and 50 mg/mL. Importantly, the EGCG-AMP combination showed enhanced efficacy compared to either treatment alone, as indicated by a fractional inhibitory concentration index between 0.009 and 0.018. The most pronounced synergy was observed in 13 drug-resistant strains, where the MIC for EGCG dropped to 8 μg/mL (from 1 mg/mL alone) and that for AMP to 50 μg/mL (from 50 mg/mL alone), achieving a 125-fold and 1000-fold reduction, respectively. Time-kill assays revealed that the bactericidal effect of the EGCG-AMP combination occurred within 2 h. The mechanism of EGCG action includes the disruption of membrane permeability and biofilm eradication in a dose-dependent manner. SEM confirmed that the combination treatment consistently outperformed the individual treatments. This study underscores the potential of restoring AMP efficacy in combination with EGCG as a promising strategy for treating MDR *E. coli* infections.

Keywords: antimicrobial resistant; ampicillin; EGCG; *Escherichia coli*; synergistic

Citation: Kiddee, A.; Yosboonruang, A.; Siriphap, A.; Pook-In, G.; Suwancharoen, C.; Duangjai, A.; Praphasawat, R.; Suganuma, M.; Rawangkan, A. Restoring Multidrug-Resistant *Escherichia coli* Sensitivity to Ampicillin in Combination with (−)-Epigallocatechin Gallate. *Antibiotics* **2024**, *13*, 1211. https://doi.org/10.3390/antibiotics13121211

Academic Editors: Katarzyna Wińska, Joanna Kozłowska and Anna Duda-Madej

Received: 19 November 2024
Revised: 10 December 2024
Accepted: 11 December 2024
Published: 13 December 2024

1. Introduction

Multidrug-resistant (MDR) *Escherichia coli* has emerged as a significant threat to public health, along with MDR *Staphylococcus aureus*, *Klebsiella pneumoniae*, *Streptococcus pneumoniae*, *Acinetobacter baumannii*, and *Pseudomonas aeruginosa* [1]. The prevalence of MDR *E. coli* varies significantly across regions, with high resistance rates reported globally. In Europe, studies have shown that *E. coli* resistance to third-generation cephalosporins has reached concerning levels, with some reports indicating up to 40% of clinical isolates in certain countries exhibit resistance [2]. In the United States, the prevalence of carbapenem-resistant *E. coli* (CRE) has been rising, posing a serious challenge to treatment options [3]. Similarly, in parts of Asia and Africa, the spread of MDR *E. coli* is also a growing concern, with high rates of resistance observed in several countries [4]. The increasing prevalence of MDR *E. coli* is largely attributed to the overuse and misuse of antibiotics in both healthcare and agricultural settings, promoting the selection and spread of resistant strains [5]. In hospitals, MDR *E. coli* is a leading cause of healthcare-associated infections, often complicating the treatment of vulnerable patients [6]. Several new drugs, including ceftolozane/tazobactam, ceftazidime/avibactam, meropenem/vaborbactam, imipenem/cilastatin/relebactam, cefiderocol, plazomicin, eravacycline, and omadacycline, have recently been recommended for the treatment of Gram-negative multidrug-resistant bacterial infections [7,8]. However, ensuring their long-term effectiveness is crucial to delaying the emergence and spread of resistance to these novel agents. In response to the threat posed by MDR *E. coli*, novel therapeutic strategies are being explored, including the development of new antibiotics, and restoring sensitivity to existing drugs in combination therapies.

Ampicillin is a broad-spectrum beta-lactam antibiotic that has been widely used for a long time. The mechanisms of action of ampicillin are well known and involve interference with cell wall synthesis through attachment to penicillin-binding proteins (PBPs), inhibition of peptidoglycan synthesis, and inactivation of inhibitors of autolytic enzymes. Despite its clinical utility, the rising prevalence of antibiotic resistance has significantly reduced its effectiveness, particularly against Enterobacteriaceae like *E. coli*. Consequently, treatment with ampicillin alone is no longer recommended [9]. Sensitivity to ampicillin has been restored in the treatment of *Enterobacteriaceae*, including *E. coli*, through combination therapy with cloxacillin [10,11]. Moreover, combining ampicillin with ceftriaxone or azithromycin has demonstrated effectiveness in treating *Enterococcus faecalis* and *S. pneumoniae*, respectively [12,13]. However, attention should be given to the potential adverse effects, such as renal failure, although this is rare [14]. Additionally, these organisms are capable of preemptively adapting to antimicrobial resistance, leading to therapeutic failures. Consequently, combining ampicillin with natural products may provide viable alternative treatments for MDR *E. coli*.

Herbal medications are increasingly regarded as superior alternatives for addressing current and emerging antimicrobial-resistant bacteria and are expected to play a significant role in protecting humans against infections. Green tea (*Camellia sinensis*) is a rich natural source of polyphenols, including phenolic acids, such as caffeic acid and gallic acid, as well as flavonoids. Among these, catechins—a class of flavonoids containing flavan-3-ol units and galloylated catechins—are particularly prominent. Structurally, green tea catechins are characterized by a benzopyran framework with at least one aromatic ring [15]. Extensive research has highlighted the diverse health benefits of green tea consumption, including its antimicrobial properties against various pathogens. A 120 mL serving of green tea infusion contains approximately 150 mg of catechins, comprising 10–15% (−)-epigallocatechin gallate (EGCG), 6–10% (−)-epigallocatechin (EGC), 2–3% (−)-epicatechin gallate (ECG), and 2% (−)-epicatechin (EC) [16]. Among these catechins, EGCG, the most abundant and biologically active catechin in green tea, has gained attention in recent years not only for its antioxidant and anti-inflammatory properties but also for its potential to enhance the efficacy of anticancer, antiviral, and antibiotic therapies [16–19]. Upon consumption, these compounds and their metabolites are distributed throughout the body, aiding in both the treatment and prevention of infections. Green tea catechins exhibit antibacterial activity against both Gram-positive and Gram-negative bacteria through several mechanisms, including disruption of cell wall and membrane synthesis, inhibition of protein and nucleic acid synthesis, and interference with metabolic pathways involved in toxin production, extracellular matrix virulence factors, oxidative stress, and iron chelation [20]. Consequently, combining ampicillin with EGCG may offer a promising strategy to combat MDR *E. coli* antibiotic resistance and restore the efficacy of the conventional antibiotic treatment.

In this study, we investigated the efficacy of green tea EGCG against MDR *E. coli* strains, its combination with ampicillin, and the mechanisms underlying their antibacterial activity.

2. Results

2.1. EGCG Inhibits MDR E. coli Strains

We first investigated the susceptibility of 40 MDR *E. coli* strains to EGCG and ampicillin. Table 1 shows that all 40 isolates were sensitized to EGCG, with minimum inhibitory concentrations (MIC) ranging from 0.002 to 2 mg/mL and minimum bactericidal concentrations (MBC) ranging from 0.004 to 4 mg/mL. Notably, treatment with 1 mg/mL EGCG was effective against 87.5% (35/40) of the strains. In contrast, the MDR *E. coli* strains were resistant to ampicillin, with MIC values ranging from 1.25 to 50 mg/mL and MBC values ranging from 2.5 to 100 mg/mL. Interestingly, MDR *E. coli* strain E48, which was resistant to 13 antibiotics, including ampicillin (AMP), amoxicillin (AML), cephalothin (KF), cefotaxime (CTX), trimethoprim-sulfamethoxazole (SXT), meropenem (MEM), imipenem (IMP), gentamicin (CN), ciprofloxacin (CIP), norfloxacin (NOR), amoxicillin/clavulanic acid (AMC), ampicillin/sulbactam (SAM), and tetracycline (TE), had a MIC of 1 mg/mL for EGCG, while its MIC for AMP was 50 mg/mL. This suggests that EGCG is more effective at inhibiting MDR *E. coli* strains than ampicillin.

Table 1. Susceptibility of 40 MDR *E. coli* strains to EGCG and ampicillin.

No	Isolates	EGCG (mg/mL)		AMP (mg/mL)	
		MIC	MBC	MIC	MBC
1	E1	1	2	1.25	2.5
2	E3	1	2	12.5	25
3	E5	1	2	5	10
4	E6	1	2	5	10
5	E8	1	2	5	10
6	E9	1	2	5	10
7	E10	1	2	3.125	6.25
8	E11	1	2	5	10
9	E14	1	2	5	10
10	E15	1	2	5	10
11	E16	1	2	5	10
12	E18	1	2	5	10
13	E19	1	2	5	10
14	E20	1	2	5	10
15	E21	1	2	1.25	2.5
16	E24	1	2	5	10
17	E25	1	2	2.5	5
18	E26	1	2	1.25	2.5
19	E27	1	2	2.5	5
20	E28	1	2	5	10
21	E29	1	2	2.5	5
22	E30	1	2	5	10
23	E31	0.002	0.004	5	10
24	E32	1	2	1.25	2.5
25	E34	1	2	2.5	5
26	E36	1	2	2.5	5
27	E37	1	2	2.5	5
28	E38	1	2	2.5	5
29	E39	1	2	2.5	5
30	E40	2	4	5	10
31	E41	1	2	12.5	25
32	E42	1	2	25	50

Table 1. *Cont.*

No	Isolates	EGCG (mg/mL)		AMP (mg/mL)	
		MIC	MBC	MIC	MBC
33	E47	2	4	5	10
34	E48	1	2	50	100
35	E49	1	2	5	10
36	E50	1	2	2.5	5
37	E52	1	2	1.25	2.5
38	E62	0.5	1	2.5	5
39	E65	1	2	5	10
40	E66	2	4	25	50
41	ATCC 25922	2	4	0.3125	0.625

Abbreviations: EGCG, (−)-epigallocatechin gallate; AMP, ampicillin; MIC, minimum inhibitory concentration; MBC, minimum bactericidal concentration.

2.2. EGCG-AMP Combination Treatment Improved Antimicrobial Effectiveness Against MDR *E. coli*

The synergistic effects of EGCG, in combination with ampicillin against 40 MDR *E. coli* strains, were next investigated. As expected, the combination treatment produced a greater synergistic response compared to treatment with EGCG or ampicillin alone, as indicated by a fractional inhibitory concentration index (FICI) ranging from 0.009 to 0.018. This synergistic effect was observed in 92.5% (37/40) of the strains, as shown in Table 2. The combination treatment reduced the required EGCG dose by up to 125-fold and the ampicillin dose by up to 1250-fold. The most pronounced synergy was observed in the 13 drug-resistant strain, E48 (No. 34), where the MIC for EGCG dropped to 8 μg/mL (from 1 mg/mL alone) and the MIC for ampicillin decreased to 50 μg/mL (from 50 mg/mL alone), achieving 125-fold and 1000-fold reductions, respectively. These findings suggest that the EGCG-AMP combination could be a more effective treatment for MDR *E. coli* than either EGCG or ampicillin alone.

Table 2. Synergistic effect of EGCG in combination with ampicillin against 40 isolates of MDR *E. coli*.

No.	Isolates	Alone MIC (mg/mL)		Combination MIC (mg/mL)		FIC [a]	FIC [b]	FICI	Outcome	Dose Reduction (Fold)	
		EGCG [a]	AMP [b]	EGCG [a]	AMP [b]					EGCG	AMP
1	E1	1	1.25	0.5	0.001	0.5	0.001	0.501	Additive	2	1250
2	E3	1	1.25	0.008	0.012	0.008	0.010	0.018	Synergy	125	104.2
3	E5	1	5	0.008	0.005	0.008	0.001	0.009	Synergy	125	1000
4	E6	1	5	0.008	0.005	0.008	0.001	0.009	Synergy	125	1000
5	E8	1	5	0.008	0.005	0.008	0.001	0.009	Synergy	125	1000
6	E9	1	5	0.008	0.005	0.008	0.001	0.009	Synergy	125	1000
7	E10	1	5	0.008	0.005	0.008	0.001	0.009	Synergy	125	1000
8	E11	1	5	0.008	0.005	0.008	0.001	0.009	Synergy	125	1000
9	E14	1	5	0.008	0.005	0.008	0.001	0.009	Synergy	125	1000
10	E15	1	5	0.008	0.005	0.008	0.001	0.009	Synergy	125	1000
11	E16	1	5	0.008	0.005	0.008	0.001	0.009	Synergy	125	1000
12	E18	1	5	0.008	0.005	0.008	0.001	0.009	Synergy	125	1000
13	E19	1	5	0.008	0.005	0.008	0.001	0.009	Synergy	125	1000
14	E20	1	5	0.008	0.005	0.008	0.001	0.009	Synergy	125	1000
15	E21	1	1.25	0.008	0.001	0.008	0.001	0.009	Synergy	125	1250
16	E24	1	5	0.008	0.005	0.008	0.001	0.009	Synergy	125	1000
17	E25	1	2.5	0.008	0.002	0.008	0.001	0.009	Synergy	125	1250
18	E26	1	1.25	0.008	0.001	0.008	0.001	0.009	Synergy	125	1250
19	E 27	1	2.5	0.008	0.002	0.008	0.001	0.009	Synergy	125	1250
20	E 28	1	5	0.008	0.005	0.008	0.001	0.009	Synergy	125	1000
21	E29	1	2.5	0.008	0.002	0.008	0.001	0.009	Synergy	125	1250
22	E30	1	5	0.008	0.005	0.008	0.001	0.009	Synergy	125	1000
23	E31	0.002	5	1.484	0.001	742.19	0.000	742.19	Antagonism	0	5000
24	E32	1	1.25	0.008	0.001	0.008	0.001	0.009	Synergy	125	1250
25	E34	1	2.5	0.008	0.002	0.008	0.001	0.009	Synergy	125	1250
26	E36	1	5	0.008	0.005	0.008	0.001	0.009	Synergy	125	1000
27	E37	1	2.5	0.008	0.002	0.008	0.001	0.009	Synergy	125	1250
28	E38	1	2.5	0.008	0.002	0.008	0.001	0.009	Synergy	125	1250
29	E39	1	2.5	0.008	0.002	0.008	0.001	0.009	Synergy	125	1250
30	E40	2	5	0.016	0.005	0.008	0.001	0.009	Synergy	125	1000

Table 2. *Cont.*

No.	Isolates	Alone MIC (mg/mL)		Combination MIC (mg/mL)		FIC [a]	FIC [b]	FICI	Outcome	Dose Reduction (Fold)	
		EGCG [a]	AMP [b]	EGCG [a]	AMP [b]					EGCG	AMP
31	E41	1	12.5	0.008	0.012	0.008	0.001	0.009	Synergy	125	1041.7
32	E42	1	25	0.008	0.024	0.008	0.001	0.009	Synergy	125	1041.7
33	E47	2	5	2.000	5.0	1.000	1.000	2.000	Additive	1	1
34	E48	1	50	0.008	0.05	0.008	0.001	0.009	Synergy	125	1000
35	E49	1	5	0.008	0.005	0.008	0.001	0.009	Synergy	125	1000
36	E50	1	2.5	0.008	0.002	0.008	0.001	0.009	Synergy	125	1250
37	E52	1	1.25	0.008	0.001	0.008	0.001	0.009	Synergy	125	1250
38	E62	0.5	2.5	0.008	0.002	0.016	0.001	0.017	Synergy	62.5	1250
39	E65	1	5	0.008	0.005	0.008	0.001	0.009	Synergy	125	1000
40	E66	2	25	0.016	0.024	0.008	0.001	0.009	Synergy	125	1041.7
41	ATCC	2	0.3125	0.016	0.0003	0.008	0.001	0.009	Synergy	125	1041.7

Abbreviations: EGCG, (−)-epigallocatechin gallate; AMP, ampicillin; MIC, minimum inhibitory concentration; FIC, fractional inhibitory concentration; FICI, fractional inhibitory concentration index.

2.3. Analysis of the Bactericidal Kinetics of EGCG-AMP Combination Treatment

To explore the pharmacological activity of the EGCG-AMP combination treatment, we investigated the bactericidal effects on the kinetic growth curves of the representative MDR *E. coli* strain E48 under treatment with EGCG or ampicillin alone, or the EGCG-AMP combination. Figure 1 shows that treatment with 1 mg/mL of EGCG suppressed bacterial growth, exhibiting bactericidal activity within 18 h, while the control group displayed robust exponential growth, increasing to approximately 12 log units within 24 h. Interestingly, the EGCG-AMP combination treatment demonstrated bactericidal activity within 2 h, whereas ampicillin alone showed this effect within 4 h, indicating a 2-h reduction in the time to achieve bactericidal activity. These results clearly demonstrate that the combination of EGCG and ampicillin is bactericidal against MDR *E. coli*.

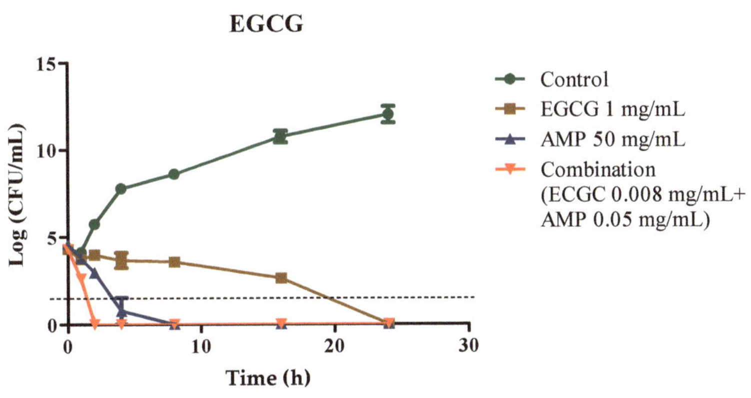

Figure 1. Effect of EGCG-AMP combination on MDR *E. coli* E48 strain. Time-kill kinetics for EGCG, ampicillin (AMP), and their combination were analyzed. Bacterial samples were taken at 1, 2, 4, 8, 16, and 24 h to measure viable cell counts. Dashed bars indicate the bactericidal threshold.

It is important to note that faster bactericidal kinetics are crucial in clinical settings to ensure rapid bacterial clearance, reduce infection duration, and minimize the likelihood of further resistance development.

2.4. EGCG-AMP Combination Treatment Disrupts the Membrane Permeability of MDR E. coli

Disruption of the cell membrane was next investigated by examining the leakage of nucleotides and proteins from bacterial cells following treatment with EGCG or ampicillin alone, or with their combination. The EGCG-AMP combination disrupts the membrane permeability of the MDR *E. coli* strain E48, as shown in Figure 2. After 1 h of treatment with various concentrations of EGCG (1, 2, and 4 mg/mL), DNA and protein leakage occurred in a dose-dependent manner. The EGCG-AMP combination treatment resulted in a higher rate of DNA release from the cells compared to treatments with either EGCG or ampicillin alone, demonstrating a strong effect on membrane permeability akin to that of Triton X-100 (Figure 2a). Similar effects were observed for protein leakage (Figure 2b). Additionally, the EGCG-AMP combination altered membrane permeabilization by increasing the relative fluorescence intensity (RFI) of NPN and decreasing the RFI of Rh123 more than treatments with EGCG or ampicillin alone (Figure 2c,d). Notably, EGCG exhibited a dose-dependent effect on outer membrane permeabilization. These results suggest that the EGCG-AMP combination modifies membrane potential and enhances membrane permeability, leading to the leakage of intracellular contents and subsequent cell death.

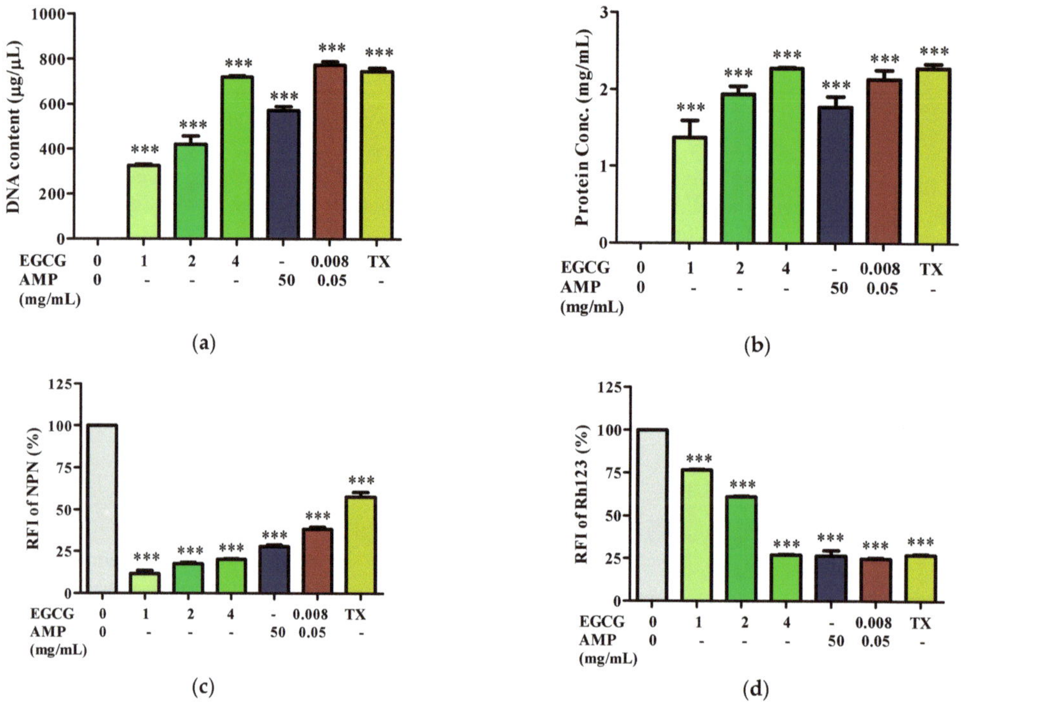

Figure 2. Effect of EGCG-AMP combination on membrane permeability. The *E. coli* E48 strain was exposed to EGCG alone (at 1×, 2×, and 4× MIC), AMP alone, or the EGCG-AMP combination for 1 h at 37 °C. DNA (**a**) and protein (**b**) levels were measured. The relative fluorescence intensity (RFI) of NPN (**c**) and Rh123 (**d**) was also assessed. Triton X-100 (0.1%) served as the positive control (TX). Significant differences compared to untreated controls are indicated by asterisks (*** $p < 0.001$).

It is important to note that this disruption is a critical mechanism, as compromised membranes facilitate the entry of antibiotics like ampicillin, enhancing their bactericidal action. This finding underscores the potential of EGCG to act as a permeabilizer, aiding the antibiotic in bypassing traditional resistance barriers.

2.5. EGCG-AMP Combination Treatment Overcomes the Biofilm Recalcitrance of MDR E. coli

Biofilms are a key mechanism through which *E. coli* develops drug resistance. Therefore, the impact of the EGCG-AMP combination was explored on the disruption or elimination of biofilms. MDR *E. coli* E48 cells were subjected to the previously described treatment conditions. Figure 3a illustrates that EGCG effectively inhibited biofilm formation in a dose-dependent manner, achieving biomass inhibition rates of 75.7 ± 8.7%, 45.9 ± 12.4%, and

42.5 ± 11.9% at concentrations of 1, 2, and 4 mg/mL, respectively, compared to untreated cells. In contrast, treatment with 50 mg/mL of ampicillin resulted in only an 18.4 ± 4.3% reduction in biofilm biomass. Notably, the EGCG-AMP combination nearly eliminated biofilm biomass, resulting in a significant 93.3% reduction to 6.7 ± 1.5%.

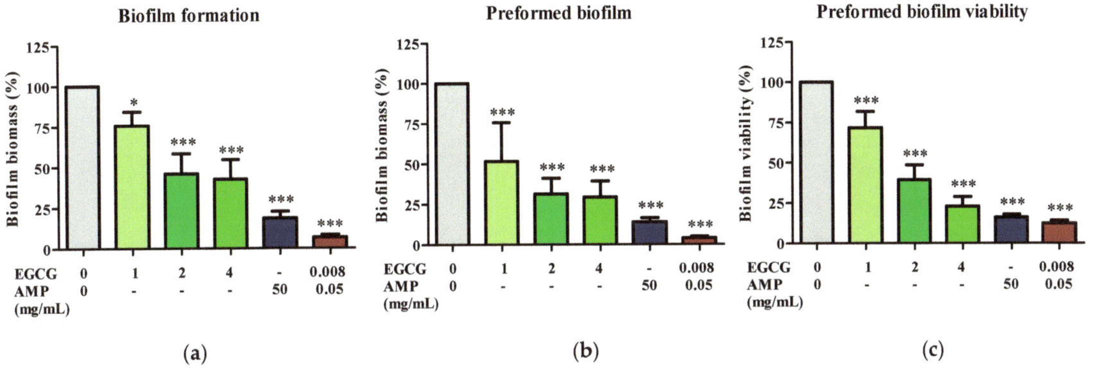

Figure 3. Effect of EGCG-AMP combination on biofilm formation. The *E. coli* E48 strain was exposed to EGCG alone (at 1×, 2×, and 4× MIC), AMP alone, or the EGCG-AMP combination for biofilm formation assays, as well as assays measuring preformed biofilm biomass and viability. Biofilm formation (**a**) and preformed biofilm (**b**) were evaluated using crystal violet staining. The viability of preformed biofilm (**c**) was assessed via the MTT assay. Significant differences compared to untreated controls are indicated by asterisks (* $p < 0.05$, *** $p < 0.001$).

The impact of the EGCG-AMP combination was also examined on the elimination of biofilms. One-day-old biofilms were treated with EGCG alone, ampicillin alone, or the combination for 24 h. Biofilm biomass was assessed using crystal violet staining and biofilm viability was evaluated through the MTT assay. Treatment with EGCG at concentrations of 1, 2, and 4 mg/mL resulted in decreases in biofilm biomass by 51.6 ± 24.3%, 31.5 ± 9.6%, and 29.4 ± 9.8%, respectively, compared to untreated biofilms (Figure 3b). The metabolic activity of the biofilms indicated that EGCG reduced the viability of MDR *E. coli* biofilm cells in a dose-dependent manner, with reductions of 71.38 ± 10.4%, 39.2 ± 9.3%, and 23.0 ± 5.9% at concentrations of 1, 2, and 4 mg/mL, respectively (Figure 3c). The EGCG-AMP combination was more effective at inhibiting both preformed biofilm biomass (3.92 ± 1.0% reduction) and biofilm viability (12.5 ± 1.8% reduction) than either treatment alone. These findings indicate that the EGCG-AMP combination can inhibit biofilm formation and eliminate existing biofilms of MDR *E. coli*, which is particularly relevant in treating chronic infections where biofilms play a critical role in persistence and resistance.

2.6. EGCG-AMP Combination Treatment Alters the Morphological Features of MDR E. coli *and Impairs Biofilm Formation*

To enhance our understanding of the effects of the EGCG-AMP combination treatment, scanning electron microscopy (SEM) was employed to examine the cellular morphology following a 4 h treatment (twice the bactericidal duration) with the individual agents and the EGCG-AMP combination. Figure 4 presents SEM images of *E. coli* E48 cells at 25,000× magnification. Untreated cells exhibited a smooth surface indicative of biofilm formation, characterized by an intact cell membrane and the absence of surface ruptures (Figure 4a). In contrast, treatment with EGCG alone resulted in minor membrane damage and a reduction in biofilm development (Figure 4b). Treatment with ampicillin alone led to significant membrane disruption and a substantial decrease in biofilm formation (Figure 4c). Notably, bacterial cells subjected to the EGCG-AMP combination exhibited pronounced membrane corrugations, cellular shrinkage, and rupture (Figure 4d). These findings indicate that the EGCG-AMP combination compromised cell membrane integrity and effectively eliminated biofilms, leading to morphological abnormalities that facilitated intracellular leakage, membrane retraction, and ultimately, cell death.

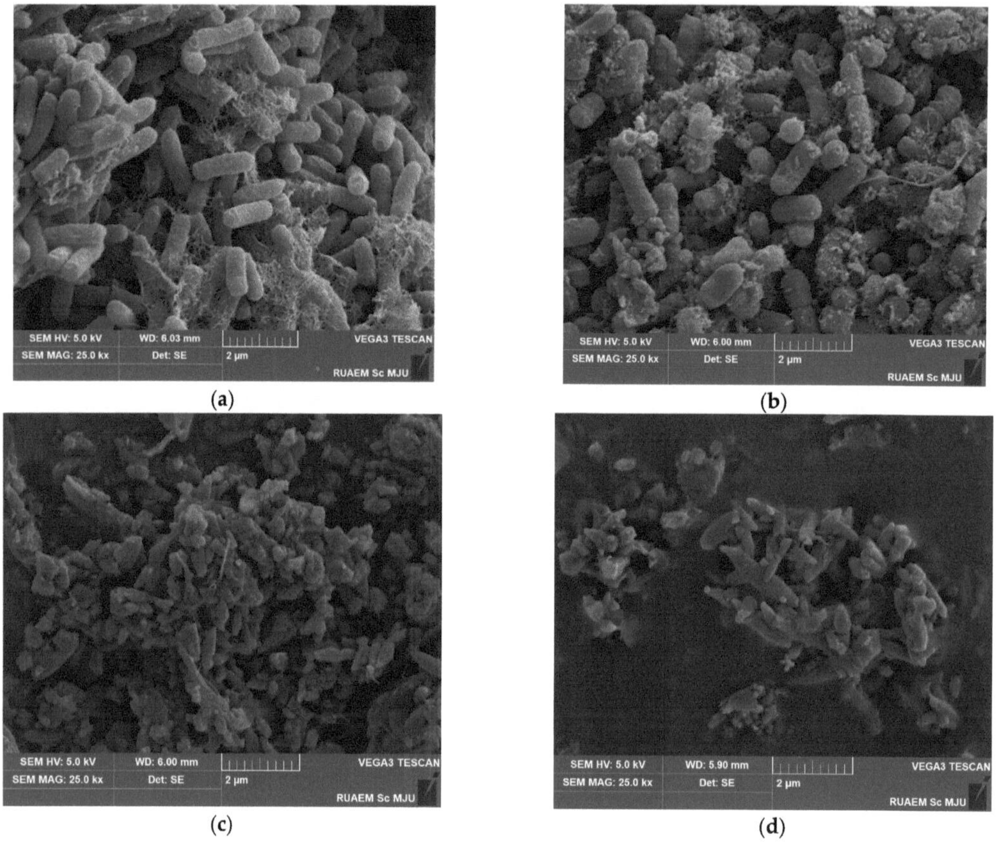

Figure 4. Effect of EGCG-AMP combination on bacterial cell morphology. The *E. coli* E48 strain was exposed for 4 h at 37 °C to either EGCG or AMP alone, and to their combination. Scanning electron microscopy (SEM) images at a magnification of 25,000× illustrate (**a**) the control; (**b**) EGCG at a concentration of 1 mg/mL; (**c**) AMP at a concentration of 50 mg/mL; and (**d**) treatment with the EGCG-AMP combination (EGCG 0.008 mg/mL + AMP 0.05 mg/mL).

It is important to note that morphological changes, like cell wall deformation and shrinkage, suggest that the bacteria were stressed or damaged by the treatment. Impaired biofilm formation indicates that the combination disrupts the bacterial community's ability to establish itself, which is crucial for preventing recurrent infections and biofilm-associated resistance.

3. Discussion

The WHO has highlighted MDR *E. coli* as a serious threat in efforts to combat antimicrobial resistance (AMR) [5]. This study suggests that combining EGCG with ampicillin could effectively overcome MDR *E. coli* infections by enhancing antibiotic efficacy, bypassing resistance mechanisms, and targeting resilient biofilms. This strategy exemplifies the restoration of antibiotic sensitivity, specifically in combination with natural compounds, as a novel approach against AMR.

Green tea is rich in catechins, a type of natural polyphenol, making up approximately 10–15% of its dry weight. The primary catechins in green tea include catechin, epicatechin, epigallocatechin, epicatechin gallate, and EGCG. Among these, EGCG is the predominant catechin, accounting for about 60% of the total, and has demonstrated the most potent antibacterial activity [21]. This aligns with our findings, which show that EGCG is effective in inhibiting all MDR *E. coli* strains. Furthermore, it is more effective than ampicillin, a drug that has largely fallen out of use. EGCG is well known for exhibiting multiple mechanisms of antibacterial action. It can disrupt bacterial cell membranes, increasing permeability and leading to the loss of essential cellular contents, which ultimately results in cell death [21]. Additionally, EGCG inhibits key bacterial enzymes, such as DNA gyrase and dihydrofolate reductase, which are crucial for DNA replication and cell survival [22]. EGCG also induces reactive oxygen species (ROS) formation by activating the Cpx system, leading to cell death [23]. Furthermore, EGCG induces oxidative stress within bacterial cells by generating ROS, which damages DNA, proteins, and lipids, leading to cell death [21,24,25]. EGCG also interferes with the formation of biofilms, protective barriers used by bacteria that evade antibiotics and immune responses, thereby reducing bacterial adhesion and biofilm integrity and rendering the bacteria more vulnerable to treatment [26,27]. Some studies suggest that EGCG may inhibit bacterial efflux pumps, mechanisms that bacteria use to expel antibiotics and other harmful compounds, thereby enhancing antibiotic retention and effectiveness within the cell [28,29].

This study found that the lowest EGCG concentration required to inhibit MDR *E. coli* was 2 μg/mL, as observed in strain E31. Conversely, the highest EGCG concentration needed for inhibition was 2 mg/mL, as observed in strains E40, E47, and E66. The most effective concentration across the tested strains was 1 mg/mL, which inhibited 35 out of 40 strains. Variations in susceptibility to EGCG among different MDR *E. coli* strains can be attributed to several factors, including differences in bacterial resistance mechanisms, such as efflux pumps, alterations in membrane permeability, or the presence of specific enzymes capable of degrading polyphenols like EGCG, as mentioned in the mechanisms above. Additionally, the genetic diversity of *E. coli* strains might play a significant role in their response to EGCG [30]. Strains exhibiting higher levels of intrinsic or acquired resistance mechanisms may require higher EGCG concentrations to achieve effective inhibition. Resistance to EGCG may also be linked to specific mutations in bacterial targets with which EGCG interacts, such as the bacterial cell wall or membrane-associated proteins. These mutations could result in reduced binding affinity for EGCG, thereby necessitating higher concentrations to achieve comparable inhibitory effects.

Notably, EGCG shows synergistic effects when used with certain antibiotics, enhancing their efficacy against drug-resistant bacteria, which is particularly useful against strains that have developed resistance mechanisms. Various studies indicate that combining EGCG with ampicillin enhances its effectiveness against methicillin-resistant *S. aureus* (MRSA) [31,32]. EGCG also demonstrates synergistic effects against *Vibrio cholerae* when used with tetracycline [21]. Additionally, combinations of EGCG with β-lactam antibiotics, such as carbapenem and meropenem, or carbenicillin, have been found to restore antibacterial activity against *A. baumannii* [33]. However, no synergistic effect had previously been reported between EGCG and ampicillin against *E. coli*. This study represents the first investigation of the combination of EGCG and ampicillin against MDR *E. coli*, demonstrating that the EGCG-AMP combination could be a more effective treatment for MDR *E. coli* than either EGCG or ampicillin alone.

A previous study reported that the combination of EGCG with gentamicin demonstrated a synergistic effect against MDR *E. coli*. Treatment with EGCG alone resulted in an MIC value of 1.25 mg/mL, while its combination with gentamicin reduced the MIC to 0.156 mg/mL, representing an 8-fold reduction. Similarly, gentamicin alone showed an MIC of 32 mg/mL, which decreased to 6.4 mg/mL when combined with EGCG, indicating a 5-fold reduction [34]. Therefore, in this study, the combination of EGCG with ampicillin may demonstrate even greater efficacy against MDR *E. coli* compared to gentamicin.

Although the MDR *E. coli* strain E31, which is resistant to six drugs (AMP, AML, KF, CTX, SXT, and SAM), exhibited good inhibitory activity with EGCG (2 μg/mL), the combination treatment of EGCG and ampicillin resulted in antagonism. This antagonism may be attributed to the strain's high sensitivity to both agents. EGCG is known to interact with various biological molecules due to its phenolic structure, which enables it to bind to proteins, enzymes, and even antibiotics. This interaction could prevent these antibiotics from reaching their target sites or alter their functional efficacy, thereby demonstrating an antagonistic effect [35]. Consequently, we chose to focus on strain E48, which exhibits resistance to 13 drugs and demonstrates higher MIC values for EGCG (1 mg/mL) and ampicillin (50 mg/mL). The combination of EGCG and ampicillin in strain E48 produced clearer and more consistent results, enabling a significant reduction in the required concentrations of both agents (125-fold for EGCG and 1000-fold for ampicillin). This substantial improvement highlights strain E48 as a more suitable candidate for further investigation, as it better reflects the potential efficacy of the combination therapy in addressing high-level multidrug resistance.

Since we found that EGCG modifies membrane potential and enhances membrane permeability, leading to the leakage of intracellular contents, structural damage, and subsequent cell death, this represents a critical mechanism. Compromised membranes facilitate the entry of ampicillin, enhancing its bactericidal action and restoring ampicillin sensitivity. Therefore, this study supports the conclusion that bactericidal EGCG damages the cell membrane, decreasing its protective function, including protection from external stress [36,37].

Biofilm formation is a critical mechanism contributing to drug resistance in *E. coli*. Biofilms are complex, multicellular communities of bacteria encased in an extracellular polymeric substance (EPS) that confers resistance to antibiotics and environmental stresses [38]. Numerous studies have demonstrated that EGCG exerts significant effects on biofilm formation and disruption by interfering with the assembly of amyloid fibers and the production of phosphoethanolamine-modified cellulose fibrils [27,30,39]. Consistent with these findings, our study showed that EGCG reduced the viability of MDR *E. coli* biofilm cells in a dose-dependent manner and significantly decreased biofilm biomass. Notably, the combination of EGCG with ampicillin enhanced biofilm eradication more effectively than either treatment alone. These findings highlight the potential of EGCG as an adjunctive agent in addressing biofilm-associated resistance, offering a promising strategy for preventing recurrent infections and combating multidrug-resistant *E. coli*.

Together, these findings highlight EGCG as a promising candidate for use in antimicrobial therapies, especially in combination with traditional antibiotics to combat multidrug-resistant bacteria. Specifically, the combination treatment with ampicillin compromised cell membrane integrity and effectively eliminated biofilms, leading to morphological abnormalities that facilitated intracellular leakage, membrane retraction, and ultimately, cell death. Further studies could explore optimizing the dosage and delivery mechanisms for the EGCG-AMP combination to maximize its therapeutic efficacy and safety. Additionally, investigating the effectiveness of this combination against other MDR pathogens may broaden its clinical applicability.

4. Materials and Methods

4.1. Green Tea EGCG and Bacterial Strains

Epigallocatechin gallate (EGCG), a green tea catechin with a purity greater than 99% as determined by high-performance liquid chromatography (HPLC), was purified from Japanese green tea leaves (*Camellia sinensis* (L.) Kuntze, Theaceae) and prepared by the Saitama Prefectural Tea Institute, Saitama, Japan, as described in the previous report [40].

Multidrug-resistant (MDR) *E. coli* strains were isolated from houseflies at the Phayao Hospital in Phayao Province (Thailand), as previously reported [41]. These strains were tested against 15 antibiotics: chloramphenicol (C), amikacin (AK), AMP, AML, KF, CTX, SXT, MEM, IMP, CN, CIP, NOR, AMC, SAM, and TE. The resistance profiles for these MDR *E. coli* strains are presented in Table S1. *E. coli* ATCC 25922 served as the reference strain.

4.2. Antimicrobial Activity Assay

The MIC of EGCG against MDR *E. coli* strains was assessed using the microdilution technique, following Clinical and Laboratory Standards Institute (CLSI) guidelines [42]. EGCG was serially diluted in Mueller-Hinton broth (HiMedia, Mumbai, Maharashtra, India) to obtain concentrations ranging from 1.95 μg/mL to 4.0 mg/mL. Bacterial cultures were prepared at a concentration of 5×10^5 CFU/mL and incubated at 37 °C for 24 h. The MIC was determined using resazurin (ALPHA CHEMIKA, Mumbai, Maharashtra, India), as an indicator of bacterial viability. A solution of resazurin (1 mg/mL) was added at 10 μL per well, and the plates were further incubated for 4 h to observe any color change. Wells where the resazurin color remained blue, indicating no metabolic activity, were recorded as having concentrations above the MIC value [43]. The MBC was established based on the absence of colony growth on Mueller-Hinton agar (HiMedia, Mumbai, Maharashtra, India) in a drop test, as outlined in previous studies. Ampicillin (PanReac AppliChem, Barcelona, Spain), which is no longer recommended by the Centers for Disease Control and Prevention as a first-line treatment for *E. coli* infections, was also included in the testing.

4.3. Checkerboard Assays

Checkerboard assays were performed to analyze the potential synergistic effects of EGCG in combination with ampicillin, following previous reports [44,45]. The MDR *E. coli* E48 strain, which exhibited antibiotic resistance to up to 13 drugs, was used as the representative strain. The fractional inhibitory concentration index (FICI) was calculated using the following formulae:

$$\text{FIC (a)} = \text{MIC of EGCG in the combination/MIC of EGCG alone;} \quad (1)$$

$$\text{FIC (b)} = \text{MIC of AMP in the combination/MIC of AMP alone;} \quad (2)$$

$$\text{FICI} = \text{FIC (a)} + \text{FIC (b)} \quad (3)$$

A synergistic effect is indicated by an FICI of <0.5, an additive effect by an FICI of 0.5–4, and an antagonistic effect by a FICI > 4 [46,47].

4.4. Time-Kill Kinetics Assay

The pharmacokinetic effects of EGCG, in combination with ampicillin, were evaluated through a time-kill assay, as described in previous studies [48,49]. Concentrations of 1× MIC for EGCG alone, ampicillin alone, and the EGCG-AMP combination were tested to assess bacterial cell growth at intervals of 0, 1, 2, 4, 8, 16, and 24 h. The kill curve indicated a bactericidal effect with a reduction of $\geq 3 \log_{10}$ CFU/mL compared to the initial inoculum (5×10^5 CFU/mL), whereas a decrease of $<3 \log_{10}$ CFU/mL indicated a bacteriostatic effect.

4.5. Outer Membrane Disruption Analysis

The MDR *E. coli* E48 strain was evaluated for DNA and protein leakage from the cell membrane. EGCG at concentrations of 1×, 2×, and 4× MIC, 1× MIC ampicillin alone, and the EGCG-AMP combination were administered for 1 h at 37 °C. The amount of DNA released from the cytoplasm was measured at 260 nm using a NANO-400A Micro Spectrophotometer (Hangzhou Allsheng Instruments Co., Ltd., Hangzhou, China) to determine DNA concentration. Protein content was determined using a Bio-Rad DC Protein Assay Kit (Bio-Rad Laboratories, Inc., Hercules, CA, USA). Subsequently, the extent of outer membrane disruption was assessed through cell staining with N-phenyl-1-naphthylamine (NPN) (TCI, Tokyo, Japan) and rhodamine 123 (Rh123) fluorescence dyes (Sigma Aldrich, St. Louis, MO, USA) [50,51]. The fluorescence intensity of NPN (measured at OD 350/420 nm) and Rh123 (measured at OD 480/530 nm) was calculated using the formula:

$$\text{Relative fluorescence intensity (\%)} = [F1/F0] \times 100,$$

where F0 represents the fluorescence intensity of untreated cells, and F1 denotes the fluorescence intensity of treated cells. A positive control consisting of 0.1% Triton X-100 (Sigma Aldrich, St. Louis, MO, USA) was also utilized.

4.6. *Biofilm Formation Assay*

The MDR *E. coli* E48 strain was inoculated at a concentration of 1×10^8 CFU/mL in a 96-well microtiter plate, with a final volume of 200 μL, containing either EGCG alone (at 1×, 2×, and 4× MIC), ampicillin alone, or the EGCG-AMP combination. After 24 h of incubation, non-adherent cells were removed by gently washing with PBS, followed by heat-drying at 60 °C. Biofilm formation was assessed by staining with 200 μL of 0.1% crystal violet (ALPHA CHEMIKA, Mumbai, Maharashtra, India), for 15 min, followed by two washes with PBS. The crystal violet was then dissolved in 200 μL of 95% ethanol for 20 min at room temperature, and the optical density was measured at 595 nm. The percentage of biofilm mass was calculated using the formula:

$$\text{Biofilm mass (\%)} = [\text{At}/\text{Ac}] \times 100,$$

where Ac represents the OD595 for the control wells and At represents the OD595 in the presence of EGCG [45].

4.7. *Preformed Biofilm Biomass and Viability Assay*

Biofilms of the MDR *E. coli* E48 strain were grown for 24 h. Next, the planktonic cells were discarded, and the biofilm was rinsed with PBS. The appropriate treatment was then applied at a final volume of 200 μL and incubated for an additional 24 h at 37 °C. Biofilm biomass was evaluated using 0.1% crystal violet staining. The viability of the preformed biofilm was determined by incubating with 5 mg/mL of 3-(4,5-dimethylthiazol-2-yl)-2,5-diphenyltetrazolium bromide (MTT) (Sigma Aldrich, St. Louis, MO, USA) for 30 min at 37 °C. After the staining solution was discarded, dimethyl sulfoxide (DMSO) (Sigma Aldrich, St. Louis, MO, USA) was added to the wells and incubated for 30 min, after which the optical density was measured at 570 nm [45,52].

4.8. *Scanning Electron Microscopy*

The MDR *E. coli* E48 strain at a concentration of 1×10^8 CFU/mL was treated with 1× MIC of both EGCG and ampicillin, as well as their combination, at 37 °C for 4 h. Following treatment, bacterial cells were collected by centrifugation at 3000× *g* for 5 min. The pellet was then resuspended in 1 mL of 2.5% glutaraldehyde (Sigma Aldrich, St. Louis, MO, USA) and incubated for 3 h at 4 °C. After washing with PBS, the bacterial cells underwent dehydration through a series of ethanol gradients (30%, 50%, 70%, and 90%) for 30 min each and were prepared for scanning electron microscopy (SEM; TESCAN, Vega III, Brno, Czech Republic), following established protocols [45].

4.9. *Statistical Analysis*

All experiments were conducted in at least triplicate. The data are expressed as mean ± standard deviation (SD). A one-way ANOVA followed by Dunnett's multiple comparison test was performed using GraphPad Prism 5.01 (GraphPad Software, Inc., La Jolla, CA, USA). Statistical significance was defined as $p < 0.05$.

5. Conclusions

These findings highlight the significance of natural compounds like EGCG in restoring the efficacy of existing antibiotics and addressing the growing challenge of antimicrobial resistance. EGCG demonstrated dose-dependent effectiveness against MDR *E. coli* by disrupting membrane permeability and eradicating biofilms. A concentration of 1 mg/mL of EGCG was effective against 87.5% of the tested strains. Furthermore, EGCG acts as a potent synergist when combined with ampicillin, significantly enhancing its activity against MDR *E. coli*. This combination reduced the required dose of ampicillin by up to

1250-fold and EGCG by 125-fold, depending on the strain, offering a promising strategy to reintroduce ampicillin as a valuable tool in combating antimicrobial resistance.

Future research should focus on optimizing the dosage and delivery mechanisms of the EGCG-AMP combination and evaluating its efficacy against other MDR pathogens to maximize its therapeutic potential. Conducting clinical trials to assess the safety and efficacy of this combination therapy in human populations will be critical for translating these findings into viable treatment options for resistant infections. Additionally, the source of green tea and the extraction method of EGCG may significantly influence its antimicrobial activity and the outcomes of biological assays. As purified analytical-grade EGCG is relatively expensive, future experiments, including in vivo studies or clinical trials, should explore the use of commercially available EGCG as a cost-effective alternative. These insights contribute to the development of novel, natural compound-based strategies to counteract the global threat of multidrug-resistant pathogens.

Supplementary Materials: The following supporting information can be downloaded at: https://www.mdpi.com/article/10.3390/antibiotics13121211/s1, Table S1: Antimicrobial resistance patterns toward 15 antimicrobial agents in *E. coli* isolates from house flies in the hospital area.

Author Contributions: Conceptualization, A.R. and A.K.; methodology, A.K., A.Y., A.S., G.P.-I., C.S., A.D., R.P., M.S. and A.R.; writing—original draft preparation, A.R. and A.K.; writing—review and editing, A.R.; project administration, A.R.; funding acquisition, A.R. and A.K. All authors have read and agreed to the published version of the manuscript.

Funding: This research was supported by the University of Phayao and Thailand Science Research and Innovation Fund (Fundamental Fund 2025, Grant no. 5031/2567 and 5136/2567) and the School of Medical Sciences, University of Phayao (Grant no. MS 241004), Thailand.

Institutional Review Board Statement: Not applicable.

Informed Consent Statement: Not applicable.

Data Availability Statement: The corresponding author will provide the data used in this study upon reasonable request.

Acknowledgments: We gratefully acknowledge Chatchaya Sumana from Chulalongkorn University and Surasak Kuimalee from the Science and Technology Service Center, Faculty of Science, Maejo University, Thailand, for their technical assistance.

Conflicts of Interest: The authors declare no conflicts of interest.

References

1. Antimicrobial Resistance Collaborators. Global burden of bacterial antimicrobial resistance in 2019: A systematic analysis. *Lancet* **2022**, *399*, 629–655. [CrossRef] [PubMed]
2. European Centre for Disease Prevention and Control & World Health Organization. *Antimicrobial Resistance Surveillance in Europe 2023—2021 Data*; European Centre for Disease Prevention and Control & World Health Organization: Stockholm, Sweden, 2023.
3. Duffy, N.; Li, R.; Czaja, C.A.; Johnston, H.; Janelle, S.J.; Jacob, J.T.; Smith, G.; Wilson, L.E.; Vaeth, E.; Lynfield, R.; et al. Trends in Incidence of Carbapenem-Resistant Enterobacterales in 7 US Sites, 2016–2020. *Open Forum Infect. Dis.* **2023**, *10*, ofad609. [CrossRef]
4. Ingle, D.J.; Levine, M.M.; Kotloff, K.L.; Holt, K.E.; Robins-Browne, R.M. Dynamics of antimicrobial resistance in intestinal *Escherichia coli* from children in community settings in South Asia and sub-Saharan Africa. *Nat. Microbiol.* **2018**, *3*, 1063–1073. [CrossRef] [PubMed]
5. Salam, M.A.; Al-Amin, M.Y.; Salam, M.T.; Pawar, J.S.; Akhter, N.; Rabaan, A.A.; Alqumber, M.A.A. Antimicrobial Resistance: A Growing Serious Threat for Global Public Health. *Healthcare* **2023**, *11*, 1946. [CrossRef] [PubMed]
6. Walker, M.M.; Roberts, J.A.; Rogers, B.A.; Harris, P.N.A.; Sime, F.B. Current and Emerging Treatment Options for Multidrug Resistant *Escherichia coli* Urosepsis: A Review. *Antibiotics* **2022**, *11*, 1821. [CrossRef] [PubMed]
7. Bassetti, M.; Peghin, M.; Vena, A.; Giacobbe, D.R. Treatment of Infections Due to MDR Gram-Negative Bacteria. *Front. Med.* **2019**, *6*, 74. [CrossRef]
8. Matlock, A.; Garcia, J.A.; Moussavi, K.; Long, B.; Liang, S.Y. Advances in novel antibiotics to treat multidrug-resistant gram-negative bacterial infections. *Intern. Emerg. Med.* **2021**, *16*, 2231–2241. [CrossRef]

9. Uddin, T.M.; Chakraborty, A.J.; Khusro, A.; Zidan, B.R.M.; Mitra, S.; Emran, T.B.; Dhama, K.; Ripon, M.K.H.; Gajdács, M.; Sahibzada, M.U.K.; et al. Antibiotic resistance in microbes: History, mechanisms, therapeutic strategies and future prospects. *J. Infect. Public Health* **2021**, *14*, 1750–1766. [CrossRef]
10. Kondampati, K.D.; Saini, S.P.S.; Sidhu, P.K.; Anand, A.; Kumar, D.; Beesam, S.; Bedi, J.S.; Kaur, R.; Bhardwaj, R. Pharmacokinetic-Pharmacodynamic Study of Ampicillin-Cloxacillin Combination in Indian Thoroughbred Horses (Equus caballus) and Safety Evaluation of the Computed Dosage Regimen. *J. Equine Vet. Sci.* **2022**, *115*, 104020. [CrossRef]
11. Sayer, B.; Bortone, B.; Sharland, M.; Hsia, Y. Fixed-dose combination antibiotics: The search for evidence using the example of ampicillin-cloxacillin. *Br. J. Clin. Pharmacol.* **2021**, *87*, 2996–2999. [CrossRef]
12. Marino, A.; Munafò, A.; Zagami, A.; Ceccarelli, M.; Campanella, E.; Cosentino, F.; Moscatt, V.; Cantarella, G.; Di Mauro, R.; Bernardini, R.; et al. Ampicillin plus ceftriaxone therapy against Enterococcus faecalis endocarditis: A case report, guidelines considerations, and literature review. *IDCases* **2022**, *28*, e01462. [CrossRef] [PubMed]
13. Majhi, A.; Kundu, K.; Adhikary, R.; Banerjee, M.; Mahanti, S.; Basu, A.; Bishayi, B. Combination therapy with ampicillin and azithromycin in an experimental pneumococcal pneumonia is bactericidal and effective in down regulating inflammation in mice. *J. Inflamm.* **2014**, *11*, 5. [CrossRef] [PubMed]
14. Peterson, S.C.; Lau, T.T.Y.; Ensom, M.H.H. Combination of Ceftriaxone and Ampicillin for the Treatment of Enterococcal Endocarditis: A Qualitative Systematic Review. *Ann. Pharmacother.* **2017**, *51*, 496–503. [CrossRef] [PubMed]
15. Zhao, T.; Li, C.; Wang, S.; Song, X. Green Tea (*Camellia sinensis*): A Review of Its Phytochemistry, Pharmacology, and Toxicology. *Molecules* **2022**, *27*, 3909. [CrossRef] [PubMed]
16. Fujiki, H.; Suganuma, M.; Imai, K.; Nakachi, K. Green tea: Cancer preventive beverage and/or drug. *Cancer Lett.* **2002**, *188*, 9–13. [CrossRef] [PubMed]
17. Suganuma, M.; Rawangkan, A.; Wongsirisin, P.; Kobayashi, N.; Matsuzaki, T.; Yoshikawa, H.Y.; Watanabe, T. Stiffening of Cancer Cell Membranes Is a Key Biophysical Mechanism of Primary and Tertiary Cancer Prevention with Green Tea Polyphenols. *Chem. Pharm. Bull.* **2020**, *68*, 1123–1130. [CrossRef] [PubMed]
18. Singh, B.N.; Shankar, S.; Srivastava, R.K. Green tea catechin, epigallocatechin-3-gallate (EGCG): Mechanisms, perspectives and clinical applications. *Biochem. Pharmacol.* **2011**, *82*, 1807–1821. [CrossRef]
19. Steinmann, J.; Buer, J.; Pietschmann, T.; Steinmann, E. Anti-infective properties of epigallocatechin-3-gallate (EGCG), a component of green tea. *Br. J. Pharmacol.* **2013**, *168*, 1059–1073. [CrossRef]
20. Renzetti, A.; Betts, J.W.; Fukumoto, K.; Rutherford, R.N. Antibacterial green tea catechins from a molecular perspective: Mechanisms of action and structure-activity relationships. *Food Funct.* **2020**, *11*, 9370–9396. [CrossRef]
21. Siriphap, A.; Kiddee, A.; Duangjai, A.; Yosboonruang, A.; Pook-In, G.; Saokaew, S.; Sutheinkul, O.; Rawangkan, A. Antimicrobial Activity of the Green Tea Polyphenol (−)-Epigallocatechin-3-Gallate (EGCG) against Clinical Isolates of Multidrug-Resistant Vibrio cholerae. *Antibiotics* **2022**, *11*, 518. [CrossRef]
22. Gradisar, H.; Pristovsek, P.; Plaper, A.; Jerala, R. Green tea catechins inhibit bacterial DNA gyrase by interaction with its ATP binding site. *J. Med. Chem.* **2007**, *50*, 264–271. [CrossRef] [PubMed]
23. Nie, T.; Zhang, C.; Huang, A.; Li, P. Epigallocatechin Gallate-Mediated Cell Death Is Triggered by Accumulation of Reactive Oxygen Species Induced via the Cpx Two-Component System in *Escherichia coli*. *Front. Microbiol.* **2018**, *9*, 246. [CrossRef] [PubMed]
24. Ouyang, J.; Zhu, K.; Liu, Z.; Huang, J. Prooxidant Effects of Epigallocatechin-3-Gallate in Health Benefits and Potential Adverse Effect. *Oxid. Med. Cell Longev.* **2020**, *2020*, 9723686. [CrossRef] [PubMed]
25. Li, R.; Lu, J.; Duan, H.; Yang, J.; Tang, C. Biofilm inhibition and mode of action of epigallocatechin gallate against Vibrio mimicus. *Food Control* **2020**, *113*, 107148. [CrossRef]
26. Shinde, S.; Lee, L.H.; Chu, T. Inhibition of Biofilm Formation by the Synergistic Action of EGCG-S and Antibiotics. *Antibiotics* **2021**, *10*, 102. [CrossRef]
27. Hengge, R. Targeting Bacterial Biofilms by the Green Tea Polyphenol EGCG. *Molecules* **2019**, *24*, 2403. [CrossRef]
28. Sudano Roccaro, A.; Blanco, A.R.; Giuliano, F.; Rusciano, D.; Enea, V. Epigallocatechin-gallate enhances the activity of tetracycline in staphylococci by inhibiting its efflux from bacterial cells. *Antimicrob. Agents Chemother.* **2004**, *48*, 1968–1973. [CrossRef]
29. Kanagaratnam, R.; Sheikh, R.; Alharbi, F.; Kwon, D.H. An efflux pump (MexAB-OprM) of Pseudomonas aeruginosa is associated with antibacterial activity of Epigallocatechin-3-gallate (EGCG). *Phytomedicine* **2017**, *36*, 194–200. [CrossRef]
30. Serra, D.O.; Mika, F.; Richter, A.M.; Hengge, R. The green tea polyphenol EGCG inhibits inhibits *E. coli* biofilm formation by impairing amyloid curli fibre assembly and downregulating the biofilm regulator CsgD via the σE-dependent sRNA RybB. *Mol. Microbiol.* **2016**, *101*, 136–151. [CrossRef]
31. Hu, Z.Q.; Zhao, W.H.; Hara, Y.; Shimamura, T. Epigallocatechin gallate synergy with ampicillin/sulbactam against 28 clinical isolates of methicillin-resistant Staphylococcus aureus. *J. Antimicrob. Chemother.* **2001**, *48*, 361–364. [CrossRef]
32. Zhao, W.H.; Hu, Z.Q.; Okubo, S.; Hara, Y.; Shimamura, T. Mechanism of synergy between epigallocatechin gallate and beta-lactams against methicillin-resistant Staphylococcus aureus. *Antimicrob. Agents Chemother.* **2001**, *45*, 1737–1742. [CrossRef] [PubMed]
33. Lee, S.; Razqan, G.S.; Kwon, D.H. Antibacterial activity of epigallocatechin-3-gallate (EGCG) and its synergism with β-lactam antibiotics sensitizing carbapenem-associated multidrug resistant clinical isolates of Acinetobacter baumannii. *Phytomedicine* **2017**, *24*, 49–55. [CrossRef] [PubMed]

34. Parvez, M.A.K.; Saha, K.; Rahman, J.; Munmun, R.A.; Rahman, M.A.; Dey, S.K.; Rahman, M.S.; Islam, S.; Shariare, M.H. Antibacterial activities of green tea crude extracts and synergistic effects of epigallocatechingallate (EGCG) with gentamicin against MDR pathogens. *Heliyon* **2019**, *5*, e02126. [CrossRef] [PubMed]
35. Hu, Z.Q.; Zhao, W.H.; Yoda, Y.; Asano, N.; Hara, Y.; Shimamura, T. Additive, indifferent and antagonistic effects in combinations of epigallocatechin gallate with 12 non-beta-lactam antibiotics against methicillin-resistant Staphylococcus aureus. *J. Antimicrob. Chemother.* **2002**, *50*, 1051–1054. [CrossRef]
36. Ikigai, H.; Nakae, T.; Hara, Y.; Shimamura, T. Bactericidal catechins damage the lipid bilayer. *Biochim. Biophys. Acta* **1993**, *1147*, 132–136. [CrossRef]
37. Kitichalermkiat, A.; Katsuki, M.; Sato, J.; Sonoda, T.; Masuda, Y.; Honjoh, K.-i.; Miyamoto, T. Effect of epigallocatechin gallate on gene expression of Staphylococcus aureus. *J. Glob. Antimicrob. Resist.* **2020**, *22*, 854–859. [CrossRef]
38. Flemming, H.C.; Wingender, J. The biofilm matrix. *Nat. Rev. Microbiol.* **2010**, *8*, 623–633. [CrossRef]
39. Stenvang, M.; Dueholm, M.S.; Vad, B.S.; Seviour, T.; Zeng, G.; Geifman-Shochat, S.; Søndergaard, M.T.; Christiansen, G.; Meyer, R.L.; Kjelleberg, S.; et al. Epigallocatechin Gallate Remodels Overexpressed Functional Amyloids in Pseudomonas aeruginosa and Increases Biofilm Susceptibility to Antibiotic Treatment. *J. Biol. Chem.* **2016**, *291*, 26540–26553. [CrossRef]
40. Hirota Fujiki, T.O. (−)-Epigallocatechin gallate. *Drugs Future* **1922**, *17*, 462–464.
41. Yosboonruang, A.; Kiddee, A.; Boonduang, C.; Pibalpakdee, P. Integron Expression in Multidrug-Resistant *Escherichia coli* Isolated from House Flies within the Hospital. *Walailak J. Sci. Technol.* **2018**, *16*, 319–327. [CrossRef]
42. Lewis Ii, J.S. *Performance Standards for Antimicrobial Susceptibility Testing*, 33rd ed.; Clinical and Laboratory Standards Institute: Wayne, PA, USA, 2023.
43. Elshikh, M.; Ahmed, S.; Funston, S.; Dunlop, P.; McGaw, M.; Marchant, R.; Banat, I.M. Resazurin-based 96-well plate microdilution method for the determination of minimum inhibitory concentration of biosurfactants. *Biotechnol. Lett.* **2016**, *38*, 1015–1019. [CrossRef] [PubMed]
44. Rawangkan, A.; Siriphap, A.; Yosboonruang, A.; Kiddee, A.; Pook-In, G.; Saokaew, S.; Sutheinkul, O.; Duangjai, A. Potential Antimicrobial Properties of Coffee Beans and Coffee By-Products Against Drug-Resistant Vibrio cholerae. *Front. Nutr.* **2022**, *9*, 865684. [CrossRef]
45. Rawangkan, A.; Yosboonruang, A.; Kiddee, A.; Siriphap, A.; Pook-In, G.; Praphasawat, R.; Saokaew, S.; Duangjai, A. Restoring Ampicillin Sensitivity in Multidrug-Resistant *Escherichia coli* Following Treatment in Combination with Coffee Pulp Extracts. *J. Microbiol. Biotechnol.* **2023**, *33*, 1179–1188. [CrossRef] [PubMed]
46. Odds, F.C. Synergy, antagonism, and what the chequerboard puts between them. *J. Antimicrob. Chemother.* **2003**, *52*, 1. [CrossRef]
47. Fatsis-Kavalopoulos, N.; Sánchez-Hevia, D.L.; Andersson, D.I. Beyond the FIC index: The extended information from fractional inhibitory concentrations (FICs). *J. Antimicrob. Chemother.* **2024**, *79*, 2394–2396. [CrossRef]
48. Betts, J.W.; Wareham, D.W. In vitro activity of curcumin in combination with epigallocatechin gallate (EGCG) versus multidrug-resistant Acinetobacter baumannii. *BMC Microbiol.* **2014**, *14*, 172. [CrossRef]
49. Petersen, P.J.; Jones, C.H.; Bradford, P.A. In vitro antibacterial activities of tigecycline and comparative agents by time-kill kinetic studies in fresh Mueller-Hinton broth. *Diagn. Microbiol. Infect. Dis.* **2007**, *59*, 347–349. [CrossRef]
50. Helander, I.M.; Mattila-Sandholm, T. Fluorometric assessment of gram-negative bacterial permeabilization. *J. Appl. Microbiol.* **2000**, *88*, 213–219. [CrossRef]
51. Wang, J.; Ma, W.; Wang, X. Insights into the structure of *Escherichia coli* outer membrane as the target for engineering microbial cell factories. *Microb. Cell Factories* **2021**, *20*, 73. [CrossRef]
52. Rawangkan, A.; Wongsirisin, P.; Pook-In, G.; Siriphap, A.; Yosboonruang, A.; Kiddee, A.; Chuerduangphui, J.; Reukngam, N.; Duangjai, A.; Saokaew, S.; et al. Dinactin: A New Antitumor Antibiotic with Cell Cycle Progression and Cancer Stemness Inhibiting Activities in Lung Cancer. *Antibiotics* **2022**, *11*, 1845. [CrossRef]

antibiotics

MDPI

Article

Phytochemical Study of Ethanol Extract of *Gnaphalium uliginosum* L. and Evaluation of Its Antimicrobial Activity

Lilia Davydova *, Angelina Menshova, Georgiy Shumatbaev, Vasily Babaev and Evgeny Nikitin

Arbuzov Institute of Organic and Physical Chemistry, FRC Kazan Scientific Center of RAS, Arbuzov Str. 8, 420088 Kazan, Russia; angelina_menshova11@mail.ru (A.M.); g-shumatbaev@mail.ru (G.S.); babaev@iopc.ru (V.B.); berkutru@mail.ru (E.N.)

* Correspondence: oleym73@inbox.ru; Tel.: +7-(937)-586-02-56

Abstract: This study evaluates the antibacterial and antifungal effects of ethanol extracts from *Gnaphalium uliginosum* L. derived from freshly harvested plant biomass, including stems, leaves, flowers, and roots. The extract was analyzed using gas chromatography-mass spectrometry (GC-MS) to determine its antimicrobial activity against phytopathogenic bacteria and fungi. Two methods were used in the experiments: agar well diffusion and double serial dilution. Extraction was carried out using the maceration method with different temperature regimes (25 °C, 45 °C, and 75 °C) and the ultrasonic method at various powers (63–352 W) for different durations (5 and 10 min). It was found that the 70% ethanol extract obtained through the ultrasonic experiment at 189 W power for 10 min and at 252 W power for 5 min had the highest antimicrobial activity compared to the maceration method. The most sensitive components of the extracts were the Gram-positive phytopathogenic bacteria *Clavibacter michiganensis* and the Gram-negative phytopathogenic bacteria *Erwinia carotovora* spp., with MIC values of 156 μg/mL. Among the fungi, the most sensitive were *Rhizoctonia solani* and *Alternaria solani* (MIC values in the range of 78–156 μg/mL). The evaluation of the antimicrobial activity of extracts using the diffusion method established the presence of a growth suppression zone in the case of *C. michiganensis* (15–17 mm for flowers, leaves, and total biomass), which corresponds to the average level of antimicrobial activity. These findings suggest that *G. uliginosum* has potential as a source of biologically active compounds for agricultural use, particularly for developing novel biopesticides.

Keywords: *Gnaphalium uliginosum* L.; antimicrobial activity; phytopathogens; ethanol extract; ultrasonic extraction; diffusion method; serial dilution method; GC-MS

Citation: Davydova, L.; Menshova, A.; Shumatbaev, G.; Babaev, V.; Nikitin, E. Phytochemical Study of Ethanol Extract of *Gnaphalium uliginosum* L. and Evaluation of Its Antimicrobial Activity. *Antibiotics* **2024**, *13*, 785. https://doi.org/10.3390/antibiotics13080785

Academic Editor: Marc Maresca

Received: 12 July 2024
Revised: 15 August 2024
Accepted: 15 August 2024
Published: 22 August 2024

1. Introduction

Plant diseases caused by pathogenic microorganisms represent a significant threat to the agriculture industry, leading to a reduction in both the quality and quantity of crops [1]. Bacterial diseases, such as those caused by strains of *Clavibacter michiganensis*, *Erwinia carotovora*, *Rhizoctonia solani*, and *Alternaria solani*, pose a significant economic risk, resulting in substantial losses in crop yields and quality. *Clavibacter michiganensis* is a bacterial pathogen that causes bacterial canker in tomato plants that can result in plant death and complete crop loss. *Erwinia carotovora* causes soft rot in various vegetables, such as potatoes, carrots, and cabbages, leading to significant yield losses and decreased product quality. *Rhizoctonia solani* causes root rot in a wide range of crops, as well as potatoes, tomatoes, and cucumbers. *Alternaria solani*, the pathogen responsible for early blight, affects potatoes, tomatoes, and eggplants, leading to reduced yields and poorer quality tubers and fruits. This pathogen negatively impacts the profitability of agriculture, increasing production costs and the risk of crop failure. Finding effective and environmentally friendly methods to control this disease is, therefore, a crucial and urgent priority.

Pesticides have been the traditional method used for controlling phytopathogens. However, it has been recently discovered that pesticides based on chemical compounds can pose a potential danger to the environment and ecosystem [2]. Soil biota, which includes microorganisms and soil fauna, plays a crucial role in maintaining soil health by regulating the nutrient cycle and contributing to soil biodiversity [3,4]. Metabolites from pesticides can negatively impact soil biota by altering gene expression and enzyme activity, leading to the inhibition of natural microorganism fertility, reducing growth, and affecting survival [5–7]. The widespread use of chemical pesticides has led to the development of resistance in pathogens, despite their high effectiveness in protecting plants from infection by phytopathogenic microorganisms [8]. Therefore, there is a need for new, low-toxicity pesticides that degrade rapidly and are environmentally friendly [9].

The discovery of natural biological agents derived from microorganisms and plants represents a safer and more environmentally friendly approach to controlling plant diseases [10,11]. According to Burtscher-Schaden et al., a comparison of pesticide active substances (AS) used in conventional and organic agriculture in Europe found that 55% of those used only in conventional farming pose risks to health or the environment, while only 3% are approved for organic farming. The search for active substances derived from plant materials and their use in creating biopesticides as alternatives or adjuncts to chemical pesticides is crucial for agricultural plant production [12]. In recent years, biologicals based on plant extracts have been actively introduced as a means to control bacterial and fungal infections in agricultural crops with antimicrobial activity comparable to that of chemical pesticides [13].

The Aster family (*Asteraceae*), one of the largest families of dicotyledonous plants, includes about 300 species worldwide, including twelve species in the Russian Federation and two in Tatarstan [14]. The genus *Gnaphalium*, with its species *Gnaphalium uliginosum*, is the most studied. This annual herb is used in folk medicine due to its anti-inflammatory, antibacterial, hemostatic, and other properties, such as for treating hypertension, duodenal ulcers, stomach ulcers, burns, etc. [15]. The chemical composition of *G. uliginosum* has been analyzed, revealing various biologically active substances such as flavonoids, alkaloids, tannins, polyphenols, coumarins, terpenoids, phytoestrogens, and carotenoids [16,17]. Extracts of some species exhibit antibacterial activity against phytopathogens and human pathogens, as shown in Table 1. Most of the plants studied were found in the wild and are abundant in Central and South America, accounting for more than half of all plant species. Research on plants typical of the flora in the Russian Federation is relatively rare.

Table 1. Characteristics of antimicrobial activity of different plant species of the genus *Gnaphalium* [18–22].

Plant	Geographical Region	Type of Raw Material	Diffusion Method, mm	Method of Serial Dilutions, μg/mL
Gnaphalium attenuatum DC.	Morelos, Mexico.	Dry-air raw materials, part Above the ground	MIC 8–21.5	MIC 50–200
Gnaphalium polycaulon	La Paz, Bolivia.	Dry-air raw materials, part	MIC 10–93	MIC 40–100
Gnaphalium uliginosum L.	Zelenodolsk district, Republic of Tatarstan, Russian Federation.	Above the ground	nd	MIC 62.5–1000
Gnaphalium polycaulon	Tamil Nadu, India.	Dry-air raw materials, part	MIC 15–33	nd
Gnaphalium oxyphyllum, *Gnaphalium americanum*	Morelos, Mexico.	Above the ground	nd	MIC 2–50
Gnaphalium uliginosum L.	Zelenodolsk district, Republic of Tatarstan, Russian Federation	Dry-air raw materials, leaves	nd	MIC 78–625

nd: not detected.

This research aimed to investigate the phytochemical composition of ethanol extract from *Gnaphalium uliginosum* L., using the GC-MS technique, and to determine its antimicrobial activity against phytopathogenic microorganisms that cause diseases in plant crops. This study aims to complement existing data on the components of *G. uliginosum* found in Tatarstan.

2. Results

2.1. GC-MS Analysis

This study was conducted using gas chromatography-mass spectrometry on a sample of the ethyl extract from *G. uliginosum*, allowing us to identify a number of major and minor components (as shown in Table 2). This significantly expanded and clarified the list of components compared to previous studies. A total of 19 individual compounds were identified using relative match factors (RMF) generated by NIST-17 software based on a mass spectral library.

Table 2. GC-MS identification and quantification of the components of *G. uliginosum* extract (components with $\omega > 0.04\%$, mass., n = 3, $p = 0.95$).

№ [1]	tR [2], min	Component	ω [3] (%)	Class of Compounds
1	3.199	Octadecane	1.946 ± 0.006	Acyclic hydrocarbon
2	5.617	2-Hexynoic acid	0.130 ± 0.002	Alkyl carboxylic acids
3	6.331	Bicyclo [2.2.1]heptan-2-ol, 1,7,7-trimethyl-, (1S-endo)	0.185 ± 0.002	Monoterpenoid
4	6.514	Dodecanoic acid	0.935 ± 0.003	
5	6.943	Eicosanoic acid, phenylmethyl ester	0.139 ± 0.002	Fatty acid ester
6	7.080	Folic Acid	0.171 ± 0.002	Vitamins
7	8.092	9-Octadecenoic acid, (2-phenyl-1,3-dioxolan-4-yl)methyl ester, trans	0.296 ± 0.003	Carboxylic acids ester
8	10.789	6-Hydroxy-4,4,7a-trimethyl-5,6,7,7a-tetrahydrobenzofuran-2(4H)-one	0.740 ± 0.003	Tetrahydrobenzofuran
9	10.984	3,7,11,15-Tetramethyl-2-hexadecen-1-ol	0.828 ± 0.003	Acyclic diterpene alcohol
10	11.607	n-Hexadecanoic acid	1.646 ± 0.004	Carboxylic acid
11	12.275	9,12-Octadecadienoic acid, methyl ester, (E,E)	1.060 ± 0.004	Carboxylic acid ester
12	12.578	9,12-Octadecadienoic acid (Z,Z)-	8.949 ± 0.012	Carboxylic acid
13	12.887	2-[4-methyl-6-(2,6,6-trimethylcyclohex-1-enyl)hexa-1,3,5-trienyl]cyclohex-1-en-1-carboxaldehyde	3.451 ± 0.008	Carboxaldehyde
14	14.567	3-(1,1-Dimethylethenyl)-7-hydroxy-6-methoxy-2H-1-benzopyran-2-one	3.227 ± 0.007	Coumarins
15	15.882	(9Z,12Z)-octadeca-9,12-dienoic acid 2-[(2-hydroxyethoxy)methyl] ester	2.356 ± 0.007	Fatty acid ester
16	25.043	Campesterol	1.979 ± 0.006	Phytosterol
17	25.924	Stigmasterol	5.965 ± 0.010	Phytosterol
18	27.644	Sitosterol	5.624 ± 0.010	Phytosterol
19	29.324	Cyclodecasiloxane, eicosamethyl	4.620 ± 0.009	Siloxane

[1] Peak number; [2] t_R—retention time; [3] ω—mass fraction of the component as area in % of 100.00% of all identified peaks.

According to the GC-MS results (Table 2), the ethanol extract of *G. uliginosum* L. contains a significant amount of phytosterols, including campesterol, stigmasterol, and sitosterol, which account for 13.56% of the total detected compounds. Among carboxylic acids, n-Hexadecanoic acid and 9,12-Octadecadienoic acid (Z, Z) dominate, accounting for 11.79%. Carboxylic acid esters account for 1.35%, including 9,12-Octadecadienoic acid, methyl ester, (E, E) and 9-Octadecenoic acid, (2-phenyl-1,3-dioxolan-4-yl)methyl ester, trans. The aromatic aldehyde is represented by 2-[4-methyl-6-(2,6,6-trimethylcyclohex-1-enyl)hexa-1,3,5-trienyl]cyclohex-1-en-1-carboxaldehyde at 3.45 (wt. % of extract). The total content of fatty acid esters is 2.49 wt %, including (9Z,12Z)-octadeca-9,12-dienoic acid 2-[(2-hydroxyethoxy)methyl] ester and Eicosanoic acid, phenylmethyl ester. In addition to the above compounds, the following were detected in the extract by GC-MS: coumarins (3-(1,1-Dimethylethenyl)-7-hydroxy-6-methoxy-2H-1-benzopyran-2-one) 3.23%; siloxane (Cyclodecasiloxane, eicosamethyl) 4.62%; alkane class hydrocarbon (Octadecane) 1.95%. The contents of diterpene alcohol (0.83%), vitamin B9 (0.17%), bicyclic monoterpenoids (0.18%), and benzofuran (0.74%) were found to be relatively low.

2.2. Antibacterial and Antifungal Activity

2.2.1. Serial Dilution Methods

To determine the antimicrobial potential of *G. uliginosum* growing in the Republic of Tatarstan against phytopathogenic microorganisms, the collected and frozen plants were extracted using maceration and ultrasound treatment methods. The effect of the solvent was controlled by concentrating all extracts in a 1% solution. The antimicrobial activity of water–ethanol extracts of *G. uliginosum* was tested against phytopathogenic bacteria causing bacterial cancer of tomato (*Clavibacter michiganensis*), the causative agent of potato blackleg, tomato, and wet rot of fruits in the plants of the *Solanaceae* family (*Erwinia carotovora* spp), as well as phytopathogenic fungi causing rhizoctoniosis (*Rhizoctonia solani* VKM F-895) and alternariasis (*Alternaria solani* K-100054) [23–25]. The results on the antimicrobial activity of the water–ethanol extracts of *G. uliginosum* obtained using the maceration method at temperatures of 25 °C, 45° C, and 75 °C are presented in Table 3.

Table 3. Indicators of the antimicrobial activity of water–ethanol extracts of *G. uliginosum* obtained by maceration under various temperature conditions.

Strains of Microorganisms	Maceration at t = 25 °C		Maceration at t = 45 °C		Maceration at t = 75 °C		Chloramphenicol * /Tebuconazole **	
	MIC, μg/mL	MFC, μg/mL	MIC, μg/mL	MFC, μg/mL	MIC, μg/mL	MFC, μg/mL	MIC, μg/mL	MBC/MFC, μg/mL
Clavibacter michiganensis	625 ± 40	>2500 ± 180	312 ± 20	>2500 ± 180	625 ± 40	>2500 ± 180	1 ± 0.3	1.9 ± 0.2
Erwinia carotovora spp.	1250 ± 80	1250 ± 80	625 ± 40	625 ± 40	312 ± 20	312 ± 20	250 ± 20.6	250 ± 19.5
Rhizoctonia solani	625 ± 40	2500 ± 160	625 ± 40	1250 ± 80	312 ± 20	1250 ± 80	31.25 ± 2.4	125 ± 10.3
Alternaria solani	>2500 ± 180	>2500 ± 180	>2500 ± 180	>2500 ± 180	>2500 ± 180	>2500 ± 180	15.62 ± 1.3	62.5 ± 5.6

* Reference drug for phytopathogenic bacteria; ** Reference drug for phytopathogenic fungi.

The antimicrobial activity of the extracts against the tested phytopathogens was 312 μg/mL or higher. The Gram-positive bacterium *C. michiganensis* was most sensitive to the water–ethanol extracts of *G. uliginosum*, obtained by maceration at 45 °C, with MIC values of 312 μg/mL. Extracts obtained at 25 °C and 75 °C were less active with MICs two times higher, up to 625 μg/mL. In relation to C. michiganensis, no MBC values were observed in any of the concentrations studied. Gram-negative bacteria *E. carotovora* and the fungus *R. solani* were more sensitive to *G. uliginosum* extracts obtained by macerating at 75 °C, with MIC 312 μg/mL. When extraction temperatures decreased, MICS increased to 625 μg/mL at 45 °C and 1250 μg/mL at 25 °C. MBC values for *E. carotovora* matched MIC, and for *R. solani* they were two to four times higher. Among the tested phytopathogens, A. solani was the most resistant to extracts of *G. uliginosum* obtained by maceration. The values of MIC and MBC exceeded 2500 μg/mL.

Currently, extraction by maceration is the simplest and most economical way to extract biologically active compounds from plant biomass. This method does not require special equipment. However, ultrasonic extraction can break down the cell walls of plant materials and make bioactive substances more accessible to the extractant (bacteria). Ultrasonic extraction significantly reduces processing time and ensures a more complete extraction of substances. The boundary diffusion layer is disrupted under the influence of ultrasonic waves, and the penetration of the extractant into the material is improved [26]. It is important to note that maceration at different temperatures can also influence the antimicrobial activity of the extract. Our study showed that maceration at higher temperatures (45 °C and 75 °C) demonstrated lower antibacterial activity than ultrasonic extraction. This may be due to the degradation of thermolabile bioactive components. At low temperatures (25 °C), the diffusion of bioactive compounds from plant cells into a solvent occurs more slowly, which can reduce antibacterial activity.

To determine the optimal conditions for extracting compounds with bacteriostatic and antifungal activity, ultrasound extraction was performed at varying power levels ranging from 63 W to 315 W with a gradual increase of 63 watts per step. Exposure times were 5 and 10 min. The results of evaluating the antimicrobial activity of *G. uliginosum* water–ethanol extracts using ultrasound are presented in Table 4.

Table 4. Indicators of antibacterial and antimycotic activity of the ethanol extract of Gnaphalium uliginosum obtained using the ultrasound method.

Power, Extraction Time	*Rhizoctonia solani*		*Alternaria solani*		*Clavibacter michiganensis*		*Erwinia carotovora* spp.	
	MIC, µg/mL	MFC, µg/mL	MIC, µg/mL	MFC, µg/mL	MIC, µg/mL	MBC, µg/mL	MIC, µg/mL	MBC, µg/mL
63 W, 5 min	625 ± 40	>2500 ± 180	>2500 ± 180	>2500 ± 180	>2500 ± 180	>2500 ± 180	1250 ± 80	2500
63 W, 10 min	625 ± 40	>2500 ± 180	1250 ± 70	>2500 ± 180	>2500 ± 180	>2500 ± 190	1250 ± 80	2500
126 W, 5 min	312 ± 30	312 ± 30	>2500 ± 180	>2500 ± 180	>2500 ± 175	>2500 ± 180	625 ± 40	2500
126 W, 10 min	312 ± 25	312 ± 25	1250 ± 70	>2500 ± 180	>2500 ± 180	>2500 ± 175	312 ± 25	625 ± 40
189 W, 5 min	156 ± 10	312 ± 20	625 ± 30	>2500 ± 180	1250 ± 80	1250 ± 80	312 ± 25	625 ± 40
189 W, 10 min	78 ± 10	78 ± 10	156 ± 10	312 ± 30	156 ± 10	312 ± 30	156 ± 10	312 ± 25
252 W, 5 min	78 ± 10	78 ± 10	156 ± 10	312 ± 30	156 ± 10	312 ± 30	156 ± 10	312 ± 25
252 W, 10 min	78 ± 10	156 ± 20	625 ± 30	>2500 ± 180	1250 ± 80	1250 ± 70	312 ± 25	625 ± 40
315 W, 5 min	312 ± 30	312 ± 30	625 ± 30	>2500 ± 180	1250 ± 80	1250 ± 70	625 ± 40	1250 ± 80
315 W, 10 min	625 ± 30	625 ± 30	1250 ± 70	>2500 ± 180	625 ± 40	625 ± 40	2500 ± 160	>2500
Tebuconazole */ Chloramphenicol **	31.25 ± 2.4	125 ± 10.3	1.9 ± 0.3	1.9 ± 0.2	250 ± 20.6	250 ± 19.5	15.62 ± 1.3	62.5 ± 5.6

* Reference drug for phytopathogenic bacteria; ** Reference drug for phytopathogenic fungi.

The antimicrobial activity of the water–ethanol extract of *G. uliginosum* against bacterial and fungal phytopathogens, obtained using ultrasound, was found to be 78 µg/mL or higher. Among all the tested microorganisms, the fungus *R. solani* was the most sensitive to the action of *G. uliginosum* extracts, regardless of the strength of the ultrasonic unit and the extraction time. MIC values of the extracts ranged from 78 µg/mL to 625 µg/mL. Increasing the ultrasonic power from 63 W to 315 W with increments of 63 W during the extraction of *G. uliginosum* resulted in a consistent decrease of the MIC value by a factor of two. The fungistatic concentration of the extract obtained by ultrasound treatment at 63 W for 5 min started at 625 µg/mL. The MIC values during the extraction were 126–312 µg/mL at 129 W, 189–156 µg/mL at 189 W, and 252–156 µg/mL at 252 W. Consequently, increasing the power of the ultrasonic unit during extraction up to 252 W led to more complete extraction of antifungal compounds against *R. solani*. A further increase in the power, along with the extraction of compounds with antifungal activity, led to the isolation of inert compounds. At the same time, an MIC of 625 µg/mL was observed.

The duration of the extraction process under different conditions had a significant effect on the antimicrobial activity of *G. uliginosum* extracts. During extraction at 63 W, 126 W, and 252 W, there was no noticeable difference in antimicrobial activity against *R. solani*. At 189 W, a 10-min exposure allowed for more antimicrobial compounds to be extracted than at a 5-min extraction: MIC at 5 min was 312 µg/mL, and at 10 min, it was two times lower than 156 µg/mL. At 315 W, an increase in extraction time led to a more intensive extraction of ballast compounds, which was reflected in a rise in MIC from 312 to 625 µg/mL. The concentrations of the extract causing death of the fungus at 63 W coincide with the MIC values. IFC was not observed at the studied concentrations during extraction in modes 189 W 5 min and 252 W 10 min. During extraction in the modes of 189 W 5 min and 252 W 10 min, the values of the IFC were two times higher than the MIC.

Phytopathogenic fungus *A. solani* showed greater resistance to the action of the *G. uliginosum* extract than *R. solani*, with an average increase in MIC values by two to four times. In extracts obtained at 5 min and 63 W or 126 W, the MIC values in the studied concentrations were not observed. The effect of extraction power and time on *A. solani* was also similar. From 63 to 252 W, there was a decrease in the MIC values, whereas above 252 W, an increase in the MIC values was observed. IFC was not seen in all extract concentrations except for 189 W 10 min and 252 W 5 min.

MIC values for *C. michiganensis* and *E. carotovora* ssp bacteria started from 156 μg/mL, with MBC from 312 μg/mL. Similar to fungi, the best antibacterial activity was achieved at 189 W for 10 min and 252 W for 5 min. The Gram-positive bacterium *C. michiganensis* was resistant to extracts produced at 63 W and 126 W, regardless of the duration of extraction. A further increase in power to 189 W led to the production of more active extracts, with an MIC value beginning at 156 μg/mL.

At the same time, increasing the extraction time from 5 min to 10 min reduced the MIC by an order of magnitude, from 1250 μg/mL to 156 μg/mL. In contrast, for the Gram-negative bacterium *E. carotovora* spp., the MIC values for extracts obtained using 63 W were 1250 μg/mL. A further increase in extraction power to 252 W resulted in a 2 to 4-fold reduction in MIC, while at 315 W there was a twofold decrease in antibacterial activity. The study of the antimicrobial activity of extracts obtained from *G. uliginosum* biomass revealed the greatest activity during ultrasonic extraction compared to the maceration method. The lowest minimum inhibitory concentration (MIC) values for all tested microorganisms were detected during ultrasonic extraction with 189 W for 10 min and 252 W for 5 min. At the same time, the contribution of different parts of *G. uliginosum* to antimicrobial activity may differ. The influence of leaves and flowers on antimicrobial activity has been described in the literature.

To determine the antibacterial and antifungal activity of the stems, roots, leaves, and flowers of a plant against tested phytopathogens, appropriate extracts were obtained using ultrasonic extraction at 189 W for 10 min and 252 W for 5 min. The results on the antimicrobial activity of extracts, including MIC and MBC values against phytopathogenic strains *R. solani*, *A. solani*, *C. michiganensis*, *E. carotovora*, are presented in Tables 5 and 6.

Table 5. Indicators of antifungal activity of ethanol extract of various parts of *Gnaphalium uliginosum* obtained using the ultrasound method.

Plant Parts	*Rhizoctonia solani*				*Alternaria solani*			
	189 W, 10 min		252 W, 5 min		189 W, 10 min		252 W, 5 min	
	MIC, μg/mL	MBC, μg/mL	MIC, μg/mL	MBC, μg/mL	MIC, μg/mL	MBC, μg/mL	MIC, μg/mL	MBC, μg/mL
Roots	1250 ± 70	>2500 ± 180	1250 ± 70	>2500 ± 180	1250 ± 70	>2500 ± 180	1250 ± 70	>2500 ± 180
Stems	625 ± 40	>2500 ± 180	625 ± 40	>2500 ± 180	1250 ± 70	1250 ± 70	1250 ± 70	1250 ± 70
Flowers	78 ± 5	156 ± 10	78 ± 5	312 ± 25	156 ± 10	312 ± 25	156 ± 10	312 ± 25
Leaves	156 ± 10	312 ± 25	312 ± 25	312 ± 30	312 ± 30	312 ± 30	312 ± 25	625 ± 40
Tebuconazole *	31 ± 2.4	125 ± 10.3	31.25 ± 2.4	125 ± 10.3	15.62 ± 1.3	62.5 ± 5.6	15.62 ± 1.3	62.5 ± 5.6

* Reference drug for phytopathogenic fungi.

In these experiments, the MICs of ethanol extracts from *G. uliginosum* flowers were 78 μg/mL; from leaves 156 μg/mL; and from roots and stems 625 μg/mL. There was no significant difference in antimicrobial activity between the extracts obtained using ultrasonic extraction at 189 W for 10 min or 252 W for 5 min. When extracting the total biomass, *R. solani* was the most sensitive to the extracts. Ethanol extracts of *G. uliginosum* flowers stopped the growth of the fungus at a concentration of 78 μg/mL and were minimal among all extracts. For *A. solani*, *C. michiganensis*, and *E. carotovora*, MICs were two times higher than for R. solani and amounted to 156 mg/mL. At a concentration of 156 μg/mL of *G. uliginosum* flower extract, the death of R. solani and *C. michiganensis* began; for *A. solani* and *E. carotovora*, death began at a concentration of 312 μg/mL.

Table 6. Indicators of antimicrobial activity of ethanol extract of various parts of *Gnaphalium uliginosum* obtained using the ultrasound method.

Plant Parts	*Clavibacter michiganensis*				*Erwinia carotovora* spp.			
	189 W, 10 min		252 W, 5 min		189 W, 10 min		252 W, 5 min	
	MIC, μg/mL	MBC, μg/mL	MIC, μg/mL	MBC, μg/mL	MIC, μg/mL	MBC, μg/mL	MIC, μg/mL	MBC, μg/mL
Roots	625 ± 40	625 ± 40	625 ± 40	1250 ± 80	>2500 ± 180	>2500 ± 180	1250 ± 80	2500 ± 160
Stems	625 ± 40	625 ± 40	625 ± 40	625 ± 40	>2500 ± 180	>2500 ± 180	1250 ± 80	2500 ± 160
Flowers	156 ± 10	156 ± 10	312 ± 25	312 ± 25	156 ± 10	312 ± 25	312 ± 25	312 ± 25
Leaves	312 ± 25	312 ± 25	312 ± 25	625 ± 40	156 ± 10	312 ± 25	312 ± 25	312 ± 25
Chloramphenicol *	1.9 ± 0.3	1.9 ± 0.2	1.9 ± 0.2	1.9 ± 0.3	250 ± 20.4	250 ± 19.5	250 ± 19.6	250 ± 19.6

* Reference drug for phytopathogenic bacteria.

The ethanol extract of *G. uliginosum* leaves exhibited high antimicrobial activity. Exposure to the *G. uliginosum* leaf extract inhibited the growth of tested fungi and bacteria at concentrations of 156–312 μg/mL. MIC values were, on average, two times higher than those of the flower extract, except for its action on *A. solani*, where the MIC values were similar. Ethanol extracts of the roots and stems of *G. uliginosum* showed the lowest antimicrobial activity against phytopathogens of all parts of the plant. Gram-positive bacteria *C. michiganensis* were the most sensitive to these extracts, with MICs and MBCs equal to 625 μg/mL. For phytopathogenic species *E. carotovora* and *A. solani*, MIC was 1250 μg/mL, an order of magnitude greater than that of flower extracts.

2.2.2. Diffusion Method

The diffusion method for determining antimicrobial activity is one of the most common and frequently used methods to determine the antibacterial activity of plant extracts. Additionally, we used the diffusion method to compare our results with data from the available literature. This method is widely used due to the fact that it is a reliable, simple, and inexpensive test for susceptibility [27]. For a more comprehensive evaluation of the antimicrobial activity of ethanol flower, leaf, stem, and root extracts of *G. uliginosum*, we conducted experiments to determine their antibacterial activities using the diffusion method. Figure 1 and Table 7 show the results of antimicrobial activity determination using *C. michiganensis* (a) and *E. carotovora* (b) as examples.

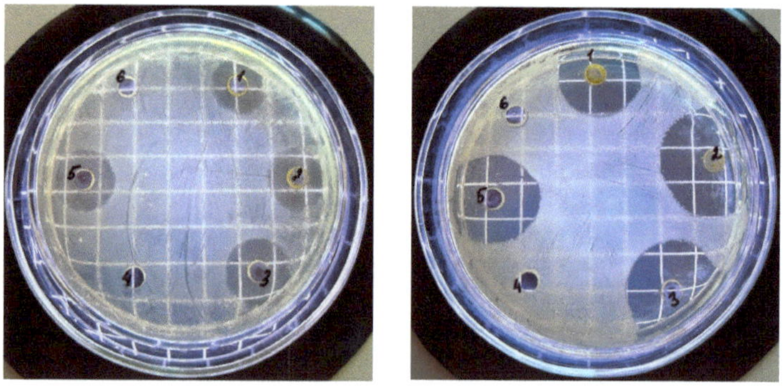

(a) (b)

Figure 1. Appearance of the Petri dish field after 24 h of inoculation of *C. michiganensis* (**a**) and *E. carotovora* (**b**) cultures when exposed to *G. uliginosum* extract: flowers (1), leaves (2), total biomass (ultrasound treatment) (3), roots (4), total biomass (maceration) (5), stems (6).

Table 7. Antimicrobial activity of extracts of individual parts of *G. uliginosum* when using the diffusion method.

Bacterial Strain	Flowers	Leaves	Roots	Stems	Total Biomass (Maceration)	Total Biomass (US)	Chloramphenicol *
	Mm						
E. carotovora	14 ± 0.42	14 ± 0.42	0	0	13 ± 0.38	18 ± 0.53	35 ± 1.1
C. michiganensis	21 ± 0.53	24 ± 0.72	0	0	22 ± 0.6	23 ± 0.69	56 ± 1.68

* Reference drug for phytopathogenic bacteria.

As a result of evaluating the antimicrobial activity using the diffusion method, we determined the average antimicrobial activity of ethanol extracts from flowers, leaves, and total biomass obtained through ultrasonic treatment or maceration. The inhibition zone of growth for these extracts against Gram-negative bacteria *E. carotovora* ranged from 13 to 18 mm, which is 2.5–3 times less than that of the control drug chloramphenicol (35 mm). The bacterium *C. michiganensis* was more sensitive to both extracts and chloramphenicol. The growth inhibitory zone for *G. uliginosum* ethanol extract was 21–24 mm and was two times lower than for chloramphenicol (54 mm), while no growth inhibitions zones were found for the tested bacteria after extracting from the roots and stem of *G. uliginosum*, indicating their low antimicrobial activities (Table 7). Comparing the results obtained using the diffusion and serial dilution methods revealed a correlation. In both cases, extracts from flower parts, leaves, and total biomass demonstrated a higher level of antimicrobial activity.

3. Discussion

Modern methods of chromatography, combined with mass spectrometry analysis, have allowed for the collection of extensive data on the phytochemical characteristics of plant extracts. Information about the phytochemical composition of *Gnaphalium* species is based on the qualitative and quantitative data presented in [28]. Studies have revealed that this plant contains more than 60 different chemical compounds, including flavonoids, sesquiterpenes, triterpenoids, phytoestrogens, anthraquinones, caffeic acid derivatives, and other substances. Phytochemical analysis of the ethanol extract using GC-MS confirmed the high content of these compounds, including phytoestrogen, carboxylic acid, aromatic aldehyde, fatty acid, coumarin, and siloxane. The obtained data are comparable to literature sources [29,30]. The high content of biologically active compounds in the extracts studied indicates potential antimicrobial activity. Coumarins and their derivatives are known to have a wide range of biological activities, including antitumor and antimicrobial properties [31,32]. In addition, the literature contains data on the antimicrobial activity of phytosteroids. For example, β-sitosterol from methanol extracts of the plant *Cissus sicyoides* has demonstrated efficacy against *Bacillus subtilis* [33]. B-sitosterol has also been shown to have antimicrobial activity against *Escherichia coli*, *Pseudomonas aeruginosa*, *Staphylococcus aureus*, and *Klebsiella pneumoniae* [34]. Another study [35] confirmed the antimicrobial activity of stigmasterol against Gram-positive and Gram-negative bacteria, as well as fungi.

The article by Sharonova et al. describes the antifungal activity of ethanol extracts from dry biomass of G. uliginosum against phytopathogenic bacteria and fungi. The extracts showed moderate activity against Gram-positive bacteria such as *S. aureus* and *B. subtilis* with a minimum inhibitory concentration of 62.5 μg/mL, but when using Gram-negative strains like *A. tumefaciens* and *X. arboricola*, a decrease in extract activity (500 μg/mL) was observed. At the same time, the extracts showed moderate antifungal activity on fungi such as *F. graminearum* and *Phytophthora* spp. (250 μg/mL). Their study, focusing on ethanolic extracts of *G. uliginosum*, reported MIC values ranging from 160 to 320 μg/mL against phytopathogenic microorganisms. These values are comparable to or slightly higher than those observed in our study (78 μg/mL for ultrasonic extracts). This similarity suggests that our extraction method, employing ultrasound, may offer potential advantages in terms of enhanced antimicrobial activity. However, it is worth noting that during the drying process of raw materials, various chemical reactions, such as oxidation, may occur and lead to a reduction in the active ingredients, including secondary metabolites with antimicrobial properties, such as phenols.

Our experiments are a continuation of previous research, which we have expanded upon. Unlike previous studies, we used freshly harvested and pre-frozen raw materials to ensure the maximum preservation of biologically active compounds. Additionally, we analyzed the influence of various plant parts, including roots, stems, leaves, and flowers, on phytopathogenic microorganisms.

Along with the maceration method, optimal extraction techniques were developed using ultrasonic equipment. The antimicrobial activity values of water–ethanol extracts from frozen plant material, including roots, stems, leaves, and flowers, obtained through maceration and ultrasound treatment, showed high levels of antibacterial and antifungal activity. According to the classification system presented in [36], plant extracts with MIC values less than 100 μg/mL should be considered very active; between 100 and 512 μg/mL—with significant activity; 512–2048 μg/mL—moderate activity; and more than 2048 μg/mL—inactive. In the available literature [37], when considering antimicrobial activity, extracts from medicinal plants with an MIC value equal to 160 μg/mL are proposed as promising. The MIC of extracts obtained by the maceration method was 312 μg/mL, which corresponds to significant activity. Using ultrasound increased the antimicrobial properties of extracts. The MIC values when using ultrasound at a power of 189 W and 252 W were 78 μg/mL, corresponding to the classification of a very active agent.

The results of our study were similar to the antimicrobial activity of plant extracts from the genus *Gnaphalium*, as described in the literature. In a paper by Diaz et al., the minimum inhibitory concentration of an extract of *Gnaphalium attenuatum* grown in Mexico against phytopathogens and pathogens of humans and animals was 100–500 μg/mL [18]. Another study [19] showed that the minimum inhibiting concentration of extracts from *Gnaphallium polycaulon* is 50–100 μL/mL. The MIC values we obtained for the *G. uliginosum* extract against *C. michiganensis* were comparable to those reported in a study investigating the antibacterial activity of *Ginkgo biloba* [38]. Both studies demonstrated low MIC values, indicating significant antibacterial activity. However, the *G. uliginosum* extract showed effectiveness against both *E. carotovora* and *R. solani*, with a broader spectrum of action compared to *Ginko biloba*, which mainly targets *C. michiganensis*. Notably, the MIC of the *G. uliginosum* extract on *E. carotovora* was lower than that reported for *Azadirachta indica*, highlighting the potential of the *G. uliginosum* extract as a promising agent for controlling soft rot caused by *E. carotovora*, demonstrating greater efficacy than other natural biopesticides [39].

The antimicrobial properties of *G. uliginosum* extracts, including ethanol extracts, depend on the combination of biologically active compounds dissolved in the extract. The alcohol extract obtained from the aboveground parts of *G. uliginosum*, using the GC/MS method, is enriched with stigmasterol, β-amyrin, sitosterol, γ-sitosterol, scopoletin, squalene, flavonoids (gnafalosides A and B, luteolin, scutellarein, glycoside scutellarin, rutin, tricin, eupafolin, quercetin), chlorogenic and caffeic acids, carotenoids, thiamine, resins, tannins, alkaloids (gnafalin), phytosterols, and ascorbic acid [17,20]. The most optimal phytochemical composition inhibiting the growth of phytopathogens *E. carotovora* and R. solani was extracted at a temperature of 75 °C using the maceration method. Lower temperatures were found to be insufficient for extracting compounds active against these bacteria. For Gram-positive bacteria *C. michiganensis* with an excellent cell wall structure, the optimal extraction temperature is 45 °C.

In our research, we demonstrated that the extract obtained using an ultrasonic generator had higher antimicrobial activity than the extract produced using the maceration method. The data we obtained correspond to those presented by Elapov et al., who found that ultrasonic extraction confirmed the high efficiency of extracting natural raw materials and contributed to a higher yield of biologically active substances [40]. The most optimal extraction of antibacterial and antifungal compounds when using an ultrasound machine also depends on the extraction mode. The extraction of active compounds from the plant was slow at low power. At low power, the extraction of active compounds was slow. However, at higher power levels, this disadvantage was compensated for by the duration of extraction. In our experiments, the optimal extraction mode was found to be 252 W for 5 min. With an increase in the extraction time in this mode, the extraction of targeted compounds was significantly lower than that of the ballast compounds, as confirmed by a reduction in antimicrobial activity. Therefore, ultrasonic extraction is more suitable for the extraction of bioactive compounds.

The results of the antimicrobial activity of ethanol extracts from the flowers, leaves, stems, and roots of *G. uliginosum* were not significantly different from those obtained using freshly harvested biomass. The contribution of each plant part to the overall antimicrobial activity was not equal. Flower extracts had the greatest contribution, with MIC values for tested microorganisms equal to those of the total extract. Leaves of *G. uliginosum* showed significant activity and were less active by one dilution, while the antimicrobial activity of leaves and roots was an order of magnitude lower. Therefore, the results suggest that the ethanol extract of *G. uliginosum* can be used as a biological fungicide and bactericide for protecting plants from various phytopathological diseases. Further studies on this plant as an antimicrobial agent and its use in herbal medicine could aim to clarify its mechanism of action and determine effective doses and application regimens.

4. Materials and Methods

4.1. Plant Material

G. uliginosum plants were collected from the experimental fields of the Tatar Scientific Research Institute of Agriculture of the Federal Research Center "Kazan Scientific Center of the Russian Academy of Sciences" of the Republic of Tatarstan in the Laishevsky district at the flowering stage in August 2022 as a weed plant. Freshly harvested plants were carefully cleaned of any impurities and dust. They were then divided into two groups. One group was left as a whole plant, while the other group was separated into its individual components, including roots, stems, leaves, flowers, and fruits. These components were then stored in a freezer at −35 °C for further analysis.

4.2. Extract Preparation by Maceration

Frozen vegetable raw materials of *G. uliginosum* L. were ground in a laboratory mill (LM 202, Easy Life LLC, Moscow, Russia) to a particle size of 0.3–2 mm. Crushed raw materials weighing 3 g were transferred to a round-bottomed flask with a condenser, and 70% ethanol was added at a biomass-to-solvent ratio of 1:30. Extraction was carried out by maceration for 1.5 h at temperatures of 25 °C, 45 °C, and 75 °C with constant stirring by an automatic magnetic stirrer (IKA RCT basic, Staufen, Germany). The resulting extracts were filtered using Whatman No. 1 filter paper, and the filtrate was concentrated using a rotary evaporator (LabTex Re 100-Pro) at 40 °C and stored at 4 °C for further study for no more than 3 days [41].

4.3. Extract Preparation by Ultrasound-Assisted Maceration

Frozen vegetable raw materials of *G. uliginosum* L. were ground in a laboratory mill (LM 202, Russia) to a particle size of 0.3–2 mm. Crushed raw materials weighing 3 g were transferred to a flat-bottomed flask, and 70% ethanol was added at a biomass-to-solvent ratio of 1:30. The extraction was performed using an ultrasonic generator (I10–0.63) with an immersion probe at a power of 63 W to 315 W for 5 and 10 min, respectively. The obtained extracts were filtered (Whatman No. 1), then the filtrate was concentrated using a rotary evaporator (LabTex Re 100-Pro, On Wing Tat Co. Ltd, Hong Kong, China) at 40 °C and stored in a laboratory refrigerator at 4 °C for no more than three days for further studies [42]. Table 8 lists the power and time settings used for the ultrasonic extraction process.

Table 8. Extraction using an ultrasonic generator (I10–0.63) with an immersion probe at different modes of power and time.

№	Power, W.	Frequency, kHz.	Time, min.
1.	63	16.9	5
2.	126	22.0	5
3.	189	22.4	5
4.	252	22.1	5
5.	315	22.1	5
6.	63	16.9	10
7.	126	22.0	10
8.	189	22.4	10
9.	252	22.1	10
10.	315	22.1	10

4.4. GS-MS Analysis

For extraction, frozen vegetable raw materials of *G. uliginosum* L. were ground in a laboratory mill (LM 202, Russia) to a particle size of 0.3–2 mm. The crushed raw materials weighing 3 g were transferred to a flat-bottomed flask, and 70% ethanol was added at a biomass-to-solvent ratio of 1:30. Extraction was performed using an ultrasonic generator (I10–0.63) with an immersion probe at a power of 189 W for 10 min. The obtained extracts were filtered (Whatman No. 1), and then the filtrate was lyophilized for 20 h (BK-FD12P, Biobase, Jinan, China). The dry extract was re-dissolved in 96% ethyl alcohol for purification. The extract was passed through a CHROMAFIL Xtra filter with a pore size of 0.45 microns (Macherey-Nagel, Duren, Germany) for further purification. GC-MS analysis of the purified extract was performed on a gas chromatograph (Agilent 6890N, Ontario, CA, USA) equipped with a mass spectrometry detector 5973N (Agilent Technologies, Ontario, CA, USA) in native form. The composition of the *G. uliginosum* extract was determined using electron ionization mass spectrometry at 70 eV, in the range of 50–550 m/i (using CH4 as a reagent gas with high purity). The ion source temperature was 230 °C. A nonpolar capillary column (HP-5 MS) was used, with a 5% diphenyl- and 95% dimethylpolysiloxane composition, 30 μm × 0.25 mm × 0.25 μm, from Thermo Fisher Scientific in the Waltham, MA, USA. Gas chromatographic conditions were as follows: an initial sample volume of 2 μL was injected. The initial temperature of the column thermostat was 75 °C, without temperature control during the first 0 min period. The temperature increased at a rate of 10 °C/min 0 until reaching 120 °C, after which the temperature control was set to 0 min for another 10 min. Then, the temperature increased from 20 °C to 280 °C at a rate of 20 °C/min, and temperature control was maintained for 20 min. After that, temperature control was held for 80 min, with the sample inlet device (injector) temperature set at 200 °C and sample injection mode without separation. Helium was used as the carrier gas at a flow rate of 0.7mL/min. The components were identified using a NIST-17 mass spectrometer (NIST, San Diego, CA, USA) with a library of 306,622 individual compound spectra.

4.5. Antimicrobial Activity

As test strains, the following microorganisms were used: phytopathogenic bacteria strains (*Clavibacter michiganensis* VKM Ac-1404, *Erwinia carotovora* spp. carotovora) and fungal strains (*Alternaria solani* K-100054, *Rhizoctonia solani* VKM F-895). The microorganisms were cultured on standard sterile nutrient media such as BTN-broth, potato dextrose agar, LB Miller broth, and potato dextrose broth. Their concentration was determined using the DEN-1B densitometer (Biosan, Riga, Latvia) according to standard protocols.

Antimicrobial activity was assessed using the diffusion method following the methodology described in [43]. The microorganisms tested were inoculated into nutrient broth and incubated at 37 °C for 24 h. A 0.1 mL culture of microorganisms was then spread over the surface of Muller–Hinton agar using a Drigalski spatula to ensure an even distribution ("continuous lawn"). After the cells of the pathogens had grown, agar blocks were removed using a sterile cork borer. A 0.1 mL solution of 1% extract was placed in the wells formed to prevent overflow. After incubation, the Petri dishes were inverted and placed on dark, matte surfaces so that light would fall on them at a 45° angle (accounting for reflected light). When measuring the inhibition zones, we focused on complete zones of inhibition of visible growth. To evaluate the antimicrobial activity of the compounds, a three-stage scale was used. The scale was based on the diameter of the zone of microbial inhibition: 1. Low level of antimicrobial activity—the diameter of the inhibition zone was less than 15 mm. 2. Average level of antimicrobial activity—from 16 to 25 mm. 3. High level of antimicrobial activity—more than 25 mm.

Antimicrobial activity was determined using the method of serial dilutions, as described in [44]. The minimum inhibitory concentration (MIC), which is the lowest concentration of juice/extract that stops the growth of a culture in a medium without causing its death, was determined. The minimum bactericidal concentration (MBC) and minimum fungicidal concentration (MFC), which are the concentrations that cause the death of a culture, were also determined.

Liquid broth with microbial spores was prepared from standard nutrient media using 24-h bacterial cultures, and 7–14-day fungal cultures were used for fungal spores. The final inoculum size was 10^5 CFU (colony-forming units)/mL for bacteria and $1.1–1.5 \times 10^2$ CFU/mL for fungi. Control tubes contained only nutrient medium. To determine the MIC, the test substance was added to the first tube or well of a tablet containing a liquid nutrient medium. Then, 1 mL of solution was taken from the first tube and transferred to the second tube. This procedure was repeated up to 10 times. Afterward, a bacterial suspension or fungal mycelial fragment was added to each tube or well on the tablet.

Microorganism incubation was carried out in a thermostat for up to 5 days at 30 °C for *Clavibacter michiganensis* VKM Ac-1404 and *Erwinia carotovora* spp. The incubation time of fungi in the thermostat at 26 °C with the corresponding substance was 7 days; the growth of bacteria and fungi was determined visually. All analyses were performed in triplicate. The reference substance for bacteria was chloramphenicol (Kazan Pharmaceutical factory), and for the fungus the reference substance was tebuconazole (obtained from the industrial preparation "Universal, KS", Registrant. LLC "Khimagro-marketing.RU").

5. Conclusions

This study investigated the composition of extracts and the effect of different extraction methods on the isolation of bioactive compounds from the entire biomass and various parts of *G. uliginosum* L. collected during flowering in the experimental fields of the Tatarstan Research Institute of Agriculture, Federal Research Center, Kazan Scientific Center of the Russian Academy of Sciences in the Republic of Tatarstan. This study assessed the antimicrobial activity of ethanolic extracts of *G. uliginosum* against several phytopathogenic microorganisms, including *Clavibacter michiganensis* VKM Ac-1404, *Erwinia carotovora* spp. carotovora, *Alternaria solani* K-100054, and *Rhizoctonia solani* VKM F-895. The assessment of antimicrobial activity for ethanol extracts from *G. uliginosum* showed significant antimicrobial potential in both the total biomass and individual parts, especially flowers and leaves. Ultrasonic extraction using 70% ethanol has been found to be the most effective method for extracting bioactive compounds from *G. uliginosum*.

The GC-MS analysis of the extracts revealed the presence of various compounds, including phytosterols, carboxylic acids, aromatic aldehydes, fatty acids, coumarins, and siloxanes. These substances are likely responsible for the observed antibacterial and antifungal properties of the ethanol extracts of *G. uliginosum*. Based on these findings, the antimicrobial activity of these extracts was moderate to high against the tested phytopathogens. This suggests that extracts from this species could potentially be used as natural biocides to protect plants from disease. Further research into the composition of these extracts and their mode of action, as well as techniques for their production, could lead to new applications in agriculture and medicine.

Author Contributions: Conceptualization, L.D. and A.M.; validation, L.D., A.M., G.S., V.B. and E.N.; writing—original draft preparation, L.D., A.M., G.S., V.B. and E.N.; formal analysis, L.D. and A.M.; investigation, L.D., A.M. and G.S.; resources, E.N. and G.S.; writing—review and editing, L.D. and A.M.; visualization, L.D. and A.M.; supervision, E.N.; project administration, E.N., funding acquisition, E.N. All authors have read and agreed to the published version of the manuscript.

Funding: This research received no external funding.

Institutional Review Board Statement: Not applicable.

Informed Consent Statement: Not applicable.

Data Availability Statement: The data presented in this study are available upon request from the corresponding author.

Conflicts of Interest: The authors declare no conflicts of interest.

References

1. Horbach, R.; Navarro-Quesada, A.R.; Knogge, W.; Deising, H.B. When and how to kill a plant cell: Infection strategies of plant pathogenic fungi. *J. Plant Physiol.* **2011**, *168*, 51–62. [CrossRef] [PubMed]
2. Burtscher-Schaden, H.; Durstberger, T.; Zaller, J.G. Toxicological comparison of pesticide active substances approved for conventional vs organic agriculture in Europe. *Toxics* **2022**, *10*, 753. [CrossRef]
3. Bhandari, G.; Kishor, A.; Vasickova, J.; Yang, X.; Geissen, V. Ecological risk assessment of pesticide residues in soils from vegetable production areas: A case study in S-Nepal. *Sci. Total Environ.* **2021**, *788*, 147921. [CrossRef] [PubMed]
4. Lavellea, P.; Decaensb, T.; Aubertb, M.; Barota, S.; Blouina, M.; Bureaub, F.; Margerieb, P.; Moraa, P.; Rossic, J.-P. Soil invertebrates and ecosystem services. *Eur. J. Soil Biol.* **2006**, *42*, S3–S15. [CrossRef]
5. Lyons, D.D.; Philibert, D.A.; Zablocki, T.; Qin, R.; Huang, R.; Gamal El-Din, M. Assessment of raw and ozonated oil sands process-affected water exposure in developing zebrafish: Associating morphological changes with gene expression. *Environ. Pollut.* **2018**, *241*, 959–968. [CrossRef] [PubMed]
6. Wang, X.; Zhu, X.; Peng, Q.; Wang, Y.; Ge, J.; Yang, G. Multi-level ecotoxicological effects of imidacloprid on earthworm (*Eisenia fetida*). *Chemosphere* **2019**, *219*, 923–932. [CrossRef] [PubMed]
7. Ye, X.; Xiong, K.; Liu, J. Comparative toxicity and bioaccumulation of fenvalerate and esfenvalerate to earthworm *Eisenia fetida*. *J. Hazard. Mater.* **2016**, *310*, 82–88. [CrossRef]
8. Mikaberidze, A.; McDonald, B.A.; Bonhoeffer, S. Can high-risk fungicides be used in mixtures without selecting for fungicide resistance? *Phytopathology* **2014**, *104*, 324–331. [CrossRef]
9. Acheuk, F.; Basiouni, S.; Shehata, A.A.; Dick, K.; Hajri, H.; Lasram, S.; Yilmaz, M.; Emekci, M.; Tsiamis, G.; Spona-Friedl, M. Status and Prospects of Botanical Biopesticides in Europe and Mediterranean Countries. *Biomolecules* **2022**, *12*, 311. [CrossRef]
10. Liu, R.; Li, J.; Zhang, F.; Zheng, D.; Chang, Y.; Xu, L.; Huang, L. Biocontrol activity of *Bacillus velezensis* D4 against apple Valsa canker. *Biol. Control* **2021**, *163*, 104–106. [CrossRef]
11. Ngegba, M.; Kamara, J.; Koroma, M. Characteristics of Rural Community Savings Systems and Impacts on Rural Livelihood in Selected Districts of Sierra Leone. *Asian J. Agric. Ext. Econ. Sociol.* **2022**, *40*, 278–298. [CrossRef]
12. Semina, Y.V.; Shcherbakova, L.A.; Slezina, M.P.; Odintsova, T.I. Investigation of the Activity of *Chenopodium album* Seed Extracts and *Fusarium sam bucinum* Culture Broth against Some Phytopathogenic Fungi. *Sel'skokhozyaistvennaya Biol.* **2016**, *51*, 739–745. [CrossRef]
13. Buono, D.D. Can biostimulants be used to mitigate the effect of anthropogenic climate change on agriculture? It is time to respond. *Sci. Total Environ.* **2021**, *751*, 141–163.
14. Efremov, A.P. *Lekarstvennye Rasteniya i Griby Srednei Polosy Rossii: Polnyi Atlas-Opredelitel (Medicinal Plants and Mushrooms of Middle Russia: A Complete Atlas-Identifier)*, 2nd ed.; Fiton XXI: Moscow, Russia, 2021; p. 504.
15. Shikov, A.N.; Kundracikova, M.; Palama, T.L.; Pozharitskaya, O.; Kosman, V.; Makarov, V.; Galambosi, B.; Kim, H.J.; Jang, Y.P.; Choi, Y.H.; et al. Phenolic constituents of *Gnaphalium uliginosum* L. *Ind. Crops Prod.* **2010**, *31*, 45–47. [CrossRef]
16. Ayad, R.; Akkal, S. Phytochemistry and biological activities of algerian *Centaurea* and related genera. *Stud. Nat. Prod. Chem.* **2019**, *63*, 357–414.
17. Lockowandt, L.J.; Pinela, C.; Roriz, C.; Pereira, R.; Abreu, R.; Calhelha, M.; Alves, L.; Barros, M.; Bredol, I. Chemical features and bioactivities of cornflower (*Centaurea cyanus* L.) capitula: The blue flowers and the unexplored non-edible part. *Ind. Crops Prod.* **2019**, *128*, 496–503. [CrossRef]
18. Diaz, E.; Javier, O.; Méndez-Gutiérrez, A.; Jesus, L.; Sharma, A.; Villarreal, M.; Cardoso Taketa, A. Antibacterial Activity of *Bougainvillea glabra, Eucalyptus globulus, Gnaphalium attenuatum,* and *Propolis* Collected in Mexico. *Pharmacol. Pharm.* **2012**, *3*, 433–438. [CrossRef]

19. Apaza Ticona, L.; Puerto Madorrán, M.J.; Hervás Povo, B.; Ortega Domenech, M.; Rumbero Sánchez, A. Isolation and characterisation of antibacterial and anti-inflammatory compounds from *Gnaphalium polycaulon* L. *J. Ethnopharmacol.* **2022**, *282*, 114. [CrossRef] [PubMed]
20. Sharonova, N.; Terenzhev, D.; Bushmeleva, K.; Gumerova, S.; Lyubina, A.; Fitsev, I.; Belov, T. Phytochemical Contents, Antimicrobial and Antioxidant Properties of *Gnaphalium uliginosum* L. Ethanolic Extract and Essential Oil for Agricultural Uses. *Asian J. Chem.* **2019**, *31*, 2672–2678.
21. Shanmugapriya, K.; Subramanian, P.; Boomi, P.; Subaskumar, R.; Sundaram, R.; Thayumanavan, T. An eco-friendly *Gnaphalium polycaulon* mediated silver nanoparticles: Synthesis, characterization, antimicrobial, wound healing and drug release studies. *J. Drug Deliv. Sci. Technol.* **2020**, *61*, 115–120. [CrossRef]
22. Rojas-Bribiesca, G.; Lévaro, J.; Tortoriello, J.; Navarro, V. Antimicrobial evaluation of certain plants used in Mexican traditional medicine for the treatment of respiratory diseases. *J. Ethnopharmacol.* **2001**, *74*, 97–101. [CrossRef]
23. Al-Zubairy, M.A.; Hussein, K.; Alkhyat, S.H.; Al-Mahdi, A.Y.; Alghalibi, S.M.; Al-Gheethi, A.A.; Al-Shaibani, M.M.; El Enshasy, H.A.; Sidik, N.M. Antibacterial Activity of a Novel Oligosaccharide from *Streptomyces californics* against *Erwinia carotovora* subsp. Carotovora. *Molecules* **2022**, *27*, 2384. [CrossRef]
24. Fitsev, I.M.; Nikitin, E.N.; Rakhmaeva, A.M.; Terenzhev, D.A.; Sakhno, T.M.; Nasybullina, Z.R. Chemical Composition of *Cupressus sempervirens* L. and *Thuja occidentalis* L. Essential Oils and Their Activity against Phytopathogenic Fungi. *Uchenye Zap. Kazan. Univ. Seriya Estestv. Nauk.* **2022**, *164*, 392–407. [CrossRef]
25. Lin, H.; Tsao, R. Antimicrobials from plants—Food preservation and shelf life extension. *Compr. Biotechnol.* **2019**, *4*, 703–714.
26. Zibareva, L.N.; Filonenko, E.S. The Influence of Ultrasonic Treatment on the Extraction of Biologically Active Compounds from Plants of the Caryophyllaceae Family. *Chem. Plant Raw Mater.* **2018**, *2*, 145–151. [CrossRef]
27. National Committee for Clinical Laboratory Standards; Barry, A.L. *Methods for Determining Bactericidal Activity of Antimicrobial Agents*; Document M26-A; National Committee for Clinical Laboratory Standards: Wayne, PA, USA, 1999; Volume 19, pp. 20–29.
28. Zhang, W.; Wu, C.Z.; Fan, S.Y. Chemical constituents from *Gnaphalium affine* and their xanthine oxidase inhibitory activity. *Chin. J. Nat. Med.* **2018**, *16*, 347. [CrossRef]
29. Platonov, V.V.; Sukhikh, G.T.; Dunayev, V.A.; Khadartsev, A.A.; Yarkova, T.A.; Datieva, F.S. Chemical composition of the ethanol extract of common tansy (*Tanacetum vulgre* L., Astrove family—*Asteraceal*). *Bull. New Med. Technol.* **2021**, *3*, 15.
30. Platonov, V.V.; Khadartsev, A.A.; Valentinov, B.G.; Sukhoi, G.T.; Dunaev, V.A.; Volochaeva, M.V. Chemical composition of hexane extract of roots of wild medicinal Dandelion (*Dandelion officinalic* wiggs., family *Asteraceae*—Compound. *Bull. New Med. Technol. Electron. Ed.* **2022**, *2*, 13.
31. Murray, R.D.; Mendez, J.; Brown, S.A. *The Natural Coumarins: Occurrence. Chemistry and Biochemistry*; Wiley: Hoboken, NJ, USA, 1982; pp. 1–12.
32. Mironov, A.; Colanzi, A.; Polishchuk, R.; Beznoussenko, G.; Mironov, A.; Fusella, A.; Di Tullio, G.; Silletta, M.; Corda, D.; De Matteis, M.; et al. Dicumarol, an inhibitor of ADP-ribosylation of CtBP3/BARS, fragments golgi non-compact tubular zones and inhibits intra-golgi transport. *Eur. J. Cell Biol.* **2004**, *83*, 263–279. [CrossRef]
33. Beltrame, F.L.; Pessini, G.L.; Doro, D.L.; Dias Filho, B.D.; Bazotte, R.B.; Cortez, D.A.G. Evaluation of the antidiabetic and antibacterial activity of *Cissus secyoides*. *Braz. Arch. Biol. Technol.* **2002**, *45*, 21–25. [CrossRef]
34. Sen, A.; Dhavan, P.; Shukla, K.K.; Singh, S.; Tejovathi, G. Analysis of IR, NMR and Antimicrobial Activity of β- Sitosterol Isolated from *Momordica charantia*. *Sci. Secur. J. Biotechnol.* **2012**, *1*, 9–13.
35. Yusuf, A.J.; Abdullahi, M.I.; Aleku, G.A.; Ibrahim, I.A.A.; Alebiosu, I.O.; Yahaya, M. Antimicrobial activity of stigmasterol from the stem bark of *Neocarya macrophylla*. *J. Med. Plants Econ. Dev.* **2018**, *2*, 38. [CrossRef]
36. Alvarez-Martinez, F.J.; Catalan, B.E.; Herranz-Lopez, M.; Micol, V. Antibacterial plant compounds, extracts and essential oils: An updated review on their effects and putative mechanisms of action. *Phytomedicine* **2021**, *90*, 153626. [CrossRef] [PubMed]
37. Van Vuuren, S.; Holl, D. Antimicrobial natural product research: A review from a South African perspective for the years 2009–2016. *J. Ethnopharmacol.* **2017**, *208*, 236–252. [CrossRef] [PubMed]
38. Sanchez-Hernandez, E.; Gonzalez-Garcia, V.; Palacio-Bielsa, A.; Lorenzo-Vidal, B.; Buzon-Duran, L.; Martin-Gil, J.; Martin-Ramos, P. Antibacterial Activity of *Ginkgo biloba* extracts against *Clavibacter michiganensis* subsp. michiganensis, *Pseudomonas* spp., and *Xanthomonas vesicatoria*. *Horticulturae* **2023**, *9*, 461. [CrossRef]
39. Bhardwaj, S.K.; Laura, J.S. Antibacterial activity of some plant extracts against pathogenic bacteria *Erwinia carotovora* subsp. Carotovora. *Potato J.* **2008**, *35*, 72–77.
40. Elapov, A.A.; Kuznetsov, N.N.; Marakhova, A.I. Application of Ultrasound in the Extraction of Biologically Active Compounds from Plant Raw Materials Used or Promising for Use in Medicine (Review). *Razrab. I Regist. Lek. Sredstv.* **2021**, *10*, 96–116.
41. Makarova, N.V.; Eremeeva, N.B. Comparative Study of the Influence of Ultrasonic Effects on the Extraction of Antioxidant Compounds from Bilberry Berries (*Vaccinium myrtillus* L.). *Khimiya Rastenitelnogo Syr'ya* **2020**, *1*, 65–67.
42. Sharma, P.; Wichaphon, J.; Klangpetch, W. Antimicrobial and antioxidant activities of defatted *Moringa oleifera* seed meal extract obtained by ultrasound-assisted extraction and application as a natural antimicrobial coating for raw chicken sausages. *Int. J. Food Microbiol.* **2020**, *332*, 97–101. [CrossRef]

43. Kim, D.S.; Kwack, Y.; Lee, J.H.; Chun, C. Antimicrobial Activity of Various Parts of Tomato Plants Varied with Different Solvent Extracts. *Plant Pathol. J.* **2019**, *35*, 149–155. [CrossRef]
44. Nikitin, E.; Fitsev, I.; Egorova, A.; Logvinenko, L.; Terenzhev, D.; Bekmuratova, F.; Rakhmaeva, A.; Shumatbaev, G.; Gatiyatullina, A.; Shevchuk, O.; et al. Five Different *Artemisia* L. Species Ethanol Extracts' Phytochemical Composition and Their Antimicrobial and Nematocide Activity. *Int. J. Mol. Sci.* **2023**, *24*, 14372. [CrossRef] [PubMed]

antibiotics

MDPI

Article

Green Synthesis of Silver Nanoparticle from *Anadenanthera colubrina* Extract and Its Antimicrobial Action against ESKAPEE Group Bacteria

Anastácia Nikolaos Deonas [1], Lucas Marcelino dos Santos Souza [1], Gabriel Jonathan Sousa Andrade [2], Jennifer Germiniani-Cardozo [2], Débora Dahmer [2], Admilton Gonçalves de Oliveira [1], Gerson Nakazato [1], José Marcelo Domingues Torezan [3] and Renata Katsuko Takayama Kobayashi [1,*]

1 Department of Microbiology, Center for Biological Sciences, State University of Londrina, Londrina 86057-970, Brazil; anastacia.nikolaos@uel.br (A.N.D.); lucas.marcelino1@uel.br (L.M.d.S.S.); admilton@uel.br (A.G.d.O.); gnakazato@uel.br (G.N.)
2 Department of Biochemistry and Biotechnology, Center for Exact Sciences, State University of Londrina, Londrina 86057-970, Brazil; gabriel.jonathan@uel.br (G.J.S.A.); jennifer.germiniani@uel.br (J.G.-C.)
3 Department of Animal and Plant Biology, Center for Biological Sciences, State University of Londrina, Londrina 86057-970, Brazil; torezan@uel.br
* Correspondence: kobayashirkt@uel.br; Tel.: +55-(43)-99182-0786

Abstract: Given the urgent need for novel methods to control the spread of multidrug-resistant microorganisms, this study presents a green synthesis approach to produce silver nanoparticles (AgNPs) using the bark extract from *Anadenanthera colubrina* (Vell.) Brenan var. colubrina. The methodology included obtaining the extract and characterizing the AgNPs, which revealed antimicrobial activity against MDR bacteria. *A. colubrina* species is valued in indigenous and traditional medicine for its medicinal properties. Herein, it was employed to synthesize AgNPs with effective antibacterial activity (MIC = 19.53–78.12 μM) against clinical isolates from the ESKAPEE group, known for causing high hospitalization costs and mortality rates. Despite its complexity, AgNP synthesis is an affordable method with minimal environmental impacts and risks. Plant-synthesized AgNPs possess unique characteristics that affect their biological activity and cytotoxicity. In this work, *A. colubrina* bark extract resulted in the synthesis of nanoparticles measuring 75.62 nm in diameter, with a polydispersity index of 0.17 and an average zeta potential of −29 mV, as well as low toxicity for human erythrocytes, with a CC_{50} value in the range of 961 μM. This synthesis underscores its innovative potential owing to its low toxicity, suggesting applicability across several areas and paving the way for future research.

Keywords: biogenic Bio-AgNPs; white angico; multidrug-resistant bacteria

Citation: Deonas, A.N.; Souza, L.M.d.S.; Andrade, G.J.S.; Germiniani-Cardozo, J.; Dahmer, D.; de Oliveira, A.G.; Nakazato, G.; Torezan, J.M.D.; Kobayashi, R.K.T. Green Synthesis of Silver Nanoparticle from *Anadenanthera colubrina* Extract and Its Antimicrobial Action against ESKAPEE Group Bacteria. *Antibiotics* **2024**, *13*, 777. https://doi.org/10.3390/antibiotics13080777

Academic Editor: Marc Maresca

Received: 10 July 2024
Revised: 7 August 2024
Accepted: 13 August 2024
Published: 16 August 2024

1. Introduction

Antimicrobial resistance has been the focus of much research, as major health organizations consider it one of the biggest global health issues. This concern arises from the emergence of microorganisms resistant to last-resort antibiotics, mainly caused by the inappropriate use of antibiotics in treating infections, such as the administration of sub-doses or treatment interruption. Other contributing factors include the excessive use of antimicrobials in domestic and hospital environments, the lack of public education policies regarding the correct use of antibiotics, and the overuse of antibiotics in livestock to increase animal productivity [1–3].

Consequently, selective pressure from the environment leads to the emergence of antimicrobial-resistant bacteria. This situation is exacerbated by the impact on various areas, including agronomy, livestock, human and animal health, and the economy, which led to the creation of the "One Health" policy. This policy aims to promote health across human, animal, and environmental contexts, as all these areas are interconnected and require joint attention [4,5].

In line with the "One Health" policy, the World Health Organization (WHO), World Organization for Animal Health (WOAH), and Food and Agriculture Organization of the United Nations (FAO) implemented, in 2015, a global program to combat antimicrobial resistance. This program, named the "Global Action Plan on Antimicrobial Resistance", aims to achieve the following objectives: (1) promote awareness and understanding of antimicrobial resistance; (2) increase scientific knowledge on antimicrobial resistance through surveillance and research; (3) reduce the incidence of new infections through preventive measures such as basic sanitation and adequate personal hygiene; (4) improve the use of antimicrobials for people and animals; and (5) advocate for sustainable investment in new medicines, diagnostic methods, and other tools to combat antimicrobial resistance [6].

However, ongoing efforts are still insufficient, as bacterial resistance to antimicrobials was directly or indirectly related to 4.95 million deaths in 2019 [7]. Moreover, these numbers are expected to escalate. According to O'Neill [8], deaths caused by antimicrobial-resistant microorganisms are estimated to reach 10 million by 2050, surpassing the combined deaths from cancer and traffic accidents. These estimates indicate the significant impact of this problem.

Many deaths caused by antimicrobial-resistant bacteria are attributed to the group of pathogens known by the acronym ESKAPEE (originally ESKAPE, composed of *Enterococcus faecium*, *Staphylococcus aureus*, *Klebsiella pneumoniae*, *Acinetobacter baumannii*, *Pseudomonas aeruginosa*, and *Enterobacter* spp. with *Escherichia coli* included in the acronym) [9,10]. The ESKAPEE group comprises bacteria that primarily cause healthcare-associated infections (HAI) and display resistance to multiple antibiotics, demanding special attention. In 2024, the WHO published a list of priority microorganisms for research and development of new antimicrobials, encouraging investment in this area. These microorganisms were classified into critical, high, and medium-priority groups, with the ESKAPEE group included in the first two categories, critical and high priority. This reinforces the urgent need for action to combat these microorganisms [11]. Among the pathogens associated with HAIs, we can highlight, in addition to those previously mentioned, the bacterial species *Escherichia coli*. Over the last few years, some authors have reported growing concern over the increase in cases of *E. coli* pathogenic strains due to their significant role in human urinary tract infections [12–14]. Reports on *E. coli* pathotypes and the prevalence of resistance to antimicrobials have led to multidrug-resistant microorganisms gaining worldwide recognition as a serious health problem affecting both humans and animals, including infections caused by ESBL-producing *E. coli* [15–17].

Given the alarming rise of multidrug-resistant bacteria, particularly the ESKAPEE group, the search for alternatives to traditional antimicrobials is intensifying, driven by the growing concern to combat resistant microorganisms. In this context, studies on nanotechnology, bacteriophages, essential oils, plant extracts, and bacteriocins are noteworthy [18]. Within the field of nanotechnology, various studies investigate the effectiveness of nanomaterials, especially metallic nanoparticles, in combating antimicrobial-resistant microorganisms [19–22]. It is important to highlight that silver nanoparticles are the most explored and studied due to their cost-effective production and potent antimicrobial action, even against multidrug-resistant strains, including those belonging to the ESKAPEE group [21–23].

These metallic nanoparticles can be produced by chemical, physical, and biological methods. Naant-based synthesis (green synthesis) is gaining great interest due to several advantages over other techniques: it offers reduced production times, eliminates the need for a culture medium, is considered safe depending on the chosen plant extract, and is considered a more sustainable and environmentally friendly technology in the synthesis of nanoparticles [24,25]. This green synthesis can be carried out using microbial agents, such as fungi and bacteria, plants and their extracts, or biological agents, such as viral DNA. The use of plants, however, has attracted significant attention due to the ability of their bioactive molecules, such as amino acids, alkaloids, and phenols, to reduce metallic ions, which is vital for the biosynthetic process of nanoparticles, and the extract used in the synthesis can be obtained from different parts of the plant, such as bark, seeds, and roots, depending on availability [26]. Furthermore, some studies have demonstrated that the formation of nanoparticles with extracts can be completed in a metal salt solution in a shorter period at room temperature, depending on the plant species used, which reduces production time and costs [27]. Some examples of plants that were used in the synthesis of nanoparticles are *Malus domestica*, *Polyalthia longifolia*, *Ficus benghalensis*, and *Ananas comosus*, among others [26,27].

Anadenanthera colubrina (Vell.) Brenan var. colubrina plant, popularly known as "Angico", is an endemic tree to South America, native to the Brazilian's biomes Cerrado, Caatinga, and Atlantic Forest. It is used frequently in folk medicine to treat infections, cough, and wound healing, and in indigenous rituals [28–30]. Its phytocomplex, which includes a lot of secondary metabolites (approximately 56 compounds), such as phenolics (gallic acid equivalents), tannins, saponins, and flavonoids, and anadanthoflavone, which is related exclusively to this species, may be related to the pharmacological properties presented by the extract, such as antimicrobial, antioxidant, and anti-inflammatory, among others [30]. Although *A. colubrina* (Vell.) Brenan metabolic composition is quite varied, and it has a wide distribution in Brazilian territory, this plant has never been used before in the synthesis of silver nanoparticles, which corresponds to a potential to be explored.

Considering the problem caused by antimicrobial resistance, the significant attention towards the ESKAPEE group, and the alternatives to combat these pathogens presented by nanotechnology, particularly biogenic silver nanoparticles, our study aimed at synthesizing silver nanoparticles using the bark extract from *Anadenanthera colubrina* (Vell.) Brenan var. colubrina plant and verifying the action of these nanoparticles against bacteria from the ESKAPEE group.

2. Results

2.1. Synthesis of Biogenic Nanoparticles

The biosynthesis of Bio-AgNPs after 24 h resulted in a yellowish, milky appearance compared to the clear, light brown appearance of the control sample (Figure 1A,B). The colloidal nature of the sample, characterized by the presence of suspended particles, was confirmed by the Tyndall effect, in which a red laser beam was visible as it passed through the sample (Figure 1C). In contrast, when the same laser beam was applied to the control, which had the same extract concentration but without the precursor salt, the light passed through the sample without any visible alteration (Figure 1D).

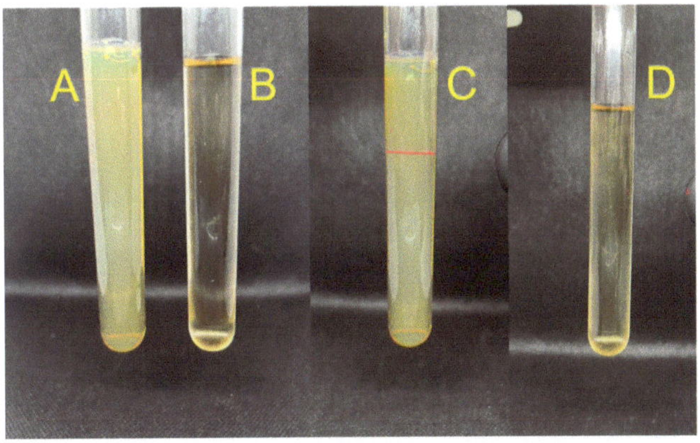

Figure 1. Comparison between the Bio-AgNPs biosynthesized using *A. colubrina* extract (**A**) and the extract without silver nitrate (**B**). Visualization of the Tyndall effect in the Bio-AgNPs solution (**C**), and absence of the Tyndall effect in the control extract sample (**D**).

2.2. Average Diameter Size, Zeta Potential and Polydispersity Index (PDI)

The polydispersity index (PDI) reveals the homogeneity of solutions, ranging from 0 to 1. A PDI value closer to 0 indicates an ideal sample with particles of the same size, while a PDI close to 1 indicates a polydisperse sample with particles of different sizes. The nanoparticles synthesized with *A. colubrina* extract had an average diameter of 75.62 ± 0.38 nm (Figure 2), with a PDI of 0.17, indicating uniform particle size and a monodisperse sample. Additionally, the zeta potential was determined at pH 5.4 from the conversion of the electrophoretic mobility using the Smoluchowski equation with a Henry factor of 1.5. The value obtained was −29 mV. The nanoparticles synthesized with *A. colubrina* extract presented readings in absorption spectra in the range between 350 nm and 420 nm, as shown in the UV–Vis spectrum (Figure 3).

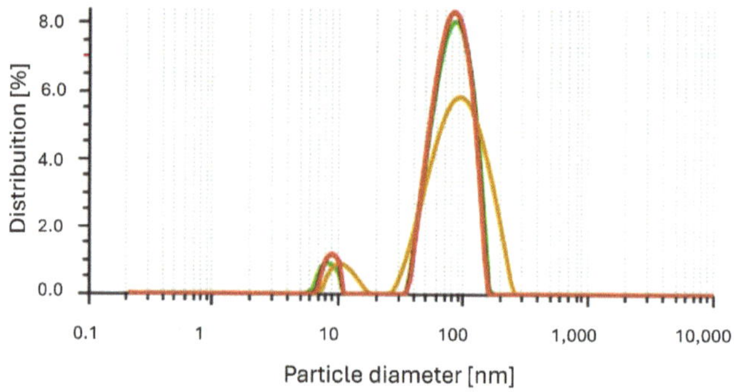

Figure 2. Histogram of the average size distribution of Bio-AgNPs (nm). The average diameter is 75.62 ± 0.38.

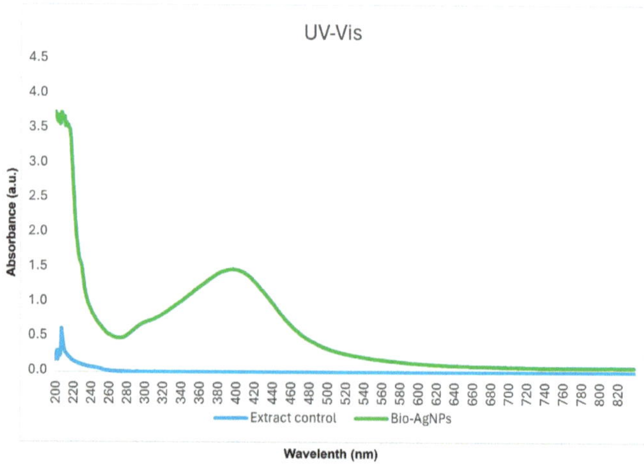

Figure 3. UV–Vis plasmonic spectrum of Bio-AgNPs and extract control.

2.3. Fourier Transform Infrared Spectroscopy (FT-IR)

FT-IR analysis of the biosynthesized silver nanoparticles (AgNPs), compared to the control (just the bark extract), did not reveal major differences in the functional groups present in the nanoparticle coating in relation to the bark extract. However, the bands associated with these groups showed a slight shift in the spectrum, and some were overlapped by neighboring bands with an increase in intensity. In the control, the band at 3266 cm^{-1} represents hydroxyl groups common in phenolic compounds such as tannins and flavonoids, which are part of the composition of *A. colubrina* extracts [31]. The band at 1616 cm^{-1} can be associated with carbon double bonds (C=C) and the presence of carbonyl groups (C=O) conjugated to aromatic systems, typical in flavonoids [32]. The peaks at 1494 cm^{-1} are characteristic of C=C stretching in aromatic rings, indicating the presence of the mentioned phenolic compounds [33]. The bands at 1282 cm^{-1} are associated with (C-O) stretching in alcohols and ethers, characteristic of tannins and flavonoids [34], while the band at 1116 cm^{-1} is also related to C-O stretching in ethers [35]. The band at 791 cm^{-1} is coherent with out-of-plane deformations of bonds (C-H) in aromatic rings, as is the band at 616 cm^{-1}, which is associated with out-of-plane deformations of C-H in benzene rings [36].

FT-IR analysis of Bio-AgNPs (Figure 4) showed the emergence of a new characteristic group (O-Ag), the maintenance of functional groups referring to (O-H), (C=C), (C-O), and the apparent absence of the band at 1494 cm^{-1}. In this case, its apparent absence can be explained by the greater intensity of the band at 1291 cm^{-1} that is overlapping the neighboring bands. The bands associated with the functional groups present in both the control and the nanoparticles are slightly shifted in the spectrum, positioning themselves at 3232, 1609, and 1291 cm^{-1}, respectively (Figure 4), indicating that these groups are part of the organic coating of the AgNPs. The displacement at 3232 cm^{-1} is attributed to stretching vibrations (O-H). The band at 1609 cm^{-1} corresponds to the stretching vibration (C=O) conjugated to aromatic systems typical of flavonoids, as well as the vibration (C=C) in aromatic rings. The bands at 1291 and 1032 cm^{-1} are typical of (C-O) stretching of ethers and alcohols. The band at 801 cm^{-1} is characteristic of out-of-plane vibrations of (C-H) bonds in aromatic rings. The band at 514 cm^{-1} is associated with metal-oxygen vibrations, in this case Ag-O, indicating interaction between the phenolic compounds tannins and flavonoids present in the *A. colubrina* extract and the silver nanoparticles [37].

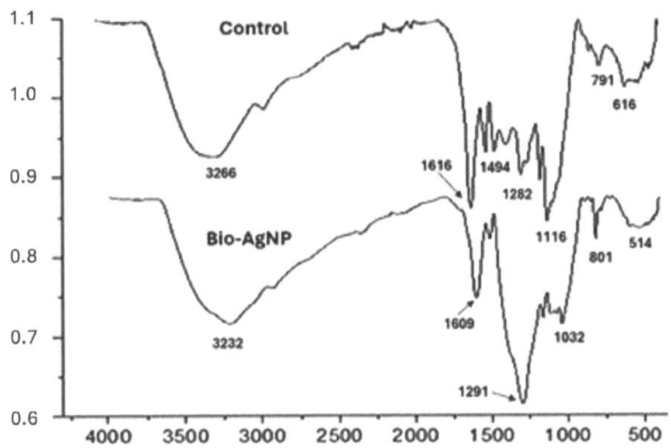

Figure 4. Comparison of Fourier transform infrared spectroscopy (FT-IR) spectra of the biosynthesized Bio-AgNPs and the control.

2.4. X-ray Diffraction (XRD)

The sample was analyzed using XRD techniques to compare the presence or absence of silver in the synthesis solution compared to the control (*A. colubrina* bark extract). The control sample did not present crystalline structures, only amorphous halos related to the phases of the plant extracts (Figure 5). The absence of organized structures in a complex sample, such as crude extracts, causes X-rays to scatter during analysis, resulting in broad peaks observed between 14° and 30° (Figure 5).

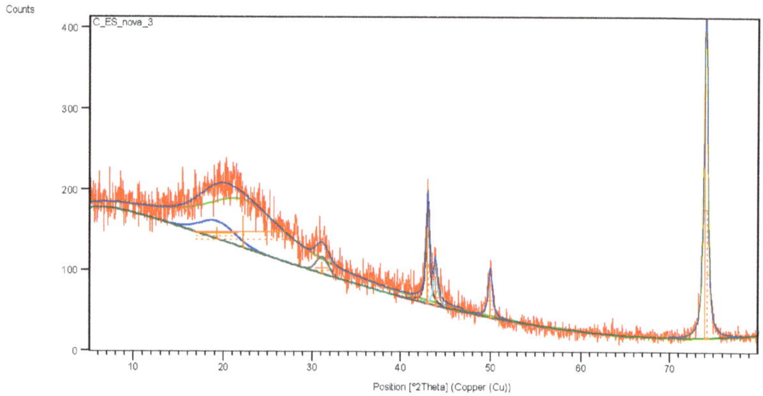

Figure 5. X-ray diffraction (XRD) of the control sample with *A. colubrina* bark extract.

In the Bio-AgNPs sample, amorphous halos were observed due to residual extract, along with peaks at 32.5° corresponding to silver and silver oxide readings, which were equivalent to 12% of the sample (Figure 6). The peaks observed in the control sample at Bragg angles of 42.9°, 43.85°, 49.9°, and 74° refer to the support where the sample was inserted to perform the analysis (Figure 5).

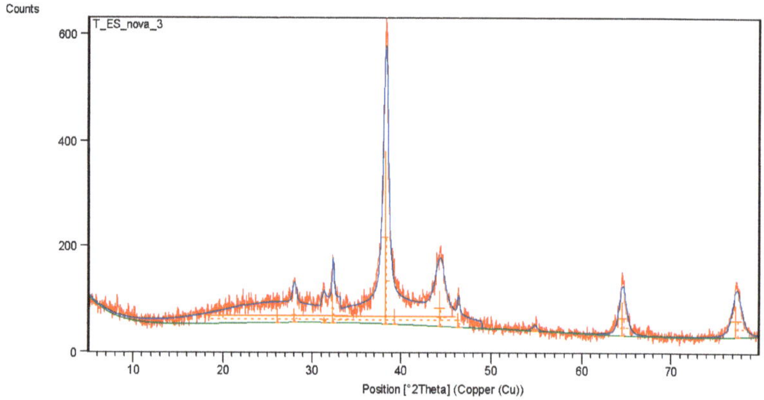

Figure 6. X-ray diffraction (XRD) of the Bio-AgNPs sample biosynthesized with *A. colubrina*.

Observing the diffraction pattern of the Bio-AgNPs particles, it is possible to measure a new crystallographic pattern, with peaks present at 38.01°, 44.3°, 64.4°, and 77.4° characteristic of the crystalline structure of silver nanoparticles, corresponding respectively to the (1 1 1), (2 0 0), (2 2 0), and (3 1 1) planes of the face-centered cubic arrangement [36].

2.5. Electron Microscopy Images

Scanning electron microscopy (SEM) images acquired 30 days after synthesis showed the presence of aggregated arrangements, suggesting potential stability loss due to clustering into larger structures, as well as the presence of organic structures from the *A. colubrina* extract within the sample and the spherical morphology of isolated Bio-AgNPs (Figure 7). The size of Bio-AgNPs was measured using the ImageJ software, yielding an average size of 65.78 ± 78.75 nm from SEM images.

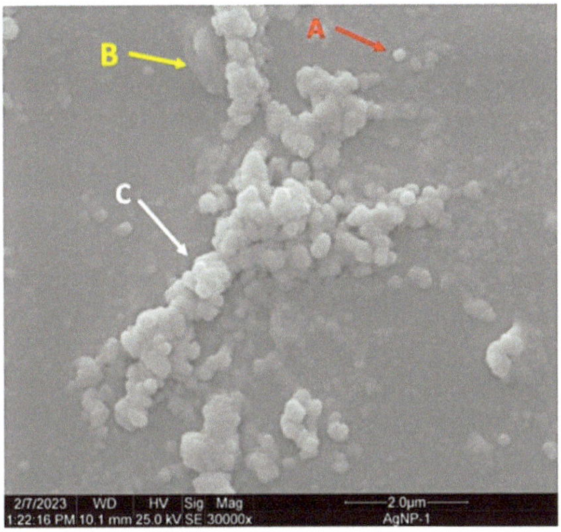

Figure 7. Morphology of Bio-AgNPs observed using scanning electron microscopy (SEM). Spherical morphology of isolated Bio-AgNPs (A). Organic structures from the *A. colubrina* extract (B). Aggregated arrangements of Bio-AgNPs and larger structures (C).

Transmission electron microscopy (TEM) analysis confirmed the presence of spherical Bio-AgNPs, with images showing aggregated Bio-AgNPs forming clusters with an average size of 200 nm. Moreover, TEM images captured some particles of approximately 50 nm and other smaller particles (Figure 8), although their distribution throughout the sample was limited.

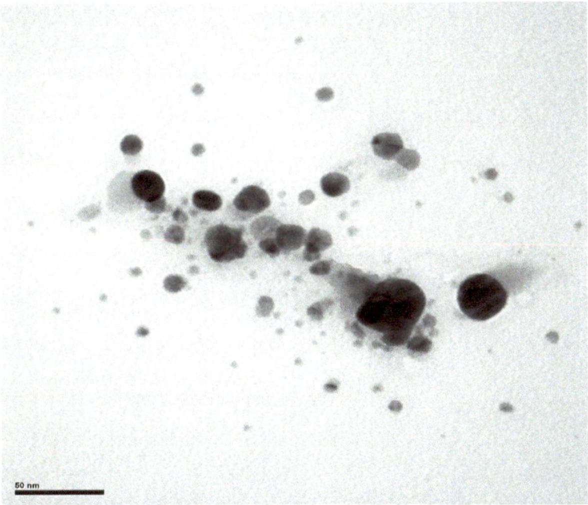

Figure 8. Morphological characterization of the Bio-AgNPs synthesized from *A. colubrina* with transmission electron microscopy (TEM). The micrograph shows spherical nanoparticles of Bio-AgNPs.

2.6. Antibacterial Activity

The antibacterial activity of Bio-AgNPs synthesized from *A. colubrina* showed significant minimal inhibitory concentration (MIC) and minimal bactericidal concentration (MBC) values, with the lowest MIC values against Gram-negative bacteria. Negative controls using culture medium Mueller–Hinton broth (MHB) and extract at equivalent synthesis concentrations (3.1 mg/mL) and superior concentrations (1250 mg/mL) were tested showing no antibacterial activity against the tested bacteria. Additionally, the antibacterial activity of silver nitrate was evaluated for comparison with Bio-AgNPs. Sensitivity tests showed that the most susceptible strains were *A. baumannii* 141, *P. aeruginosa* 3167, and *E. cloacae* 9434, with MIC and MBC values of 19.53 μM. Both *Staphylococcus aureus* strains, ATCC 25923 and BEC 9393, showed identical MIC and MBC values of 78.12 μM and 156.25 μM, respectively (Table 1).

Table 1. Antibacterial activity of Bio-AgNPs synthesized from *A. colubrina* against Gram-positive and Gram-negative bacteria, including reference and multidrug-resistant strains.

Tested Bacteria	MIC (μM)	MBC (μM)	$AgNO_3$ MIC
Enterococcus faecium 700	39.06	156.25	31.25
Staphylococcus aureus ATCC 25923	78.12	156.25	62.5
Staphylococcus aureus BEC 9393	78.12	156.25	62.5
Klebsiella pneumoniae 3978	39.06	39.06	31.25
Acinetobacter baumannii 141	≤19.53	≤19.53	7.81
Pseudomonas aeruginosa 3167	≤19.53	≤19.53	7.81
Enterobacter cloacae 9434	≤19.53	≤19.53	31.25
Escherichia coli ATCC 25922	78.12	78.12	15.62
Escherichia coli ATCC 8739	78.12	156.25	15.62
Escherichia coli 5616	39.06	78.12	15.62

MIC: minimal inhibitory concentration; MBC: minimal bactericidal concentration.

Escherichia coli strain ATCC 25922 presented an MIC value of 78.12 μM and was subjected to a survival curve assay to elucidate the kinetics and dynamics of Bio-AgNPs action over time. This assay aimed at verifying whether Bio-AgNPs exerted a bacteriostatic or bactericidal effect. Bactericidal activity is characterized by a reduction of 3 $\log_{10}$ in bacterial concentration relative to the initial concentration of the inoculum [22]. As illustrated in Figure 9, the biosynthesized Bio-AgNPs showed bactericidal kinetics. Within just over 1 h of treatment at the MIC, bacterial concentrations dropped by 1 $\log_{10}$ compared to growth control. After 2 h of treatment, bacterial concentrations were completely reduced, indicating that the biosynthesized Bio-AgNPs exhibit bactericidal action at the MIC found in this study.

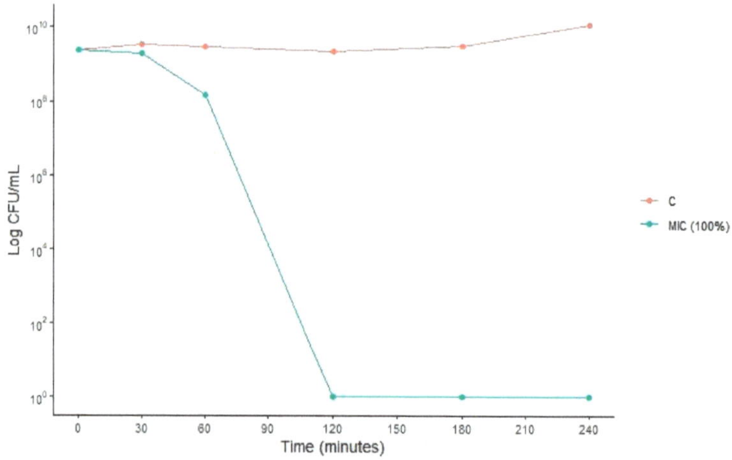

Figure 9. Growth kinetics and activity of Bio-AgNPs against *E. coli* ATCC 25922 at concentrations of MIC (78.12 μM). C (control) indicates the bacterial growth without antimicrobial.

2.7. Hemolytic Activity

In this study, a cytotoxicity test was carried out using human erythrocytes to verify the hemolytic capacity of Bio-AgNPs biosynthesized from *A. colubrina*. Figure 10 shows the CC_{50} of Bio-AgNPs determined by linear regression, as represented by the equation (Y = 50.03x + 1.914). The assay was performed with a maximum concentration of AgNP (1000 μM). The Bio-AgNPs biosynthesized using *A. colubrina* bark extract exhibited low hemolytic activity, with a 50% hemolysis rate at 961 μM (Figure 10). These findings suggest that the Bio-AgNPs synthesized in this work showed low toxicity for human erythrocytes once the cytotoxic concentration for these cells significantly exceeded the MIC range.

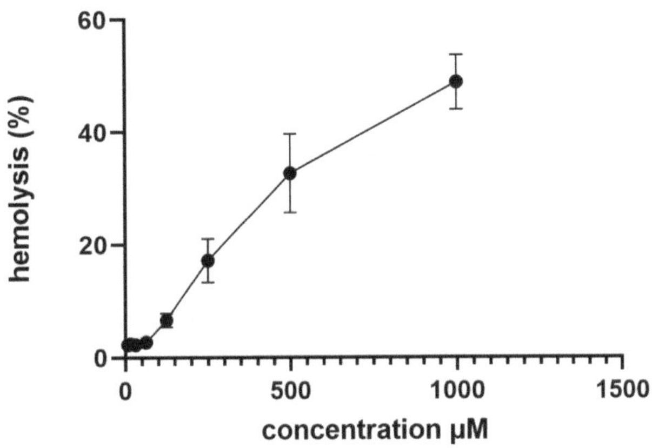

Figure 10. Hemolytic activity of Bio-AgNPs biosynthesized for *A. colubrina* through human erythrocytes. Values of hemolysis percentage are the mean ± standard deviation. The linear model equation used to predict the CC_{50} was y = 50.03x + 1.914 and the R-squared (R^2).

3. Discussion

The use of plants and plant-derived materials for treating diseases is attributed to traditional knowledge recorded over millennia by indigenous peoples and traditional communities. This practice has been encouraged and recognized by the World Health Organization since 1978 [38]. The investigation of plant-derived compounds with medicinal properties offers a basic strategy to characterize bioactive molecules and improve therapeutic approaches [39,40]. Among all the species described to date, Brazil stands out for its wide variety of plant species, distributed across six predominant biomes [41]. One of the most studied botanical families is the *Fabaceae* family, which includes *Anadenanthera colubrina*. This tree species reaches a crown height of up to 15 m, with a reddish-brown trunk and bipinnate leaves, and is found from the Caatinga to the Semideciduous Seasonal Forest [40,42,43]. The use of this plant species extends to civil construction due to the ligneous nature of its trunks [29]. *A. colubrina* is recognized for its medicinal properties, such as antimicrobial, anti-inflammatory, antioxidant, and antitumor activities [43,44]. Although this species has several therapeutic properties, some authors highlight the importance of environmental preservation owing to some beneficial properties, particularly in the seeds, in addition to warning about the necessity of replanting to minimize environmental impacts [44].

The green synthesis of Bio-AgNPs can be carried out in several ways, such as through the use of fungi, bacterial supernatant, and sand plants [45]. Studies using plants for the synthesis of Bio-AgNPs have shown nanoparticles varying from 17 nm to 50 nm, indicating that synthesis from plants can produce Bio-AgNPs of varying sizes, directly impacting their action and application [46]. Our study represents the first report of the use of *A. colubrina* bark extract to carry out Bio-AgNPs synthesis, resulting in spherical nanoparticles with an average size of 75.62 nm.

The data obtained by FT-IR analysis indicates the presence of an organic coating on the silver nanoparticles. Displacements of the bands referring to the functional groups present in the crude extract were observed, such as the bands at 3232, 1609, and 1291 cm^{-1}, attributed to stretching vibrations (O-H), (C=O), and (C-O), respectively. Functional groups (C=O) conjugated in aromatic rings and (C-O) of ethers and alcohols were also detected. These characteristics are associated with phenolic compounds, including tannins and flavonoids, present in the bark extract of *A. colubrina*, as described by [31–33]. Therefore, this suggests that tannin and flavonoid compounds interact with nanoparticles, as their characteristic functional groups undergo a shift in the spectrum after nanoparticle synthesis. These groups probably act as stabilizing and reducing agents during biosynthesis, as evidenced by the appearance of the band at 514 cm^{-1}, attributed to stretching (Ag-O). These findings are consistent with the literature, which reports the use of biocompounds for the synthesis and stabilization of silver nanoparticles [47].

Previous studies also identified the functional groups related to the *A. colubrina* bark extract as well as the interaction of biocompounds with nanoparticles, reinforcing the validity of the results presented [31,34–37]. This may have significant implications for the stability and functionality of AgNPs, potentially influencing their antibacterial and antioxidant properties, as reported in studies characterizing biosynthesized silver nanoparticles [37,48–50].

To determine the presence of metallic silver in the biosynthesized nanoparticles, an X-ray diffraction analysis was performed. The biological extracts exhibited no crystallinity pattern, resulting in a molecular network characterized by broad halos, which led to diffuse dispersion of the X-rays at Bragg angles [51,52]. This typical behavior was evident in the sample, which showed broad peaks between 14° and 30°. When comparing the control sample (*A. colubrina* bark extract) with silver nanoparticles (AgNP), it was possible to observe the amorphous characteristics of the crude extract and the emergence of new peaks indicative of metallic silver in the Bio-AgNPs. The presence of amorphous halos on the nanoparticles indicates their coating with biocompounds. However, the presence of crystallographic peaks at Bragg angles of 38.01°, 44.3°, 64.4°, and 77.4°, corresponding to the planes (111), (200), (220), and (311), are intrinsic characteristics of the face-centered cubic structure of silver nanoparticles. The XRD data indicate that the sample is composed predominantly of silver (88%) and a small amount of silver oxide (12%). The main peaks of the diffractogram are dominated by the silver structure, reflecting its high presence in the sample, while additional peaks correspond to the silver oxide phase. These results corroborate previous studies, confirming the presence of metallic silver in the biosynthesized nanoparticles, as well as their organic coating consisting of tannins and flavonoids from *A. colubrina* bark extract [53,54].

Despite being biogenically synthesized, the coating with plant extract around the silver nanoparticles confers unique characteristics and biological activities. According to the results obtained regarding the Tyndall effect, which indicates the scattering of laser light in the solution, and the PDI value of 0.17, both confirm the production of monodisperse Bio-AgNPs in a colloidal solution [55]. Previous research highlights that nanoparticles with a zeta potential ranging from 0 to $|10|$ present a tendency towards instability, while those between $|15|$ and $|20|$ show a tendency towards stability, and values between $|20|$ and $|30|$ represent strong stability [56,57]. The Bio-AgNPs synthesized using *A. colubrina* extract showed an average zeta potential of −29, indicating strong stability in solution. The size, shape, composition (structure, core, and envelope), and dispersity of Bio-AgNPs reflect their physicochemical properties, in addition to influencing their bioactivity and ability to penetrate materials [58–60]. Studies have shown that very small particles (4 to 10 nm) are capable of permeating intact skin, potentially causing histopathological changes, while larger particles (45 nm) do not have this capacity [24,61].

Furthermore, the Bio-AgNPs exhibit antimicrobial action mechanisms, decreasing the properties of protection of the bacterial cell wall and affecting the cell membrane, thereby inhibiting electron transport, changing bacterial cellular function, and causing damage to genetic material [19,62,63]. Previous works that performed Bio-AgNPs synthesis from other plants have also shown excellent results against Gram-negative bacteria [19,63]. The MIC of Bio-AgNPs tested against *E. coli* strain ATCC 25922 in this and other studies using plant-synthesized Bio-AgNPs was similar, ranging around 78.12 μM. However, this work highlights low MIC values (19.53 μM) against multidrug-resistant clinical isolates such as *A. baumannii*, *P. aeruginosa*, and *E. cloacae*.

Bacteria from the ESKAPEE group are mostly resistant to conventional drugs used to control infections, underscoring their great clinical importance once these pathogens are responsible for causing numerous healthcare-associated infections [10]. According to the WHO (2024), *Escherichia coli* is among the top three pathogens on the list of critical priority due to its resistance to third-generation cephalosporins and carbapenems [11]. The pathogen *A. baumannii*, also included in this critical group, represents a serious threat to public health [11,64,65], given its challenging treatment and high risk of transmission due to its virulence mechanisms and resistance to antimicrobial drugs, leading to patient deaths and affecting the global population [66–68]. We can highlight that the Bio-AgNPs synthesized from *A. colubrina* were successful in controlling the growth of all microorganisms within the ESKAPEE group, precisely those listed as critical priority pathogens. Despite the absence of reports on new, effective drugs against these pathogens, continued investments in research and development for novel treatment strategies are still needed [69].

The use of Bio-AgNPs synthesized using plants represents an excellent strategy for controlling microbial growth [19,70,71]. When using plants known for their medicinal characteristics, the coat surrounding Bio-AgNPs acquires unique characteristics derived from compounds that act in the control of microorganisms and plant defense resulting from the secondary metabolism of plants [72–74]. *A. colubrina* bark extract has demonstrated antibacterial properties, as described in previous works [74], with MIC and MBC values ranging from 8 μg/mL to 10 μg/mL for Gram-positive and Gram-negative bacteria. This antimicrobial activity may be related to the rich phytochemical profile presented by the plant, such as the presence of phenolic compounds, which can generate metabolic instability and eliminate the enzymatic activity of proteasomes, reducing the growth rate of the microorganism and its ability to form and mature biofilms. The presence of tannins, which are part of this phenolic composition, also contributes to this activity due to the ability of these compounds to precipitate proteins, inactivating enzymes and proteins' transportation in the cell envelope, and consequently interfering in the availability of metal ions essential to the metabolism of the microorganism [75,76]. It is important to highlight that the extraction method has a direct influence on the antibacterial compounds extracted [73,74,77]. The concentrations of the extract alone showed no activity against the tested bacteria, indicating that the antimicrobial action found in our study is exclusively attributable to the synthesized AgNPs.

Silver nitrate displayed antimicrobial action with MIC values ranging around 30 μM against *E. coli* ATCC 25922, as demonstrated in other works [78]. However, the aqueous solution of silver nitrate is reactive, presenting health risks and causing cytotoxic damage to human cells and the environment [19,22]. Optimizing AgNP synthesis using plant extracts may offer novel ways to control microorganisms, especially multidrug-resistant bacteria, directly influencing potential future applications. Thus, in this work, Bio-AgNPs were produced through green biosynthesis, which is ecologically sustainable, economically viable, and maintains exclusive characteristics justified by its synthesis. Additionally, these Bio-AgNPs possess antibacterial activity against clinically relevant Gram-positive and Gram-negative bacteria, as observed in other works [19,74,77].

The AgNP synthesis method is directly related to their cytotoxicity. Previous research using fungi for the biosynthesis of Bio-AgNPs showed lower cytotoxicity compared to other methods, such as chemical synthesis [23,25]. The CC_{50} value obtained from green synthesis using fungi was approximately 97.22 μM, indicating low toxicity [19]. Conversely, when *A. colubrina* bark extract was used for the synthesis of Bio-AgNPs, we observed even lower cytotoxicity, with a CC_{50} value of 961 μM, a concentration 12.8-fold greater than the highest MIC value (78.12 μM) obtained for the multidrug-resistant bacteria tested in this work. The highest MIC (78.12 μM) and MBC (156.25 μM) values obtained did not promote more than 5% hemolysis, which suggests that they have low toxicity and safe concentrations [79], with potential for pharmaceutical application [80]. Other studies show that the synthesis of Bio-AgNPs using plants is emerging as a promising alternative with a cost-effective approach and low impacts on human health and the environment [19,23,25,49].

4. Materials and Methods

4.1. Preparation of the Extract

The plant material used for the study was the species *Anadenanthera colubrina* (Vell.) Brenan var. colubrina. The plant sample was collected from the campus of the State University of Londrina from a single specimen located at the following geographic coordinates: 23° 19′ 26.7″ S/51° 11′ 42.0″ W, Londrina, Paraná State, Brazil. Collection was carried out using a pruning tool, and the sample was identified at the Herbarium of the State University of Londrina, under registration FUEL 56789. Moreover, registration on the National Forest Management System Platform for Genetic Heritage and Associated Traditional Knowledge (Sisgen) for recognition of genetic heritage is documented under AFCD831.

Only the outer part (the bark) was collected, avoiding pruning the specimen, ensuring minimal environmental damage or impact. The plant sample was sanitized by immersion in a solution containing distilled water and 2% sodium hypochlorite for 30 min, rinsed in running distilled water, and dried in a drying oven at 28 °C for 24 h. After drying, the sample was ground using an electric hammer mill (with Botini B55 adjustment) and stored in a glass jar with a lid, protected from light.

A decoction of the ground plant material was performed using ultrapure water at 80 °C for 1 h. The obtained extract was then sterilized by filtration using a PES syringe filter membrane (Millipore®, Burlington, MA, USA, 0.22 μm pore size) and stored at −4 °C for subsequent synthesis and testing.

4.2. Biosynthesis of Bio-AgNPs

A 100 mM stock solution of silver nitrate ($AgNO_3$; molar mass 169.87 g/mol, Sigma-Aldrich®, Waltham, MA, USA) was prepared, filtered using a PES syringe filter membrane (Millipore®, 0.22 μm pore size, 33 mm diameter), and stored at −4 °C under total light protection until synthesis.

The synthesis was carried out at 37 °C under light protection for 24 h using a stock solution of plant extract (final concentration: 3.1 mg/mL) and silver nitrate (10 mM). The resulting solution had a final pH of 5.4.

4.3. Characterization of Bio-AgNPs

Characterizations of the obtained nanoparticles included evaluations of their morphology using scanning electron microscopy (SEM), zeta potential, polydispersity index (PDI), Fourier transform infrared spectroscopy (FT-IR), and X-ray diffraction (XRD).

The absorbance of the synthesis was evaluated by ultraviolet–visible (UV–vis) molecular absorption spectroscopy for analysis of the plasmonic band, with a reading spectrum from 250 nm to 850 nm at 5 nm intervals using a UV–vis 2600 spectrophotometer (Shimadzu, Tokyo, Japan) at the Spectroscopy Laboratory of the Multiuser Center for Research Laboratories of PROPPG/UEL.

The evaluation of the electronegativity index of the samples was determined using a zetasizer (Litesizer 500, Anton Paar, Graz, Austria), through dynamic light scattering methodology, and electrophoretic dispersion at the Multiuser Laboratory of the Federal Technological University of Paraná (UTFPR), Londrina campus.

The Bio-AgNPs were subjected to FT-IR analysis using a Michelson 30° with an automatic Peltier-controlled system for polymeric identification as well as sample freeze-dried and ground for XRD analysis composition by analyzing the molecular bonds present and functional groups present in the nanoparticle coating. A 50 mL sample of Bio-AgNPs was freeze-dried (Model L101, LIOBRAS, São Carlos, Brazil) at −56 °C and 40 mmHg. The spectrum was subjected to a resolution of 2 cm^{-1} in the range of 400–4000 cm^{-1} using a Bruker Vertex 70 spectrophotometer equipped with an ATR accessory with Ge crystal at 45° (Bruker, Ettlingen, Germany). A plant extract at the same concentration without salt was used as a control. The obtained images were processed using OriginPro 9.5 software.

The structural analyses of Bio-AgNPs were performed by XRD on a PANalytical X'Pert PRO MPD diffractometer at the X-ray Diffraction Laboratory of LARX (Multiuser Laboratory of the Pro-Rectorate for Research and Postgraduate Studies—UEL) with a voltage of 40 kV and a current of 30 mA, as well as a scanning range from 8 to 80° with an angular step of 0.04°. The samples were frozen, then freeze-dried and ground for XRD analysis.

Electron microscopy images of Bio-AgNPs were obtained using SEM (FEI Quanta 200, Quanta, Minato City, Japan) and TEM (JEM-1400, JEOL, Tokyo, Japan), following well-established methodologies and protocols for samples grown in liquid media to determine the arrangements and shapes of the synthesized Bio-AgNPs. TEM imaging was conducted at the Laboratory of Technological Innovation in the Development of Drugs and Cosmetics, COMCAP-UEM Microscopy Center, State University of Maringá, Paraná, Brazil.

4.4. Evaluation of the Biological Activity of Bio-AgNPs

The analyses and evaluations of the biological activity of the nanoparticles were carried out using serial microdilution methodology in broth, according to the standards described in document M07-A10 from the Clinical and Laboratory Standards Institute (CLSI) [81]. To determine the minimum bactericidal concentration, document M26-A [82] was used as a reference, with adaptations allowing the determination of the minimum inhibitory concentration (MIC) and the minimum bactericidal concentration (MBC), as well as the evaluation of the action of nanoparticles over time, assessing bacterial survival and death after treatment with inhibitory concentrations of Bio-AgNPs. Finally, the hemolytic activity of the nanoparticles was evaluated using the methodology described elsewhere [83], with some modifications, approved by the Research Ethics Committee with human beings (CAAE47661 1 15.0.0000.5231, nº. 1.268.019-UEL).

4.4.1. Tested Strains

The microbiological material used in this study came from the bacterial collection of the Basic and Applied Bacteriology Laboratory, Department of Microbiology, Center for Biological Sciences, State University of Londrina, Londrina, PR. Strains stored at −20 °C in Brain-Heart Infusion (BHI) broth containing 30% glycerol were used.

Standard strains of microorganisms, including 3 standard strains from the American Type Culture Collection (ATCC): *Escherichia coli* ATCC 25922, *Escherichia coli* ATCC 8739, and *Staphylococcus aureus* ATCC 25923, were used. Multidrug-resistant clinical isolates from the University Hospital of Londrina, generously provided by Prof. Dr. Eliana Carolina Vespero and Prof. Dr. Floristher Elaine Carrara-Marroni (Clinical Microbiology Laboratory—Department of Pathology, Clinical and Toxicological Analysis of the University Hospital of Londrina), were also tested, including *E. coli* 5616, *Acinetobacter baumannii* 141, *Enterococcus faecium* 700, *Enterobacter cloacae* 9434, and *Klebsiella pneumoniae* 3978 samples. Additionally, tests were conducted with clinical isolates of *Pseudomonas aeruginosa* 3167, provided by Dr. Erika Kushikawa Saeki (Adolfo Lutz Institute—IAL of Presidente Prudente-SP) (Table 2).

Table 2. Antimicrobial resistance profile.

Bacterial Isolates	Resistance Profile
Enterococcus faecium 700	Vancomycin-resistant
Staphylococcus aureus BEC 9393	Methicillin-resistant
Klebsiella pneumoniae 3978	Carbapenem-resistant *
Acinetobacter baumannii 141	Carbapenem-resistant *
Pseudomonas aeruginosa 3167	Carbapenem-resistant *
Enterobacter cloacae 9434	Carbapenem-resistant *
Escherichia coli 5616	Carbapenem-resistant *

* Carbapenemase-producing bacteria.

4.4.2. Antibacterial Activity

To determine the minimum bactericidal concentrations, subcultures were prepared after exposure to treatments at different dilutions (ranging from 2500 µM to 19.5 µM/mL of Bio-AgNPs) in a 96-well polystyrene plate. Bacteria were exposed to treatments in each well at a final concentration of 7.5×10^5 CFU/mL and incubated at 37 °C for 24 h. MIC values were determined by visual reading, checking the turbidity. MBC was assessed after an incubation period of 24 h, involving plating 10 µL from each well onto Mueller–Hinton agar (MHA), followed by another incubation for 18 to 24 h at 37 °C for CFU counting. MBC was determined as the concentration corresponding to 99.9% bacterial death compared to the bacterial growth control. All assays were performed in triplicate.

Bacterial time-kill assays were performed according to the National Committee for Clinical Laboratory Standard (NCCLS) [82], with some modifications. The *Escherichia coli* ATCC 25922 strain was cultivated in culture broth until reaching the exponential bacterial growth phase after 24 h. The bacterial suspension was adjusted to approximately 1.5×10^8 CFU/mL and added to the microtubes containing Bio-AgNPs at 78.12 µM (MIC), then incubated at 37 °C. Aliquots of cultures after treatment were diluted and plated on MHA for CFU counting at different times of exposure to treatment (0.5 h, 1 h, 2 h, 3 h, 4 h, 5 h, 12 h, and 24 h). A kinetic curve of AgNP action over time was constructed as a function of incubation times.

4.4.3. Hemolytic Activity Assay of Bio-AgNPs

The hemolytic activity assay of Bio-AgNPs was carried out using human erythrocytes in heparinized tubes, according to previous studies [19,83] approved by the Research Ethics Committee (CAAE 47661115.0.0000.5231, n°. 1.268.019-UEL). Erythrocytes were separated from plasma by centrifugation at 5000 rpm, 4 °C for 5 min, and subsequently diluted in phosphate-buffered saline (PBS) to a concentration of 6% (*v*/*v*). The tests were carried out in a 96-well polystyrene plate with varying concentrations of Bio-AgNPs, with the highest concentration tested being 1000 µM. PBS was used as the negative control, while 1% Triton X-100 (Sigma-Aldrich®) was used as the positive control, representing 100% hemolytic activity. The plate was incubated at 37 °C for 3 h, and the supernatant absorbance was read at 570 nm.

5. Conclusions

This work demonstrated for the first time the green synthesis of Bio-AgNPs from the bark extract of *Anadenanthera colubrina*, a tree species endemic to South America. The bark extract of *A. colubrina* resulted in the synthesis of stable spherical silver nanoparticles measuring 75.62 nm. These nanoparticles presented bactericidal action against bacterial strains and multidrug-resistant clinical isolates of the ESKAPEE group. The extract effectively coated the nanoparticles, resulting in Bio-AgNPs with unique characteristics, including excellent antibacterial activity and low toxicity to human erythrocytes (CC_{50} = 961 µM). The antibacterial action was observed against Gram-positive and Gram-negative bacteria, with the clinical isolates *Acinetobacter baumannii*, *Pseudomonas aeruginosa*, and *Enterobacter cloacae* displaying heightened susceptibility, with MIC values ranging approximately 19.53 μM, highlighting the effectiveness of plant-based synthesis for producing Bio-AgNPs.

6. Patents

This work has been granted a patent by the National Institute of Industrial Property under registration number BR 10 2024 003295 0.

Author Contributions: Conceptualization, A.N.D. and R.K.T.K.; methodology, A.N.D., L.M.d.S.S. and J.M.D.T.; formal analysis, A.N.D., L.M.d.S.S., G.J.S.A., J.G.-C. and D.D.; investigation, A.N.D., A.G.d.O. and G.N.; resources, J.M.D.T.; writing—original draft, A.N.D.; writing—review and editing, L.M.d.S.S., G.N. and R.K.T.K.; supervision, A.G.d.O., G.N. and R.K.T.K.; funding acquisition, R.K.T.K. All authors have read and agreed to the published version of the manuscript.

Funding: Fundação Araucária: Conv. 736/2022-PDI-PESQUISA-INICIATIVA AMAZÔNIA +10/ Paraná (2022171000010); NAPI Biodiversidade. Coordenação de Aperfeiçoamento de Pessoal de Nível Superior: Code 001. National Council for Scientific and Technological Development: DT 2022 (305972/2022-7).

Institutional Review Board Statement: The study was conducted in accordance approved by the Research Ethics Committee CAAE 47661115.0.0000.5231, nº. 1.268.019-UEL.

Informed Consent Statement: Not applicable.

Data Availability Statement: The original contributions presented in the study are included in the article, further inquiries can be directed to the corresponding author.

Conflicts of Interest: The authors declare no conflict of interest.

References

1. Cuong, N.V.; Padungtod, P.; Thwaites, G.; Carrique-Mas, J.J. Antimicrobial usage in animal production: A review of the literature with a focus on low-and middle-income countries. *Antibiotics* **2018**, *7*, 75. [CrossRef]
2. Franco, B.E.; Altagracia, M.M.; Sánchez, R.M.A.; Wertheimer, A.I. The determinants of the antibiotic resistance process. *Infect. Drug Resist.* **2009**, *2*, 1–11. [PubMed]
3. Kasimanickam, V.; Kasimanickam, M.; Kasimanickam, R. Antibiotics use in food animal production: Escalation of antimicrobial resistance: Where are we now in combating AMR? *Med. Sci.* **2021**, *9*, 14. [CrossRef] [PubMed]
4. Hernando-Amado, S.; Coque, T.M.; Baquero, F.; Martínez, J.L. Defining and combating antibiotic resistance from One Health and Global Health perspectives. *Nat. Microbiol.* **2019**, *4*, 1432–1442. [CrossRef] [PubMed]
5. Pitt, S.J.; Gunn, A. The One Health Concept. *Br. J. Biomed. Sci.* **2024**, *81*, 12366. [CrossRef] [PubMed]
6. World Health Organization. Available online: https://www.who.int/publications/i/item/9789241509763 (accessed on 9 July 2024).
7. Antimicrobial Resistance Collaborators. Global burden of bacterial antimicrobial resistance in 2019: A systematic analysis. *Lancet* **2022**, *399*, 629–655. [CrossRef]
8. O'Neill, J.I.M. Tackling drug-resistant infections globally: Final report and recommendation. *Rev. Antimicrob. Resist.* **2014**, 84. [CrossRef]
9. Lin, Q.; Deslouches, B.; Montelaro, R.C.; Di, Y.P. Prevention of ESKAPE pathogen biofilm formation by antimicrobial peptides WLBU2 and LL37. *Int. J. Antimicrob. Agents* **2018**, *52*, 667–672. [CrossRef]
10. Willyard, C. Drug-resistant bacteria ranked. *Nature* **2017**, *543*, 15. [CrossRef]
11. World Health Organization. Available online: https://www.who.int/publications/i/item/9789240093461 (accessed on 9 July 2024).
12. Gomes, T.A.T.; Elias, W.P.; Scaletsky, I.C.A.; Guth, B.E.C.; Rodrigues, J.F.; Piazza, R.M.F.; Ferreira, L.C.S.; Martinez, M.B. Diarrheagenic *Escherichia coli*. *Brazilian J. Microbiol.* **2016**, *47*, 3–30. [CrossRef]
13. Terlizzi, M.E.; Gribaudo, G.; Maffei, M.E. UroPathogenic *Escherichia coli* (UPEC) infections: Virulence factors, bladder responses, antibiotic, and non-antibiotic antimicrobial strategies. *Front. Microbiol.* **2017**, *8*, 1566. [CrossRef] [PubMed]
14. Campos, J.C.; Antunes, L.C.M.; Ferreira, R.B.R. Global priority pathogens: Virulence, antimicrobial resistance and prospective treatment options. *Future Microbiol.* **2020**, *15*, 649–677. [CrossRef] [PubMed]
15. Hamers, V.; Huguet, C.; Bourjot, M.; Urbain, A. Antibacterial Compounds from Mushrooms: A Lead to Fight ESKAPEE Pathogenic Bacteria? *Planta Med.* **2021**, *87*, 351–367. [CrossRef] [PubMed]
16. Islam, M.S.; Rahman, A.M.M.T.; Hassan, J.; Rahman, M.T. Extended-spectrum beta-lactamase in *Escherichia coli* isolated from humans, animals, and environments in Bangladesh: A One Health perspective systematic review and meta-analysis. *One Health* **2023**, *16*, 100526. [CrossRef] [PubMed]

17. Brindangnanam, P.; Sawant, A.R.; Prashanth, K.; Coumar, M.S. Bacterial effluxome as a barrier against antimicrobial agents: Structural biology aspects and drug targeting. *Tissue Barriers* **2022**, *10*, 2013695. [CrossRef]
18. Ghosh, C.; Sarkar, P.; Issa, R.; Haldar, J. Alternatives to conventional antibiotics in the era of antimicrobial resistance. *Trends Microbiol.* **2019**, *27*, 323–338. [CrossRef]
19. Scandorieiro, S.; Camargo, L.C.; Lancheros, C.A.; Yamada-Ogatta, S.F.; Nakamura, C.V.; Oliveira, A.G.; Andrade, C.G.; Duran, N.; Nakazato, G.; Kobayashi, R.K. Synergistic and additive effect of oregano essential oil and biological silver nanoparticles against multidrug-resistant bacterial strains. *Front. Microbiol.* **2016**, *7*, 760. [CrossRef]
20. Figueiredo, E.P.; Ribeiro, J.M.; Nishio, E.K.; Scandorieiro, S.; Costa, A.F.; Cardozo, V.F.; Oliveira, A.G.; Durán, N.; Panagio, L.A.; Kobayashi, R.K.; et al. New approach for simvastatin as an antibacterial: Synergistic effect with bio-synthesized silver nanoparticles against multidrug-resistant bacteria. *Int. J. Nanomedicine* **2019**, *14*, 7975–7985. [CrossRef]
21. Sánchez-López, E.; Gomes, D.; Esteruelas, G.; Bonilla, L.; Lopez-Machado, A.L.; Galindo, R.; Cano, A.; Espina, M.; Ettcheto, M.; Camins, A.; et al. Metal-based nanoparticles as antimicrobial agents: An overview. *Nanomaterials* **2020**, *10*, 292. [CrossRef]
22. Souza, L.M.S.; Miriam, D.; Sarmiento, J.J.P.; Seabra, A.B.; Medeiros, L.P.; Lourenço, I.M.; Kobayashi, R.K.; Nakazato, G. Biosynthesis of selenium nanoparticles using combinations of plant extracts and their antibacterial activity. *Curr. Res. Green. Sustain. Chem.* **2022**, *5*, 100303. [CrossRef]
23. Fatima, M.; Zaidi, S.; Amraiz, D.; Afzal, F. In vitro antiviral activity of *Cinnamomum cassia* and its nanoparticles against H7N3 influenza A virus. *J. Microbiol. Biotechnol.* **2016**, *26*, 151–159. [CrossRef]
24. Khan, F.; Shariq, M.; Asif, M.; Siddiqui, M.A.; Malan, P.; Ahmad, F. Green nanotechnology: Plant-mediated nanoparticle synthesis and application. *Nanomaterials* **2022**, *12*, 673. [CrossRef] [PubMed]
25. Liaqat, N.; Jahan, N.; Khalil-Ur-Rahman; Anwar, T.; Qureshi, H. Green synthesized silver nanoparticles: Optimization, characterization, antimicrobial activity, and cytotoxicity study by hemolysis assay. *Front. Chem.* **2022**, *10*, 952006. [CrossRef]
26. Mustapha, T.; Misni, N.; Ithnin, N.R.; Daskum, A.M.; Unyah, N.Z. A Review on Plants and Microorganisms Mediated Synthesis of Silver Nanoparticles, Role of Plants Metabolites and Applications. *Int. J. Environ. Res. Public Health* **2022**, *19*, 674. [CrossRef]
27. Alabdallah, N.M.; Hasan, M.M. Plant-Based Green Synthesis of Silver Nanoparticles and Its Effective Role in Abiotic Stress Tolerance in Crop Plants. *Saudi J. Biol. Sci.* **2021**, *28*, 5631–5639. [CrossRef]
28. de Almeida Maia, C.M.; Vasconcelos, P.G.S.; Pasetto, S.; Godwin, W.C.; Silva, J.P.R.; Tavares, J.F.; Pardi, V.; Costa, E.M.M. de B.; Murata, R.M. Anadenanthera Colubrina Regulated LPS-Induced Inflammation by Suppressing NF-KB and P38-MAPK Signaling Pathways. *Sci. Rep.* **2024**, *14*, 16028.
29. Delices, M.; Muller, J.d.A.I.; Arunachalam, K.; Martins, D.T.O. Anadenanthera Colubrina (Vell) Brenan: Ethnobotanical, Phytochemical, Pharmacological and Toxicological Aspects. *J. Ethnopharmacol.* **2023**, *300*, 115745. [CrossRef]
30. Nascimento, J.P.B.; Bispo, J.S.; Dantas, B.F. Angico Anadenanthera colubrina varo cebil (Vell.) Brenan. Nota Técnica 10 Angico. *Inf. Abrates* **2019**, *29*, 47–51.
31. Mota, G.S.; Sartori, C.J.; Miranda, I.; Quilhó, T.; Mori, F.A.; Pereira, H. Bark anatomy, chemical composition and ethanol-water extract composition of *Anadenanthera peregrina* and *Anadenanthera colubrina*. *PLoS ONE* **2017**, *12*, e0189263. [CrossRef]
32. Krysa, M.; Szymańska-Chargot, M.; Zdunek, A. FT-IR and FT-Raman fingerprints of flavonoids—A review. *Food Chem.* **2022**, *393*, 133430. [CrossRef]
33. Araújo, T.S.L.; de Oliveira, T.M.; de Sousa, N.A.; Souza, L.K.M.; Sousa, F.B.M.; de Oliveira, A.P.; Nicolau, L.A.D.; da Silva, A.A.V.; Araújo, A.R.; Magalhães, P.J.C.; et al. Biopolymer extracted from *Anadenanthera colubrina* (Red angico gum) exerts therapeutic potential in mice: Antidiarrheal activity and safety assessment. *Pharmaceuticals* **2020**, *13*, 17. [CrossRef] [PubMed]
34. Ortuño, J.; Stergiadis, S.; Koidis, A.; Smith, J.; Humphrey, C.; Whistance, L.; Theodoridou, K. Rapid tannin profiling of tree fodders using untargeted mid-infrared spectroscopy and partial least squares regression. *Plant Methods* **2021**, *17*, 14. [CrossRef]
35. Omran, A.M.E. Green route synthesis of silver nanoparticles driven by Cassia fistula flower extract: Characterization, antioxidant, antibacterial, anticancer, and photocatalytic assessment. *Biomass Conv. Bioref* **2023**. [CrossRef]
36. Feria-Reyes, R.; Ramírez-Cruz, S.O.; Ruiz-Aquino, F.; Robledo-Taboada, L.H.; Sánchez-Medina, M.A.; Mijangos-Ricárdez, O.F.; Gabriel-Parra, R.; Suárez-Mota, M.E.; Puc-Kauil, R.; Porcallo-Vargas, J. Pine Bark as a Potential Source of Condensed Tannin: Analysis through Fourier Transform Infrared Spectroscopy (FTIR), Scanning Electron Microscopy (SEM), and Energy Dispersive X-ray (EDX). *Forests* **2023**, *14*, 1433. [CrossRef]
37. Alzubaidi, A.K.; Al-Kaabi, W.J.; Al Ali, A.; Albukhaty, S.; Al-Karagoly, H.; Sulaiman, G.M.; Asiri, M.; Khane, Y. Green Synthesis and Characterization of Silver Nanoparticles Using Flaxseed Extract and Evaluation of Their Antibacterial and Antioxidant Activities. *Appl. Sci.* **2023**, *13*, 2182. [CrossRef]
38. World Health Organization. *Report of the International Conference on Primary Health Care, Jointly Sponsored by the World Health Organization and the United Nations Children's Fund, Alma-Ata, URSS*; World Health Organization: Geneva, Switzerland, 1978.
39. Rodrigues, A.R.D.S.P.; Alencar, C.D.C. Ação antioxidante de espécies vegetais nativas do Brasil: Uma revisão integrativa. *Revista Fitos* **2023**, *17*, 551–560. [CrossRef]
40. Silva, A.C.D. Fungos Endofíticos de *Anadenanthera colubrina* (Vell.) Brenan de área de Caatinga: Diversidade e Potencial Antifúngico Contra Espécies de *Sporothrix*. Master's Thesis, Universidade Federal de Pernambuco, Recife, Brazil, 2023.
41. Flora do Brasil. Available online: https://floradobrasil.jbrj.gov.br/FB126378 (accessed on 22 December 2023).
42. Nakaoka Sakita, M.; Vallilo, M.I. Preliminary phytochemical studies of forest tree species of Morro do Diabo State Park. *Rev. Do Inst. Florest.* **1990**, *2*, 215226.

43. Lorenzi, H.; Matos, F.J. *Plantas Medicinais No Brasil: Nativas e Exóticas*, 2nd ed.; Nova Odessa, S.P., Ed.; Instituto Plantarum: Nova Odessa, Brazil, 2008; p. 576.
44. Luna, E.M.; Lopes, H.T.O.; Rodrigues, F.A.Á.; Coutinho, H.D.M.; de Oliveira, L.C.C. Antioxidant potential of the Caatinga flora. *Phytomedicine Plus* **2022**, *2*, 100240. [CrossRef]
45. Duran, N.; Seabra, A.B. Biogenic synthesized Ag/Au nanoparticles: Production, characterization, and applications. *Current Nanosci.* **2018**, *14*, 82–94. [CrossRef]
46. Karatoprak, G.Ş.; Aydin, G.; Altinsoy, B.; Altinkaynak, C.; Koşar, M.; Ocsoy, I. The Effect of *Pelargonium endlicherianum* Fenzl. root extracts on formation of nanoparticles and their antimicrobial activities. *Enz. Microbial. Technolo.* **2017**, *97*, 21–26. [CrossRef]
47. Bar, H.; Bhui, D.K.; Sahoo, G.P.; Sarkar, P.S.P.; Misra, A. Green synthesis of silver nanoparticles using latex of Jatropha curcas. *Colloids Surf. A Physicochem. Eng. Asp.* **2009**, *339*, 134–139. [CrossRef]
48. Bezza, F.A.; Tichapondwa, S.M.; Chirwa, E.M.N. Synthesis of biosurfactant stabilized silver nanoparticles, characterization and their potential application for bactericidal purposes. *J. Hazard. Mater.* **2020**, *393*, 122319. [CrossRef]
49. Poinern, G.E.J.; Chapman, P.; Shah, M.; Fawcett, D. Green biosynthesis of silver nanocubes using the leaf extracts from *Eucalyptus macrocarpa*. *Nano. Bull.* **2013**, *2*, 1–7.
50. Danaei, M.R.M.M.; Dehghankhold, M.; Ataei, S.; Hasanzadeh Davarani, F.; Javanmard, R.; Dokhani, A.; Khorasani, S.; Mozafari, M.R. Impact of particle size and polydispersity index on the clinical applications of lipidic nanocarrier systems. *Pharmaceutics* **2018**, *10*, 57. [CrossRef]
51. Bates, S.; Zografi, G.; Engers, D.; Morris, K.; Crowley, K.; Newman, A. Analysis of amorphous and nanocrystalline solids from their X-ray diffraction patterns. *Pharm. Res.* **2006**, *23*, 2333–2349. [CrossRef]
52. Lu, Z.; Rong, K.; Li, J.; Yang, H.; Chen, R. Size-dependent antibacterial activities of silver nanoparticles against oral anaerobic pathogenic bacteria. *J. Mater. Sci. Mater. Med.* **2013**, *24*, 1465–1471. [CrossRef]
53. Bieski, I.G.C.; Santos, F.R.; Oliveira, R.M.; Espinosa, M.M.; Macedo, M.; Albuquerque, U.P.; Martins, D.T.O. Ethnopharmacology of medicinal plants of the pantanal region (Mato Grosso, Brazil). *Evi. Based Complement Altern. Med.* **2012**, *2012*, 272749. [CrossRef] [PubMed]
54. Bezerra, J.J.L.; Pinheiro, A.A.V.; Oliveira Barreto, E. Medicinal plants used in the treatment of asthma in different regions of Brazil: A comprehensive review of ethnomedicinal evidence, preclinical pharmacology and clinical trials. *Phytomedicine Plus* **2022**, *2*, 100376. [CrossRef]
55. Sena, A.E.C.; Ramos, A.L.; Faria, F.S.E.D.V. Avaliação da síntese de nanopartículas de prata sob diferentes concentrações do extrato de *Copaíba multijuga* (Heine). *Sci. Nat.* **2019**, *1*, 1.
56. Ssekatawa, K.; Byarugaba, D.K.; Kato, C.D.; Ejobi, F.; Tweyongyere, R.; Lubwama, M.; Kirabira, J.B.; Wampande, E.M. Nanotechnological solutions for controlling transmission and emergence of antimicrobial-resistant bacteria, future prospects, and challenges: A systematic review. *J. Nanoparticle Res.* **2020**, *22*, 117. [CrossRef]
57. Ssekatawa, K.; Byarugaba, D.K.; Angwe, M.K.; Wampande, E.M.; Ejobi, F.; Nxumalo, E.; Maaza, M.; Sackey, J.; Kirabira, J.B. Phyto-mediated copper oxide nanoparticles for antibacterial, antioxidant and photocatalytic performances. *Front. Bioeng Biotechnolo* **2022**, *10*, 820218. [CrossRef]
58. Filon, F.L.; Mauro, M.; Adami, G.; Bovenzi, M.; Crosera, M. Nanoparticles skin absorption: New aspects for a safety profile evaluation. *Regul. Toxicol. Pharmacol.* **2015**, *72*, 310–322. [CrossRef] [PubMed]
59. Shaikh, S.; Nazam, N.; Rizvi, S.M.D.; Ahmad, K.; Baig, M.H.; Lee, E.J.; Choi, I. Mechanistic insights into the antimicrobial actions of metallic nanoparticles and their implications for multidrug resistance. *Int. J. Mol. Sci.* **2019**, *20*, 2468. [CrossRef] [PubMed]
60. Mundekkad, D.; Cho, W.C. Nanoparticles in clinical translation for cancer therapy. *Int. J. Mol. Sci.* **2022**, *23*, 1685. [CrossRef]
61. Kumar, V.; Yadav, S.C.; Yadav, S.K. Syzygium cumini leaf and seed extract mediated biosynthesis of silver nanoparticles and their characterization. *J. Chem. Technol. Biotechnol.* **2010**, *85*, 1301–1309. [CrossRef]
62. World Health Organization. *Prioritization of Pathogens to Guide Discovery, Research and Development of New Antibiotics for Drug-Resistant Bacterial Infections, Including Tuberculosis*; World Health Organization: Genebra, Suíça, 2017.
63. Jiang, Y.; Ding, Y.; Wei, Y.; Jian, C.; Liu, J.; Zeng, Z. Carbapenem-resistant *Acinetobacter baumannii*: A challenge in the intensive care unit. *Front. Microbiol.* **2022**, *13*, 1045206.
64. Ayoub Moubareck, C.; Hammoudi Halat, D. Insights into Acinetobacter baumannii: A Review of Microbiological, Virulence, and Resistance Traits in a Threatening Nosocomial Pathogen. *Antibiotics* **2020**, *9*, 119. [CrossRef] [PubMed]
65. Yu, Z.; Tang, J.; Khare, T.; Kumar, V. The alarming antimicrobial resistance in ESKAPEE pathogens: Can essential oils come to the rescue. *Fitoterapia* **2020**, *140*, 104433. [CrossRef]
66. World Health Organization. *2021 Antibacterial Agents in Clinical and Preclinical Development: An Overview and Analysis*; World Health Organization: Genebra, Suíça, 2022.
67. Hashemi, A.; Tavafi, H.; Naseri, M.; Mojtabazadeh, H.; Abedi, M.; Tork, N. Structural and antibacterial properties of $AgFe_2O_4$ and Fe_3O_4 nanoparticles, and their nanocomposites. *Prog. Phys. Appl. Mater.* **2024**, *4*, 37–46.
68. Naikoo, G.A.; Mustaqeem, M.; Hassan, I.U.; Awan, T.; Arshad, F.; Salim, H.; Qurashi, A. Bioinspired and green synthesis of nanoparticles from plant extracts with antiviral and antimicrobial properties: A critical review. *J. Saudi Chem. Soc.* **2021**, *25*, 101304. [CrossRef]

69. Thomas, B.; Vithiya, B.; Prasad, T.; Mohamed, S.B.; Magdalane, C.M.; Kaviyarasu, K.; Maaza, M. Antioxidant and photocatalytic activity of aqueous leaf extract mediated green synthesis of silver nanoparticles using *Passiflora edulis f. flavicarpa*. *J. Nanosci. Nanotechnol.* **2019**, *19*, 2640–2648. [CrossRef]
70. Lakkim, V.; Reddy, M.C.; Pallavali, R.R.; Reddy, K.R.; Reddy, C.V.; Inamuddin; Bilgrami, A.L.; Lomada, D. Green Synthesis of Silver Nanoparticles and Evaluation of Their Antibacterial Activity against Multidrug-Resistant Bacteria and Wound Healing Efficacy Using a Murine Model. *Antibiotics* **2020**, *9*, 902. [CrossRef] [PubMed]
71. Martins, G.M.C.; da Silva, J.M.; Silva, R.B.; da Silva, H.C.H.; Silva, J.V.; Moura, F.D.B.P. Potencial alelopático de extratos aquosos de folhas de *Anadenanthera colubrina* (vell.) Brenan. *Rev. Ouricuri.* **2020**, *10*, 001–010. [CrossRef]
72. Lima, A.R.N.; Macedo, R.G.; Batista, G.G.; Câmara, G.B.; de Freitas Lima, R.; de Oliveira, T.K.B. Atividade antimicrobiana e anti-inflamatória da *Anadenanthera colubrina* (Vell.) Brenan. *Res. Soc. Dev.* **2020**, *9*, e121911770. [CrossRef]
73. Volpato, L.E.R.; Trigueiro, P.G.D.C.; Aranha, A.M.F.; Violante, I.M.P.; Silva, R.A.D.; Oliveira, R.C.D. Antimicrobial potential of plant extracts from the Brazilian Cerrado. *Braz. Dent. J.* **2022**, *33*, 96–104. [CrossRef]
74. Gomes, D.M.D.; Durán, N.; Seabra, A.B.; de Paiva Silva, L.; Prado, F.B.; de Amorim Silva, T.; Teixeira, M.F.S. Síntese verde de nanopartículas de prata intermediada por fungo anamórfico e eficácia antibacteriana e antifúngica. *Bol. Do Mus. Para. Emílio Goeldi-Ciências Nat.* **2020**, *15*, 433–443. [CrossRef]
75. Rocha, E.A.L.S.S.; de Medeiros, A.C.D.; de Castro, R.D.; Rosalen, P.L.; Saraiva, K.L.A.; Godoy, G.P.; da Silva, L.R.A.; Aleixo, C.S.S.; Silva, P.G.; de Brito Costa, E.M.M. Antifungal Activity, Phytochemical Characterization and Thermal Profile of Anadenanthera Colubrina (Vell.) Brenan. *Pesqui. Bras. Odontopediatria Clin. Integr.* **2017**, *17*, 1–14. [CrossRef]
76. Vannini, A.B.; Gallassini Tonini, R.C.; Carles Sturn, T.; Onofre, S.B.; Caovilla Follador, F.A.; Arruda, G. Evaluation of the Antibacterial Activity of Bark Extracts from Anadenanthera Colubrine (Vell.) Brenan (Fabaceae Lindl). *Int. J. New Technol. Res.* **2019**, *5*, 21–24. [CrossRef]
77. Roy, A.; Bulut, O.; Some, S.; Mandal, A.K.; Yilmaz, M.D. Green synthesis of silver nanoparticles: Biomolecule-nanoparticle organizations targeting antimicrobial activity. *RSC Adv.* **2019**, *9*, 2673–2702. [CrossRef]
78. Ssekatawa, K.; Byarugaba, D.K.; Kato, C.D.; Wampande, E.M.; Ejobi, F.; Nakavuma, J.L.; Maaza, M.; Sackey, J.; Nxumalo, E.; Kirabira, J.B. Green strategy–based synthesis of silver nanoparticles for antibacterial applications. *Front. Nanotechno* **2021**, *3*, 697303. [CrossRef]
79. Sæbø, I.P.; Bjørås, M.; Franzyk, H.; Helgesen, E.; Booth, J.A. Optimization of the hemolysis assay for the assessment of cytotoxicity. *Int. J. Mol. Sci.* **2023**, *24*, 2914. [CrossRef]
80. Gorr, S.U.; Flory, C.M.; Schumacher, R.J. In vivo activity and low toxicity of the second-generation antimicrobial peptide DGL13K. *PLoS ONE* **2019**, *14*, e0216669. [CrossRef] [PubMed]
81. *M07-A10*; Methods for Dilution Antimicrobial Susceptibility Tests for Bacteria that Grow Aerobically: 11th ed. Clinical and Laboratory Standards Institute: Wayne, PA, USA, 2018.
82. *M26-A*; Methods for Determining Bactericidal Activity of Antimicrobial Agents. Clinical and Laboratory Standards Institute: Wayne, PA, USA, 1999.
83. Izumi, E.; Ueda-Nakamura, T.; Veiga, V.F., Jr.; Pinto, A.C.; Nakamura, C.V. Terpenes from *Copaifera* demonstrated in vitro antiparasitic and synergic activity. *J. Med. Chem.* **2012**, *55*, 2994–3001. [CrossRef] [PubMed]

antibiotics

MDPI

Article

Clinacanthus nutans (Burm. f.) Lindau Extract Inhibits Dengue Virus Infection and Inflammation in the Huh7 Hepatoma Cell Line

Kanyaluck Jantakee [1,†], Suthida Panwong [2,†], Pachara Sattayawat [1,3], Ratchaneewan Sumankan [4], Sasithorn Saengmuang [4], Kiattawee Choowongkomon [5] and Aussara Panya [1,3,*]

1 Department of Biology, Faculty of Science, Chiang Mai University, Chiang Mai 50200, Thailand; kanyaluckjan@gmail.com (K.J.); pachara.sattayawat@cmu.ac.th (P.S.)
2 Doctoral of Philosophy Program in Applied Microbiology (International Program), Faculty of Science, Chiang Mai University, Chiang Mai 50200, Thailand; suthida_pan@cmu.ac.th
3 Cell Engineering for Cancer Therapy Research Group, Faculty of Science, Chiang Mai University, Chiang Mai 50200, Thailand
4 Graduate Master's Degree Program in Biology, Faculty of Science, Chiang Mai University, Chiang Mai 50200, Thailand; ratchaneewan_sum@cmu.ac.th (R.S.); sasithorn-mind@hotmail.com (S.S.)
5 Department of Biochemistry, Faculty of Science, Kasetsart University, Bangkok 10900, Thailand; kiattawee.c@ku.th
* Correspondence: aussara.pan@cmu.ac.th; Tel.: +66-53-943346
† These authors contributed equally to this work.

Abstract: Dengue virus (DENV) infection has emerged as a global health problem, with no specific treatment available presently. *Clinacanthus nutans* (Burm. f.) Lindau extract has been used in traditional medicine for its anti-inflammatory and antiviral properties. We thus hypothesized *C. nutans* had a broad-ranged activity to inhibit DENV and the liver inflammation caused by DENV infection. The study showed that treatment using *C. nutans* extract during DENV infection (co-infection step) showed the highest efficiency in lowering the viral antigen concentration to 22.87 ± 6.49% at 31.25 μg/mL. In addition, the virus–host cell binding assay demonstrated that *C. nutans* treatment greatly inhibited the virus after its binding to Huh7 cells. Moreover, it could remarkably lower the expression of cytokine and chemokine genes, including *TNF-α*, *CXCL10*, *IL-6*, and *IL-8*, in addition to inflammatory mediator *COX-2* genes. Interestingly, the activation of the NF-κB signaling cascade after *C. nutans* extract treatment was dramatically decreased, which could be the underlying mechanism of its anti-inflammatory activity. The HPLC profile showed that gallic acid was the bioactive compound of *C. nutans* extract and might be responsible for the antiviral properties of *C. nutans*. Taken together, our results revealed the potential of *C. nutans* extract to inhibit DENV infection and lower inflammation in infected cells.

Keywords: dengue virus; *Clinacanthus nutans*; antiviral agent; pro-inflammatory cytokine; anti-inflammation; NF-κB signaling

Citation: Jantakee, K.; Panwong, S.; Sattayawat, P.; Sumankan, R.; Saengmuang, S.; Choowongkomon, K.; Panya, A. *Clinacanthus nutans* (Burm. f.) Lindau Extract Inhibits Dengue Virus Infection and Inflammation in the Huh7 Hepatoma Cell Line. *Antibiotics* **2024**, *13*, 705. https://doi.org/10.3390/antibiotics13080705

Academic Editor: Marc Maresca

Received: 25 May 2024
Revised: 29 June 2024
Accepted: 8 July 2024
Published: 28 July 2024

1. Introduction

Dengue virus (DENV) infection is a globally important public health problem. It is estimated that approximately 390 million DENV cases are reported per year. Among these, people in Asia account for up to 70% of the worldwide disease burden [1,2]. DENV belongs to the family of Flaviviridae, with a positive single-stranded RNA genome. It encodes for 11 proteins including capsid, pre-membrane, and envelope (structural) proteins, in addition to non-structural proteins (NS1, NS2A, NS2B, NS3, NS4A, NS4B, and NS5) [3]. There are four serotypes, DENV-1, DENV-2, DENV-3, and DENV-4, which are co-circulated in the endemic area and transmitted via mosquito vectors in *Aedes* genus. The infections demonstrate a wide range of symptoms ranging from mild disease to dengue fever and dengue hemorrhagic fever (DHF). Dengue shock syndrome (DSS) is the most severe disease, with a life-threatening manifestation in which vascular leaking can occur [4]. Nowadays, there is no specific drug for DENV treatment, while the available vaccine, Dengvazia, cannot fully protect against all DENV serotypes. Furthermore, other developing vaccines, i.e., the Qdenga vaccine and TV005 tetravalent live-attenuated dengue vaccine, are not yet approved by the US FDA and are undergoing clinical trials [5]. Thus, the search for anti-DENV agents is still needed to lower its severity and mortality, especially in endemic countries.

DENV can infect diverse types of immune cells including macrophages, lymphocytes, dendritic cells, and organs such as the lungs, liver, and brain, which is associated with the complication of severe symptoms [6–8]. The infection can promote inflammation and antiviral responses via the production of cytokines and chemokines to eliminate viral particles. The uncontrolled response via massive cytokine/chemokine production or a cytokine storm occurs in addition to the acute inflammation that participates in the pathogenesis of DENV infection. Hepatocytes are one of the critical targets of DENV. The inflammation of liver tissue is observed in severe cases of DENV and is correlated with an accumulation of pro- and anti-inflammatory cytokines such as tumor-necrosis factor alpha (TNF-α), interferon-gamma (IFN-γ), interleukin-10 (IL-10), transforming growth factor-beta (TGF-β), and regulated upon activation, normal T-cell expressed and secreted (RANTES), produced by mononuclear cell types [9,10]. Concordantly, DENV infection in HepG2 cells showed an increase in the innate immune response via toll-like receptor 3 (TLR3), toll-like receptor 8 (TLR8), retinoic acid-inducible gene I (RIG-I), and melanoma differentiation-associated gene 5 (MDA5) and upregulated pro-inflammatory gene expression, i.e., *IL-6*, *IL-8*, and *RANTES* [11]. Moreover, in Brazil, the increases in IL-10, macrophage migration inhibitory (MIF), and C–X–C-motif chemokine 10/interferon gamma-induced protein 10 (CXCL10/IP-10) during the acute phase in severe cases have been reported to be correlated with hepatomegaly and liver dysfunctions [12]. The liver is considered a site of DENV tissue tropism. Several receptor types, such as the scavenger receptor B type 1 (SRB1), erb-b2 receptor tyrosine kinase 2 (ERBB2), and insulin-like growth factor 1 receptor (IGF1R), have been reported to respond to DENV infection in liver cells, promoting pathogenic mechanisms [13,14]. Although dengue is an infectious disease, its severity is often due to an overreaction of the immune response, resulting in a cytokine storm, which can be fatal. An antiviral agent that solely targets the virus without reducing inflammation might not be sufficient to mitigate the risk of severe disease. Taken together, a promising treatment for DENV infection must not only control DENV replication but also reduce serious symptoms such as massive cytokine production, liver inflammation, and dysfunction during the critical phase.

Bioactive compounds from natural products are used in traditional healthcare worldwide. Unlike small synthetic compounds or drugs, the herb/plant extract provides broad-spectrum pharmacological properties since it contains numerous compounds and secondary metabolites which can target different proteins through different pathways. Accordingly, herb/plant extracts provide a high possibility for dual antiviral and anti-inflammatory activities, which is promising for DENV treatment. *Clinacanthus nutans* (*C. nutans*), a member of Thailand's national list of essential medicines [15], has been reported for its pharmacological activities, i.e., antiviral, anti-inflammatory, antioxidant, anti-cancer, anti-diarrheal, anti-diabetic, and renoprotective activities. It has been used to treat herpes simplex virus-induced lesions as well as inflammatory conditions, i.e., rheumatism, sprains, injuries, contusion, and hematoma [16,17]. Its antiviral properties are also reported to inhibit infection by human papillomavirus and varicella zoster [16,18]. Recently, the antiviral effect of *C. nutans* extracts has been reported to inhibit chikungunya virus by inhibiting viral progeny release [19] and SARS-CoV-2 by binding to viral protease and angiotensin-converting enzyme 2 (ACE2) receptor proteins [20]. However, no study has reported on the antiviral and anti-inflammation activity of *C. nutans* against DENV, particularly in hepatocyte cells. In this study, the antiviral effect of *C. nutans* extract on inhibiting DENV serotype 2 (DENV-2) was demonstrated in the Huh7 hepatocellular carcinoma cell line. The anti-inflammation was evaluated according to the reduction in inflammatory cytokines, and the suppression of inflammation via the inhibition of the nuclear factor kappa B (NF-κB) signaling pathway was also studied. This study is the first to demonstrate the potential of *Clinacanthus nutans* extract and explore the mechanisms by which the extract affects DENV during the early stages of its life cycle, providing crucial information for treatment regimens.

2. Results

2.1. Inhibition of DENV-2-Infected Huh7 Hepatoma Cells Upon Treatment with Clinacanthus nutans Extract

The cytotoxicity of *C. nutans* extract was evaluated in Huh7 cells to select a non-toxic dose. The result showed that *C. nutans* extract is highly safe, with a CC50 of 632.4 µg/mL (Supplementary Materials, Figure S1). The antiviral activity was then tested at a concentration ranging from 31.25 to 250 µg/mL in Huh7 cells to inhibit DENV-2 infection. The *C. nutans* extract was added at different steps of infection: before-infection, co-infection, and post-infection conditions. The result showed that *C. nutans* extract inhibited DENV infection in all three conditions in a dose-dependent manner judged by IFA, cell-based ELISA, and an FFU assay (Figure 1A–C). Among the three conditions, addition of the extract at the co-infection step showed the greatest effectiveness in inhibiting viral infection. Adding *C. nutans* extract at a concentration of 125 µg/mL during the co-infection step could fully protect Huh7 cells from infection (Figure 1B), whereas adding the extract during the before-infection condition (Figure 1A) and post-infection condition (Figure 1C) reduced the percentage of E antigen to 31.95 ± 10.62% and 39.32 ± 18.19%, respectively. At the lowest concentration (31.25 µg/mL) of *C. nutans* extract, it reduced the percentage of E antigen to 22.87 ± 6.49% at the co-infection step (Figure 1B). Concordantly, the IFA showed a dramatic reduction in infected cell number as well as the production of new viral progeny after treatment with *C. nutans* extract during the co-infection step, greater than that observed under the other conditions. These results suggest that *C. nutans* extract inhibited DENV infection effectively at the early step of viral infection.

2.2. Inhibition of DENV-2 Infection in the Early Steps of Infection by C. nutans Extracts

Since *C. nutans* extract efficiently inhibited DENV-2 infection at the co-infection step, we also hypothesized that *C. nutans* extract might affect the early step of infection. To prove this hypothesis, we tested whether the virus could affect RNA replication using qRT-PCR after a viral binding step. The virus was allowed to bind to Huh7 at 4 °C and then treated with *C. nutans* extract for 2 h during the early step of infection (Figure 2A). At 24 h of infection, DENV-2 RNA was decreased to 0.46 ± 0.27-fold and 0.18 ± 0.06-fold after treatment with 62.5 and 125 µg/mL of *C. nutans*, respectively (Figure 2B). The levels of E antigen percentages were reduced dramatically by 48 h after treatment with *C. nutans* extract at 62.5 and 125 µg/mL by 70.19 ± 10.80 and 52.36 ± 13.62%, respectively (Figure 2C), in accompaniment to the IFA result which showed the reduction in infected cell numbers in *C. nutans* extract-treated cells (Figure 2D). These results confirmed the effect of *C. nutans* extract in inhibiting dengue virus in the early step of virus infection.

2.3. Treatment of C. nutans Extracts Inhibited the Virus Binding to Host Cells and Virus RNA Synthesis

To assess the effect of the *C. nutans* extract on the steps of infection, the direct effect of the extract on DENV infection was determined (Figure 3A,B). We hypothesized that the extract might directly affect the virus, partially contributing to its ability to reduce DENV infection. The varying concentrations of *C. nutans* extract were incubated with DENV before infection. At the lowest concentration (31.25 µg/mL), the extract reduced the infection to 22.25 ± 12.38%, and higher doses completely inhibited the infection (Figure 2B). This indicated the extract's ability to attenuate the virus. We further investigated whether the *C. nutans* extract affected the virus capability to bind to the host cells. The virus was incubated with the extract, and we allowed the complex to bind to the host cells at 4 °C.

The amount of virus on the cell surface was determined by measuring their genomic RNA using real-time PCR. Incubating the virus with the extract significantly reduced the virus binding to host cells in a dose-dependent manner. Concentrations of 62.5, 125, and 250 μg/mL significantly decreased the virus binding to 0.70 ± 0.01-fold, 0.59 ± 0.18-fold, and 0.42 ± 0.28-fold, respectively (Figure 3C). The result suggested that the treatment affected the binding of the virus to the host.

Furthermore, our previous data showed that the extract also affected the virus during the post-infection step. We thus investigated the effect of the extract on protein translation and RNA synthesis, which occur in the early time point after virus entry. The extract was added at 0, 2, or 4 h after the infection. The result showed that the highest efficiency was observed when the cells were exposed to the extract immediately after the virus entered the cells, with reduced effectiveness over time (Figure 3D,E). Treatment after infection reduced the infection to 18.69 ± 13.20%, while treatment at 2 and 4 h post-infection inhibited infection to 29.60 ± 8.64% and 44.50 ± 3.56%, respectively (Figure 3D). This suggests that the *C. nutans* extract affected protein translation and RNA synthesis.

2.4. Reduction in Inflammatory-Related Gene Expression in DENV-2 Infected Cells by C. nutans Extracts

DENV infection is known to induce pro-inflammatory cytokines/chemokines and inflammatory enzymes, which are associated with the severity of dengue fever disease [21]. According to the previous report of anti-inflammatory activity of *C. nutans*, we expected that the *C. nutans* extract would lower cytokine/chemokine production in Huh7 cells after DENV-2 infection. The expression of anti-inflammatory-related genes was investigated by real-time PCR. The results demonstrated the upregulation of pro-inflammation genes including *TNF-α*, *CXCL10*, *IL-8*, *IL-6*, and *COX-2* after DENV infection. At 48 h of infection, the gene expression of *CXCL10*, *TNF-α*, *IL-8*, and *IL-6* was increased up to 47.14-fold, 26.43-fold, 10.87-fold, and 3.45-fold, respectively, while the *COX-2* gene was expressed 70.70-fold after 72 h of dengue viral infection (Figure 4). Interestingly, *C. nutans* extract could suppress the inflammation by lowering the transcription of *CXCL10*, *TNF-α*, *IL-8*, and *IL-6* to 8.77-fold, 3.78-fold, 1.59-fold, and 1.11-fold, respectively, after treatment with the *C. nutans* extract at 62.5 μg/mL for 48 h. Moreover, after treatment with the extract at 125 μg/mL for 48 h, the transcriptions of *CXCL10*, *TNF-α*, *IL-8*, and *IL-6* were reduced to 5.63-fold, 3.21-fold, 1.82-fold, and 2.08-fold, respectively. A reduction in *COX-2* expression was observed at 72 h of infection by lowering the gene expression to 22.53-fold and 15.13-fold when treated with the extract at 62.5 and 125 μg/mL, respectively.

2.5. C. nutans Extract Suppressed the Inflammation of DENV-Infected Cells by Inhibiting the NF-κB Signaling Pathway

To investigate the mechanism of *C. nutans* extract in lowering inflammation, we determined its effect on disturbing NF-κB signaling activation during DENV-2 infection. DENV infection activated the NF-κB cascade as evidenced by NF-κB nuclear translocation, which dramatically increased after the virus infection compared to the mock control (Figure 5A). As expected, NF-κB nuclear translocation was remarkably blocked by *C. nutans* extract treatment (Figure 5A). To confirm the inhibitory effect of *C. nutans* extract, we determined the NF-κB p65 protein levels and its phosphorylation after DENV infection compared to the conditions with or without *C. nutans* extract treatment (Figures 5B and S2). In concordance with the nuclear translocation, the extract at the concentration of 125 μg/mL effectively decreased the protein level of NF-κB and its phosphorylated form by 0.74 ± 0.17-fold and 0.47 ± 0.26-fold, respectively (Figure 5B, Figure S2).

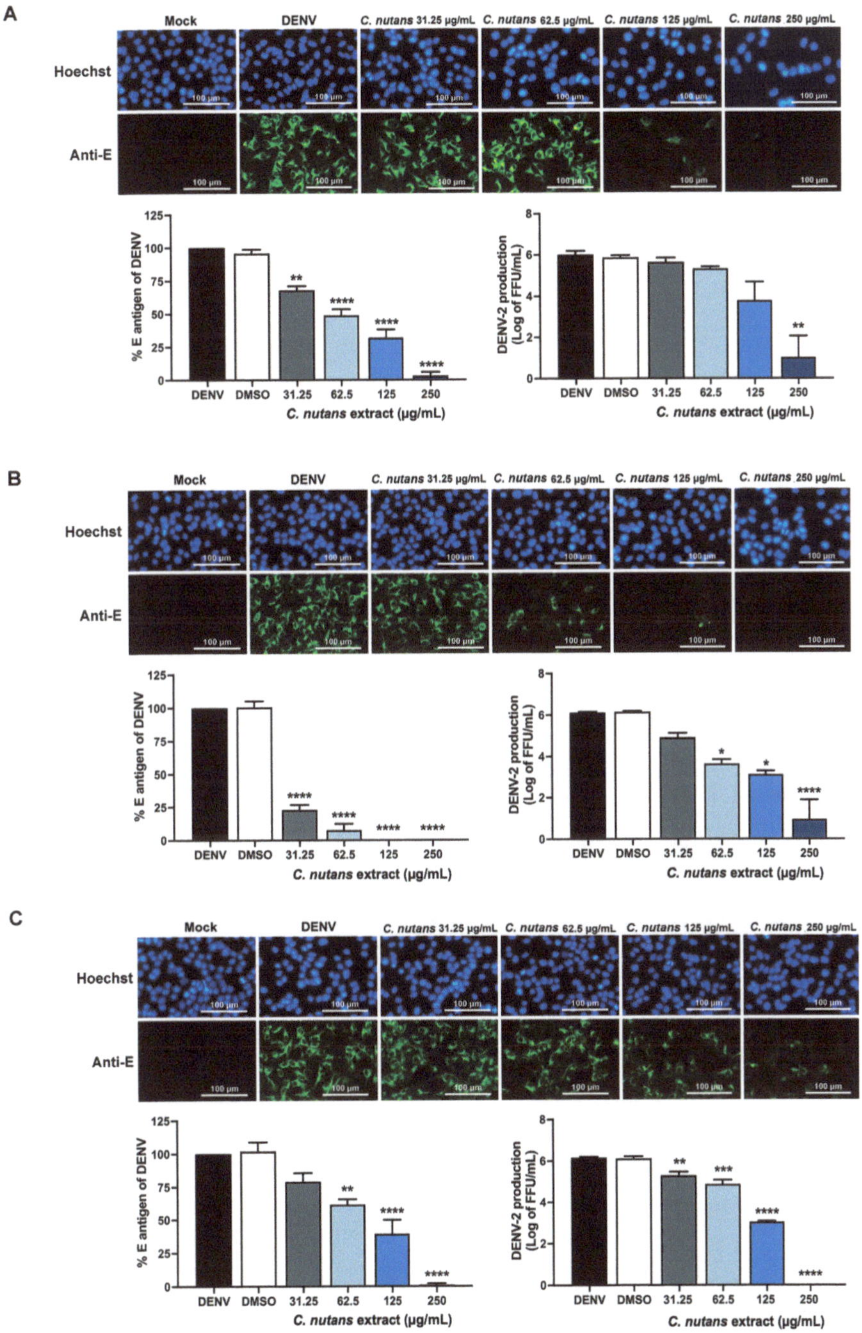

Figure 1. Antiviral activity of *C. nutans* extract against DENV-2 in Huh7 hepatoma cells. The *C. nutans* extract (31.25–250 µg/mL) was added to the cells at different steps including before infection (**A**), co-infection (**B**), and post-infection (**C**). After 48 h of infection, the infected cells were measured by using IFA, cell-based ELISA assay, and virus titration compared to that of the non-treatment control (* $p < 0.05$; ** $p < 0.01$; *** $p < 0.001$; and **** $p < 0.0001$).

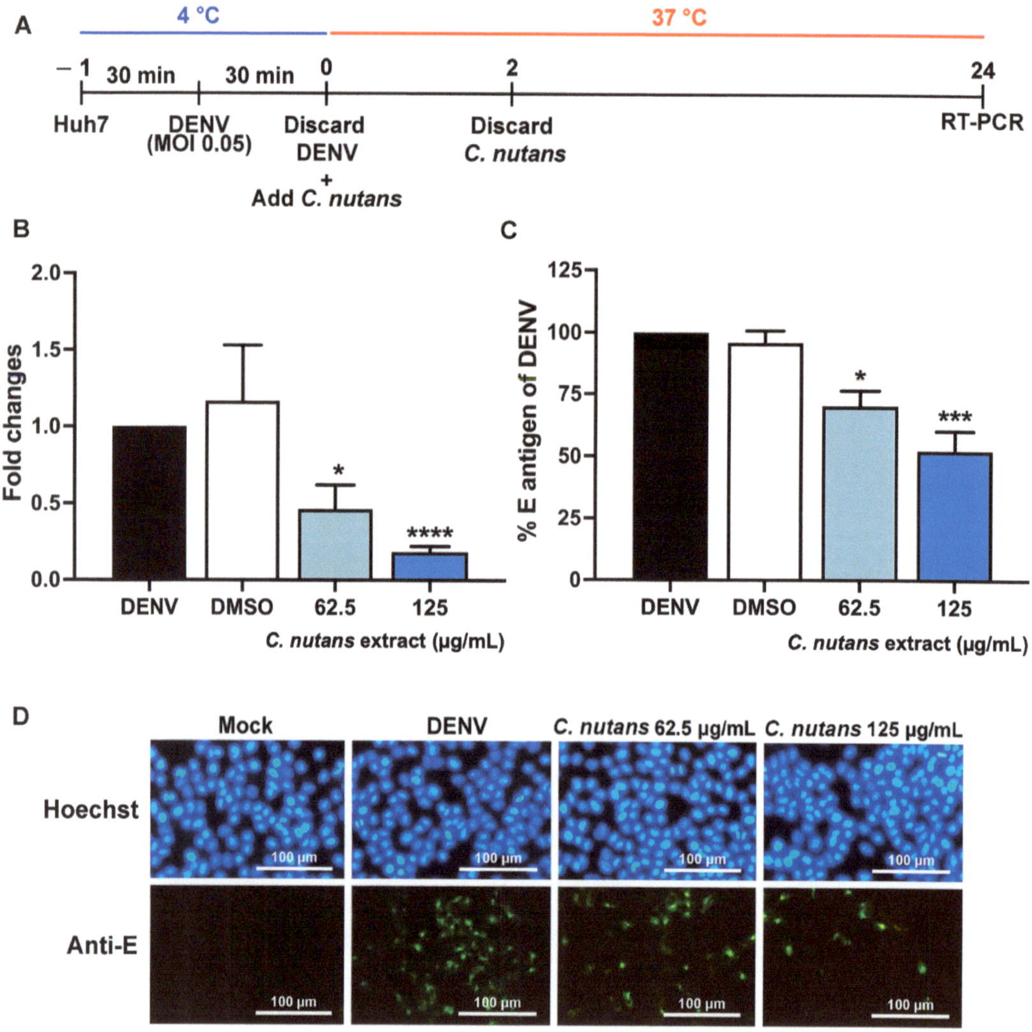

Figure 2. Efficacy of *C. nutans* extracts on an early step of infection. The scheme of extract treatment during the early step of DENV-2 infection (**A**). After 24 h of infection, the genomic RNA was measured using qRT-PCR comparing between the infected cells with or without *C. nutans* extract (62.5–125 µg/mL) (**B**). The reduction in DENV-2 was confirmed by using ELISA (**C**) and IFA assay after 48 h of infection (**D**). The results show the mean ± SEM values from three independent experiments (* $p < 0.05$, *** $p < 0.001$, **** $p < 0.0001$).

2.6. Gallic Acid Was a Bioactive Compound of C. nutans *Extract*

The bioactive compounds of *C. nutans* extract were reported in several studies. Interestingly, gallic acid found in the ethanolic extract of *C. nutans* has been reported for its antiviral activities against DENV [22]. We thus hypothesized that gallic acid is present in *C. nutans* extract and might be responsible for the anti-DENV activity of *C. nutans*. The total phenolic content of *C. nutans* extract was analyzed, which was found to be 27.44 ± 7.76 mg gallic acid/g extract. We further identified gallic acid as a representative bioactive compound in *C. nutans* extract by using HPLC. The results showed that gallic acid was a major compound that was identified by absorption spectra of relevant peaks and retention times compared with the gallic acid standard reference. The HPLC chromatograms showed a distinct peak of gallic acid at 1.503 min. Similarly, the chromatograms of the *C. nutans* extract also showed a dominant peak at 1.503 min (Figure 6).

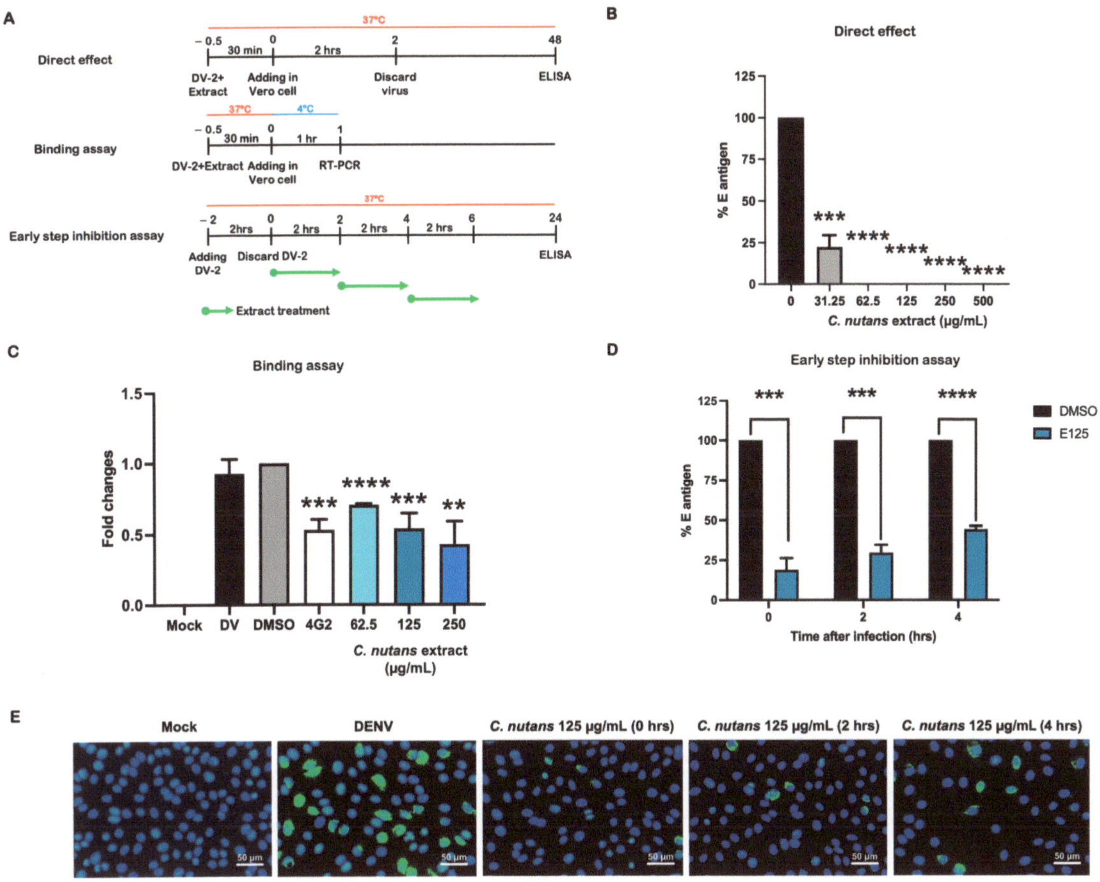

Figure 3. Effects of *C. nutans* extract on virus binding and early step of infection. (**A**) The scheme represents the protocol used to determine the direct effect, virus-receptor binding assay, and early-step inhibition assay (**B**). The direct effect of the extract on virus infectivity was determined by incubating DENV with varying concentrations of *C. nutans* extracts prior to the infection. The rate of infection was measured by cell-based ELISA (**C**). A virus–receptor binding assay was conducted to determine the extract's effect on virus binding ability. The virus–extract complexes were added to the cells at 4 °C. The amounts of bound viruses were measured by using real-time PCR (**D**,**E**). An early-step inhibition assay was conducted to monitor the effect of *C. nutans* extracts on the early step of virus life cycle. The extract was added to the infected cells at 0, 2, and 4 h post-infection. The infection rate was determined by using cell-based ELISA and IFA. The results show the mean ± SEM values from three independent experiments (** $p < 0.01$, *** $p < 0.001$, **** $p < 0.0001$).

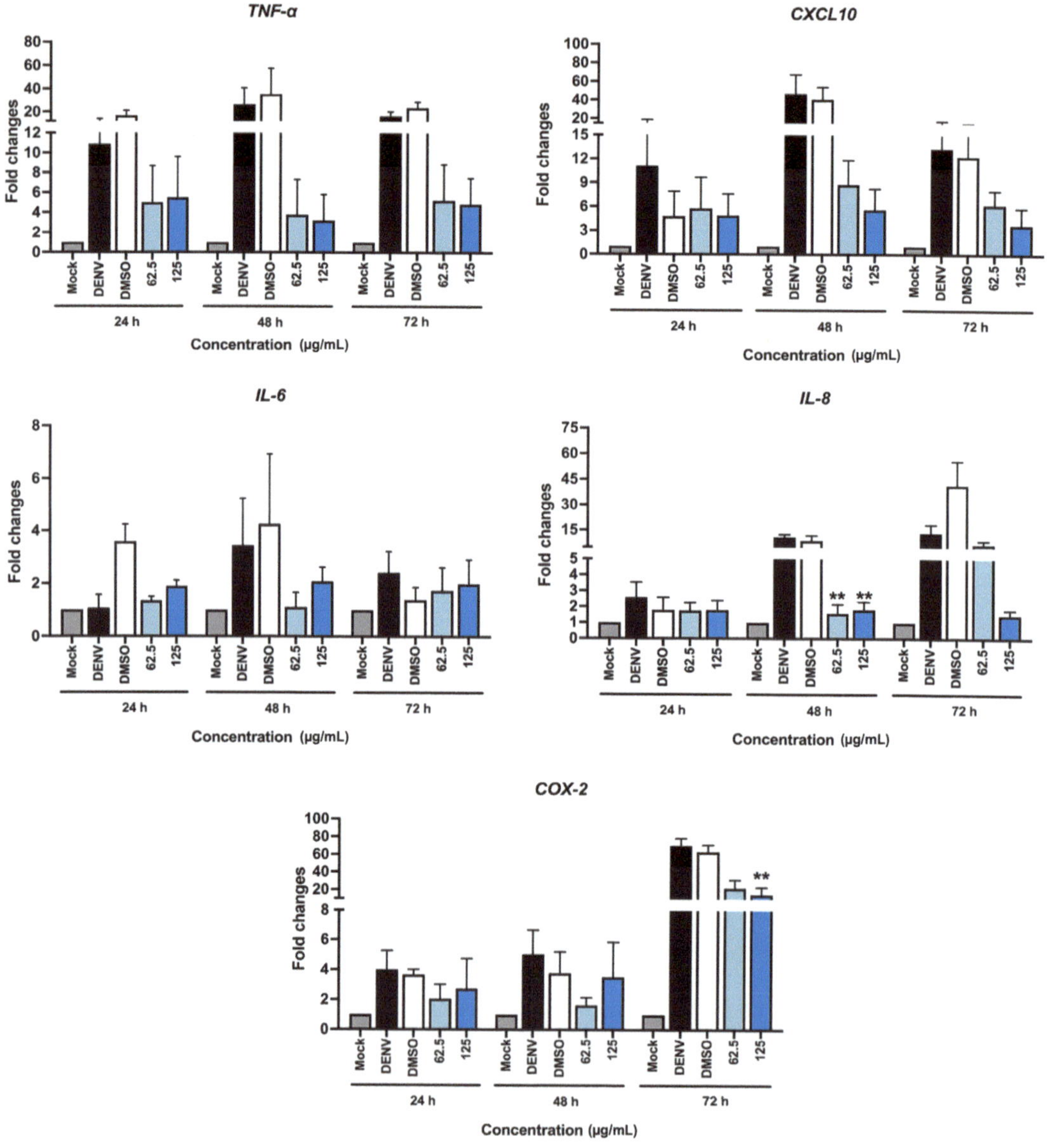

Figure 4. *C. nutans* extract reduces cytokine/chemokine production in DENV-infected Huh7 cells. Huh7 cells were infected with DENV-2 at an MOI of 0.1 and treated with the *C. nutans* extract at concentrations of 62.5 and 125 μg/mL for 24, 48, and 72 h. The transcription of *TNF-α, IL-6, IL-8, CXCL10,* and *COX-2* genes in the infected cells was determined by qRT-PCR. The expression levels were normalized with GAPDH and calculated relative to those of untreated DENV-infected cells. After 48 h of *C. nutans* extract treatment, a significant lowering of IL-8 gene expression was observed. In addition, after 72 h of *C. nutans* extract treatment, the expression of the *COX-2* gene was significantly lowered (** $p < 0.01$).

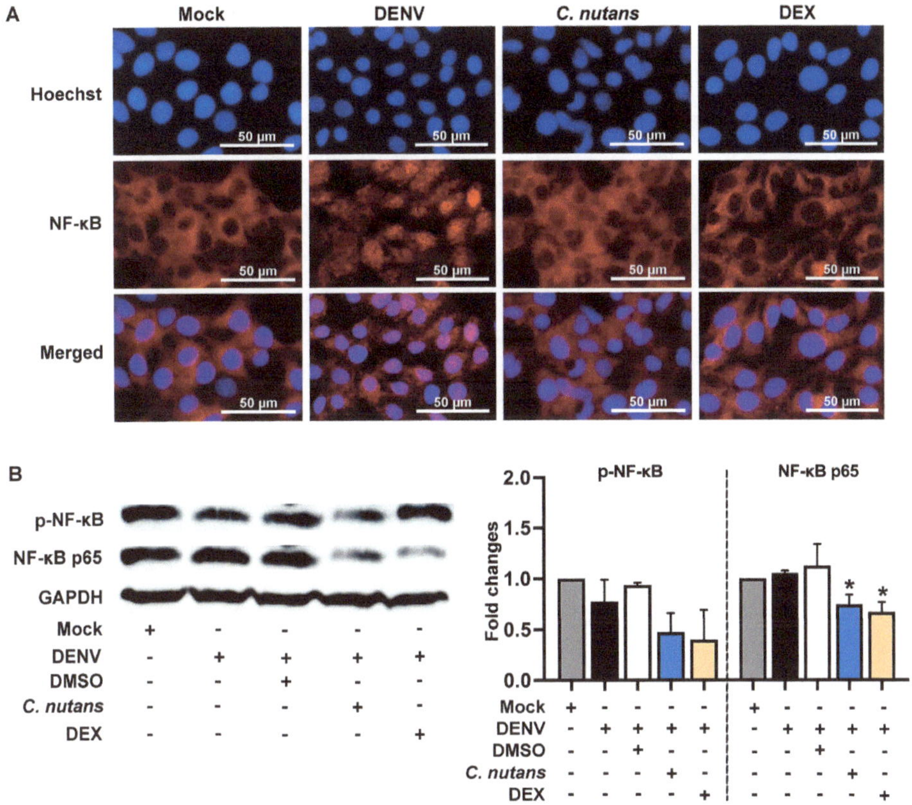

Figure 5. *C. nutans* extract suppressed NF-κB p65 signaling. Huh7 cells were infected with DENV-2 at an MOI of 0.1 and treated with 125 μg/mL of *C. nutans* extract, whereas 200 μM of DEX was used as a positive control. After 24 h of infection, the NF-κB p65 nuclear translocation was determined (red) and the nucleus was located with Hoechst staining (blue) (**A**). The protein expression of NF-κB p65 and its phosphorylated form was determined by immunoblotting (**B**). Protein band intensity was normalized to GAPDH by ImageJ software and calculated as a fold-change relative to the mock control (* $p < 0.05$).

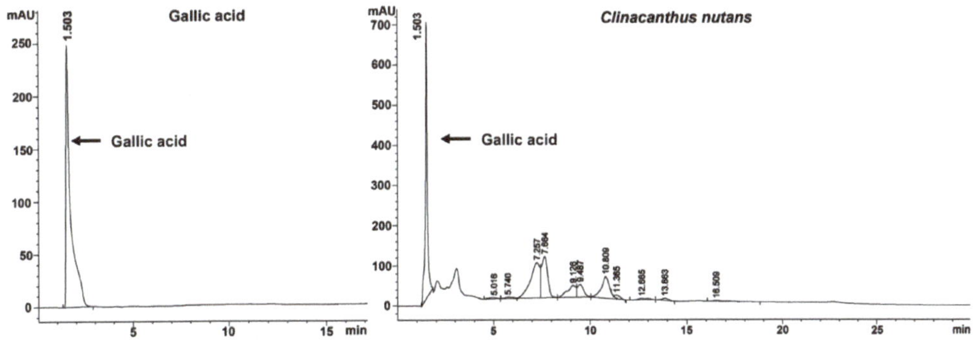

Figure 6. The gallic acid constituent in *C. nutans* extract. HPLC chromatograms of both the gallic acid standard and *C. nutans* extract were detected at 267 nm. A peak at the retention time of 1.503 min was observed in both the gallic acid standard and the *C. nutans* extract.

2.7. Gallic Acid Inhibited DENV Infection

The antiviral activity of gallic acid against DENV infection has been reported in several studies. To confirm the activity of gallic acid and investigate the possible mechanism, we performed a time-of-addition assay in which gallic acid was applied to the cell in different conditions; before infection, co-infection, post-infection (Figure 7). In conjugation with *C. nutans* extract, gallic acid exhibited a stronger inhibitory activity on DENV infection during co-infection and post-infection conditions compared to the before-infection condition (Figure 7A). The greatest potential activity of gallic acid was found during the post-infection condition by reducing the relative E antigen to 32.71 ± 9.23% at the concentration of 12.5 μM, while the treatments during co-infection and before infection resulted in reductions of 67.76 ± 6.31% and 74.2 ± 20.58%, respectively. The inhibitory effect of gallic acid was confirmed by using IFA, which showed the highest reduction in infected cells during the post-infection condition, followed by the co-infection and before-infection treatments (Figure 7B). Additionally, we performed the early-step inhibition assay to further investigate the effect of gallic acid at various post-infection time points (0, 2, and 4 h post-infection). Although significant reduction in DENV infection was observed, there were no significant differences in the inhibitory activities among different time points (Figure 7C,D).

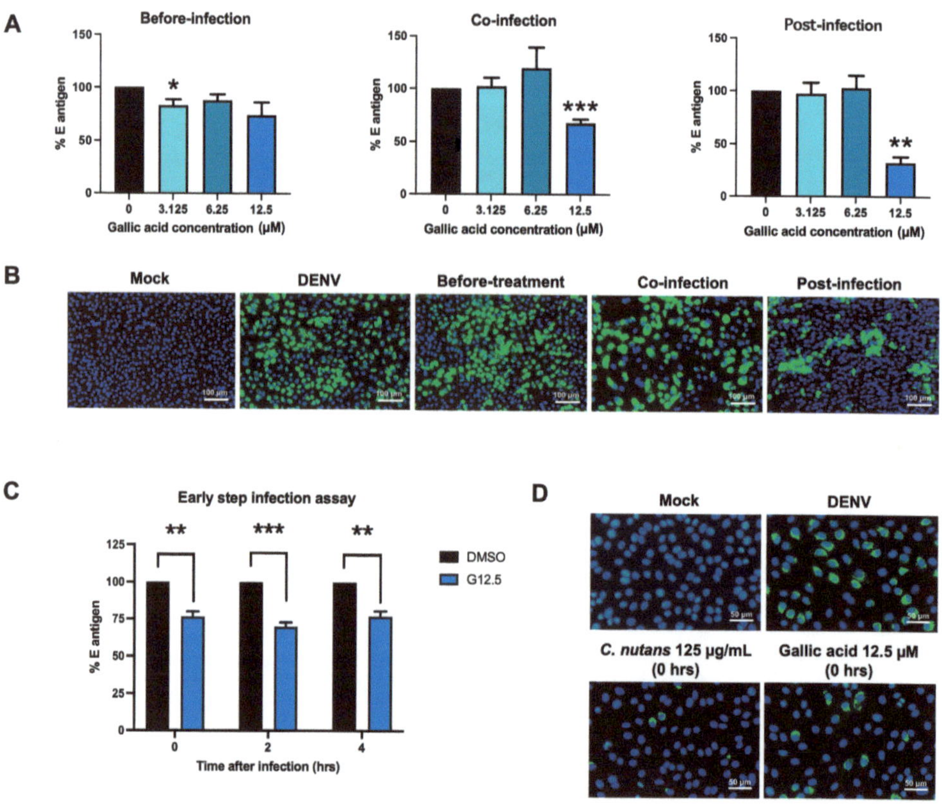

Figure 7. Gallic acid inhibited DENV infection. Gallic acid at concentrations of 3.125, 6.25, and 12.5 μM was applied to DENV-2 infected cells under different conditions: before-infection, co-infection, and post-infection conditions. The antiviral activity of gallic acid on reducing DENV infection was determined by cell-based ELISA (**A**) and IFA (**B**). An early-step infection assay was performed to investigate the effect of gallic acid after DENV entry by using cell-based ELISA (**C**) and IFA (**D**). The results show the mean ± SEM values from three independent experiments (* $p < 0.05$, ** $p < 0.01$, *** $p < 0.001$).

Molecular docking was performed to further investigate gallic acid antiviral activity to target potential viral proteins, and DENV envelope proteins and NS3 protease were selected as they are crucial viral proteins involved in attachment and protein synthesis. Based on the analysis, gallic acid could bind to envelope proteins and NS3 protease as shown in Figure 8. The fitness scores for 1OAN, 1OKE, and 2M9P were 38.72, 41.98, and 41.71, respectively. The envelope protein 1OAN interacted with gallic acid on VAL151, GLY152, ASP154, and THR155 residuals with hydrogen bonds (Figure 8A). The other target site of the envelope protein, the hydrophobic pocket [23], interacted using the hydrogen bonds with THR48, GLU49, ALA50, GLN271, and SER274 residues. In the case of NS3 protease, the interaction was on PRO193 using Pi-alkyl and VAL216 using a conventional hydrogen bond (Figure 8C). These results suggest that gallic acid could interact with key viral proteins and could potentially be one of the bioactive compounds from *C. nutans* extracts. Certainly, this further supports the findings from the time-of-addition assay that demonstrated the antiviral activity of gallic acid, and the mechanism could be that gallic acid bound to the envelope proteins and viral protease.

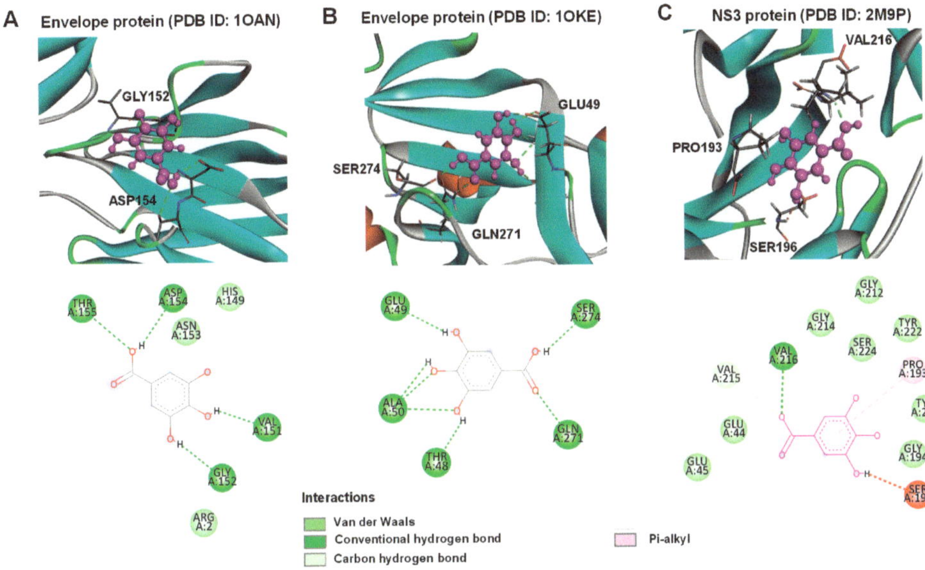

Figure 8. The interactions of dengue 2 virus envelope proteins and NS3 protease when docked with gallic acid. (**A**) Dengue 2 virus envelope protein [23]; (**B**) Dengue 2 virus envelope protein [24]; (**C**) Dengue NS3 protease [25] were docked with gallic acid using GOLD Protein–ligand docking software. Gallic acid is represented in magenta (**top**) and the detailed interacting residuals are presented in 2D diagram (**bottom**). All figures were visualized using Discovery Studio software.

3. Discussion

Reported cases of DENV infection are increasing continuously, and a large part of the world's population is affected. Approaches to preventing the spread of DENV, as well as effective therapeutic treatments, are still needed to control the situation and reduce the risk of death. In the present study, the effects of *C. nutans* extract on decreasing viral replication and reducing the inflammation induced by DENV-2 infection in Huh7 cells were elucidated.

The efficiency of *C. nutans* extract in inhibiting DENV infection was determined during different steps of infection; before infection, co-infection, and post-infection (Figure 1A–C). The treatment with *C. nutans* extract at the co-infection step was the most effective in inhibiting DENV infection (Figure 1B), suggesting that the *C. nutans* extract might act during the early step of infection, i.e., viral binding and internalization. Moreover, considering the treatment before viral infection where the extract was applied to the cell prior to infection, the *C. nutans* extract had less effectiveness to inhibit the viral infection, which suggested the extract could directly impact on the virus rather than the cell receptor. Allowing DENV to bind to the host cells and then treating those cells with *C. nutans* extract after the viral infection for 2 h caused a significant reduction in viral RNA assessed at 24 h after infection (Figure 2B). This reflected that the extract did not only inhibit the viral binding step but also the steps after internalization.

We further conducted experiments to investigate whether the extract could directly affect DENV by incubating the virus with extracts shortly before the infection. The results demonstrated that the virus–extract incubation attenuated the infectivity of DENV (Figure 3B). This finding supports our previous hypothesis that the extract did not block the host cell receptor but instead lowered the virus's ability to bind to the cells. To further confirm this phenomenon, we conducted a virus–receptor binding assay to assess the virus's capability to bind to the host cells after treatment with the extract (Figure 3C). In alignment with our previous results, treatment with the *C. nutans* extract led to a significant reduction in virus binding, which would impact the ability of DENV to enter host cells. The entry step of the virus is a promising target of antiviral inhibitors. The binding of the virus to the host cell receptor promotes receptor-mediated endocytosis, allowing the virus to enter the host cells. According to the broad-ranged tropism and host cells of DENV, a variety of host receptors have been reported at present, i.e., glycosaminoglycans (GAGs), mannose receptors, DC-SIGN, heat shock protein 90/heat shock protein 70, and TIM/TAM [26,27], etc. Virus entry through the clathrin-mediated endocytosis pathway has been reported as the main route of DENV to hepatic cells [28]. Apart from the classical entry mechanism, the protein claudin-1 has been shown to interact with the DENV prM/M protein and is necessary for the viral entry step [29]. Blocking virus entry through interfering with the virus and host cell receptor binding is a well-known strategy for antiviral drugs or inhibitors. Previously, our group reported peptide inhibitors [30–32] and single chain variable fragment (ScFv) antibodies [33] to inhibit the interaction of the DENV envelope protein and host cell receptors, leading to a significant reduction in the DENV infection rate. Our result emphasized the effect of *C. nutans* on interfering with the virus–host binding. Previously, the effect of isolated bioactive compounds from *C. nutans* in disturbing the viral entry steps has been reported in other virus types. The purified monogalactosyl diglyceride (MGDG) and digalactosyl diglyceride (DGDG), extracted from *C. nutans* leaves, had the potential to inhibit HSV plaque formation [34]. Furthermore, two bioactive compounds called 136C and 136D from *C. nutans* leaves were shown to reduce the binding between protein receptors and human papillomavirus, as well as internalization [35]. DENV infection is a dynamic process. The virus enters the human body through a mosquito bite, replicates, and releases new virions into the bloodstream. DENV can circulate and infect various cell types depending on tissue tropism, initiating its life cycle by entering new host cells and releasing new virions. Although the *C. nutans* extract was most effective at the stage before viral entry, it is crucial to inactivate the virus and block its entry, thereby reducing the overall viral titer. This reduction can lower the risk of disease progression and severity. Moreover, decreasing the active virus in the bloodstream helps to interrupt the transmission cycle of dengue virus, which spreads among humans through mosquito bites, underscoring the approach's impact on controlling disease transmission.

On the other hand, the effect of the *C. nutans* extract was also observed in the post-infection conditions; thus, we cannot exclude the possibility that the extract partially affected the other later steps such as RNA replication, protein translation, virus assembly, and virus maturation [36]. We conducted an early-step inhibition assay to analyze the potential impact of the *C. nutans* extract on viral infection. The extract was applied to cells at 0, 2, or 4 h post-infection, coinciding with the time frame of viral protein translation and RNA synthesis. Our results demonstrated a significant inhibitory effect of the extract on virus infection, which decreased over time (Figure 3D). The most effective inhibition occurred when cells were treated immediately after infection, while treatment at 4 h post-infection showed reduced effectiveness. Considering the correlation between extract efficiency and treatment timing, we hypothesize that the extract may inhibit early events, such as protein translation or RNA synthesis, rather than late-stage processes.

The uncontrolled response of the host cells after viral infection leads to massive cytokine/chemokine production, resulting in a cytokine storm, which is a well-known hallmark of severe DENV infection [37]. Clinical studies have shown a relationship between disease severity and increased pro-inflammatory cytokine/chemokines in the patient's blood circulation [38]. Imad and colleagues studied the cytokine profile of 96 hospitalized adult patients admitted to the Hospital for Tropical Diseases, Bangkok, Thailand, during 2015–2016, which showed the increased expression of protein interleukins (IL-4, -6, -8, -10), TNFα, and IFNγ during infection [38]. Apart from the cytokine production, signs of severe hepatitis have been observed in these patients with elevated liver enzymes with an AST/ALT ratio over 400 U/L [38]. Interestingly, those with an AST/ALT ratio over 1000 U/L have a strong association with an increase in IL-8 during the acute phase and defervescence [38]. Liver injury including hepatomegaly, jaundice, increased liver aminotransferase enzymes, and acute liver failure has been reported as the warning sign for severe DENV and found in up to 60%–90% of dengue hemorrhagic fever (DHF) patients [39]. In accompaniment to the reported data, we demonstrated the direct effect of DENV infection in causing increases in *TNF-α*, *IL-6*, *IL-8*, and *CXCL10* expression in Huh7 cells (Figure 3). Importantly, the treatment with the *C. nutans* extract could remarkably lower the expression of these pro-inflammatory cytokines, especially *IL-8* which was significantly reduced at the mRNA level (Figure 4) and in protein secretion (Table S1). Moreover, the *C. nutans* extract treatment could reduce the expression of *COX-2*, which is responsible for the inflammation pathway (Figure 4). We thus expected that the treatment with *C. nutans* extract would help to ameliorate liver injury after DENV infection.

C. nutans extract treatment was shown to suppress the signaling of the NF-κB pathway by lowering the protein expression of NF-κB and phosphorylated NF-κB and inhibiting its activation by blocking the nuclear localization (Figure 5). NF-κB signaling was reported to contribute to the inflammation in DENV-infected cells and those correlated with the DENV NS5 translocation to the nucleus [40]. Inhibition of DENV and NS5 nuclear translocation by alpha-mangostin showed to suppress NF-κB signaling resulting in downregulation of the NF-κB responsive inflammatory genes including *RANTES*, *IP-10*, *TNF-α*, and *IL-6* production [40]. It is possible that the *C. nutans* extract treatment lowered the viral infection and corresponded to viral proteins which decreased the magnitude of inflammation activation in DENV infection maybe by reducing DENV NS5 protein level and its translocation. On the other hand, *C. nutans* extract could directly block the inflammation pathway since many phytochemical compounds from *C. nutans* ethanolic extracts have been reported for their anti-inflammation activities such as schaftoside, isoorientin, orientin, isovitexin, and vitexin, etc. [41]. The anti-inflammation activity of schaftoside was evidenced by its ability to reduce the gene expression of *IL-1β* and *IL-6* and *TNF-α* at the mRNA and protein levels in microglia [42,43]. The vitexin compound was reported to suppress the expression of IL-1β, IL-6, IL-33, TNF-α, and IL-10 in peripheral tissues and reduce MCP-1, IL-6, IL-8, TNF-α, CXCL1, and CX3CL1 in endothelial cells through the NF-κB and Nrf2 pathways [44–47]. Moreover, isovitexin and orientin were shown to reduce IL-1β, IL-6, IL-18, and TNF-α, as well as the corresponding elements in the inflammation pathway in RAW 264.7 macrophages upon LPS stimulation, i.e., prostaglandin E2, nitric oxide, and COX-2 via the NF-κB pathway [48,49]. Taken together, the anti-inflammation mechanism of *C. nutans* could either be a direct effect caused by the suppression of the inflammatory pathway or caused by its ability to protect the cells from viral infection.

To date, the characterization of *C. nutans*, particularly its leaves, has been well-documented by several groups, including our own. Based on the characterization by Lin et al., seven major bioactive components in *C. nutans* ethanolic extracts were identified, including schaftoside, orientin, vitexin, isoorientin, isovitexin, 6,8-apigenin-C-α-Lpyranarabinoside, and gallic acid (Table S2). Among them, only isoorientin, isovitexin, and gallic acid have been reported for anti-DENV activity in different cell types and anti-inflammatory activity in different disease models, not specifically in DENV infection. Indeed, phenolic compounds of medicinal plants have been reported for their capacity to inhibit DENV infection and exert anti-inflammatory effects [22]. Previously, we reported gallic acid from the Triphala Ayurvedic formulation comprising of *Phyllanthus emblica*, *Terminalia chebula*, and *Terminalia bellirica* to inhibit DENV infection [22]. Interestingly, Khoo et al. showed gallic acid was the bioactive ingredient in a *C. nutans* ethanolic extract [50], highlighting the possibility of gallic acid being found in our *C. nutans* extract and exerting anti-DENV function. As expected, the HPLC profile showed gallic acid as the major compound in *C. nutans* extract with a similar retention time to the pure gallic acid reference (Figure 6). Our previous study showed gallic acid had the capacity to bind with DENV NS5 proteins (MTase and RNA-dependent RNA polymerase/RdRp), which later resulted in inhibition of the RNA synthesis of DENV [22]. In this study, the molecular docking of gallic acid with crucial viral proteins, the envelope proteins and NS3 protease, confirmed that these proteins may be potential inhibitory targets of gallic acid (Figure 8). In fact, gallic acid is not the only substance contributing to anti-dengue activity. When compared with the crude extract, gallic acid showed a higher toxicity to cells and lower effectiveness in reducing dengue infection, suggesting that the use of gallic acid alone as a drug should be concerning. This result supports the use of *C. nutans* extract over a single compound treatment. Notably, although gallic acid is a component of many extracts, the proportion of gallic acid in those herbs and its natural combination with other bioactive substances are crucial. These combinations play significant roles in the effectiveness of *C. nutans* extract due to their synergistic effects. This suggested that the use of traditional crude extracts might offer greater benefits compared to the isolated bioactive compounds, which can broadly impact the treatment of public health problems like dengue fever in low- and middle-income countries. However, to evaluate the potential of *C. nutans* extract, an in vivo study in an animal model is needed to demonstrate its effectiveness in controlling viral titer and preventing liver injury, which is a critical symptom in severe DENV cases.

4. Materials and Methods

4.1. Plant and Extraction

Leaves of *C. nutans* were collected from Chiang Mai, Thailand, and were classified by Dr. Narin Printarakul, Faculty of Science, Chiang Mai University (deposited voucher: Specimen No. CN001-003). After drying at 60 °C for 24 h in a hot air oven, they were extracted using 70% ethanol (1:20 g/v) by ground, macerated, and shaken at 160 rpm for 12 h at room temperature. Then, the solvent was evaporated using a rotary evaporator. After drying in a water bath, the dried extract was kept at −20 °C until use and dissolved in DMSO before analysis.

4.2. Cell Cultures and DENV Propagation

Vero and Huh7 cells were purchased from the American Type Culture Collection (ATCC) (Manassas, VA, USA). DMEM-F12 media and MEM media supplemented with 2 mM L-glutamine, and 10% (*v*/*v*) FBS (Gibco, Thermo Fisher Scientific, Waltham, MA, USA) were used to grow Huh7 and Vero cells, respectively. All cell types were incubated in 5% CO_2 at 37 °C.

The DENV serotype 2 (DENV-2) strain 16681 was kindly gifted from Prof. Pa-thai Yenchitsomanus, Division of Molecular Medicine, Faculty of Medicine Siriraj Hospital, Mahidol University, Thailand. The propagation of DENV was conducted in the C6/36 cell line using Leibowitz-15 culture medium supplemented with 1% FBS and 10% tryptose phosphate broth (TPB) (Gibco, Thermo Fisher Scientific, Waltham, MA, USA) for 5 days. After that, the amount of DENV titer was investigated by using a focus-forming units (FFU) assay.

4.3. Cell Viability Assay

The toxicity of the *C. nutans* extract was evaluated using a cell viability assay which was performed using PrestoBlue™ Cell Viability Reagent (Invitrogen, Carlsbad, CA, USA). Briefly, Huh7 cells at 1.5×10^4 cells per well were cultivated in 96-well plates overnight. After that, *C. nutans* extract (62.5–1,000 μg/mL) was added and further incubated for 48 h. DMSO was used as a negative vehicle control. Then, PrestoBlue™ reagent was added to determine cell viability by measuring the absorbance at 560 nm and 590 nm using a microplate reader. The cell viability was calculated relative to that of the non-treatment control using the following equation.

$$\% \text{ Cell viability} = \left(\frac{\text{Absorbance of tested condition}}{\text{Absorbance of non-treatment control}} \right) \times 100 \quad (1)$$

4.4. DENV Infection and Time-of-Addition Assay

To determine the effects of *C. nutans* extract on the various steps of DENV infection in Huh7-infected cells, we performed a time-of-addition assay. The treatment of *C. nutans* was divided into three conditions including (1) before-infection, (2) co-infection, and (3) post-infection steps. Briefly, Huh7 cells (1.5×10^4 cells/well) were plated in a 96-well plate overnight before the experiment. DENV-2 at an MOI of 0.01 was used to infect the cells with or without *C. nutans* extract (a concentration ranging from 31.25 to 250 μg/mL). In the before-infection step, *C. nutans* extract was added to the cells for 2 h before infecting with DENV-2 for another 2 h. In the co-infection step, DENV-2 and *C. nutans* extract were added at the same time to Huh7 cells for 2 h at 37 °C. In the post-infection step, DENV-2 was added and incubated for 2 h before treating with *C. nutans* extract. After the infection and treatment, the excess unbound virus was discarded, and the cells were washed with 1X PBS. After 48 h of infection, the amount of virus in supernatants was determined by FFU assay and the infected cells were detected by a cell-based enzyme-linked immunosorbent assay and immunofluorescence assay (cell-based ELISA).

4.5. Focus Forming Unit (FFU) Assay

The amount of virus was titrated and stained using FFU assay. Briefly, Vero cells (1.5×10^4 cells/well) were plated in a 96-well plate overnight before the experiment. The virus in the culture supernatants was diluted using a 10-fold serial dilution and then added to the cells. The plate was incubated at 37 °C in a 5% CO_2 incubator for 2 h. After that, infected cells were covered with 1.5% carboxymethylcellulose in 2% FBS MEM media. After 72 h of infection, cells were fixed and permeabilized with 3.6% formaldehyde in 1X PBS and with 0.2% Triton X-100 in 1X PBS, respectively, for 15 min at room temperature. After washing with PBS twice, the mouse monoclonal anti-DENV-E antibody (4G2) was added to the cells for at least 3 h. After washing with 1X PBS containing 0.1% Tween-20 (PBST) three times, rabbit anti-mouse IgG conjugated with HRP (Dako, Santa Clara, CA, USA) was added to the cells at a 1:2000 dilution. The plate was incubated at room temperature for 1 h in a dark place. The foci formation of infected cells was indicated after adding the 3,30-diaminobenzidine (DAB) substrate (Sigma-Aldrich Corporation, St. Louis, MO, USA). The number of foci were counted manually under a light microscope (Drawell BDS400; Shanghai Drawell Scientific Instrument Co., Ltd., Shanghai, China) and expressed as FFU/mL.

4.6. Cell-Based ELISA

The cell-based ELISA assay was performed to determine the effect of the extract on lowering the intracellular viral envelope (E) protein. The detection of DENV-2 infection was performed as described previously in the presence or absence of *C. nutans* extract. The viral infected cells were harvested after 48 h of incubation and fixed with 3.6% formaldehyde followed by permeabilization with 0.2% Triton X-100 for 15 min. The 4G2 monoclonal antibody was added to the cells for at least 3 h. After washing with 1X PBS containing

0.1% Tween-20 (PBST) three times, rabbit anti-mouse IgG conjugated with HRP (Dako, Santa Clara, CA, USA) was added to the cells at a 1:2000 dilution. The plate was incubated at room temperature for 30 min in a dark place. After washing with 0.1% PBST, the 3,3′,5,5′-tetramethylbenzidine substrate (Invitrogen, Carlsbad, CA, USA) was used as a detection reagent where 2 N sulfuric acid was used to stop the reaction. The absorbance was measured at 450 nm and the mock control was set as blank. The intracellular E antigen was calculated as a percentage following this equation:

$$\%\ \text{E antigen} = \left(\frac{\text{Absorbance of treatmemt condition}}{\text{Absorbance of non-treatment control}}\right) \times 100 \quad (2)$$

4.7. Immunofluorescence Study (IFA)

The number of infected cells was determined after *C. nutans* treatment using IFA. Huh7 cells (1.5×10^4 cells/well) were plated in a 96-well plate overnight before the experiment. DENV-2 infection was performed as described previously with or without *C. nutans* extract. The infected cells were harvested after 48 h of incubation and fixed with 3.6% formaldehyde followed by permeabilization with 0.2% Triton X-100 for 15 min. The cells were stained with 4G2 monoclonal antibody for at least 3 h at 37 °C. After washing with 0.1% PBST three times, goat anti-mouse IgG conjugated with Alexa Fluor 488 (dilution 1:2000) (Invitrogen, Carlsbad, CA, USA) and Hoechst® 33342 (dilution 1:10,000) (Invitrogen, Carlsbad, CA, USA) were added and incubated for 30 min in dark conditions at room temperature. After washing with 0.1% PBST, the stained cells were investigated under a fluorescence microscope (Nikon Instruments, Inc., Melville, NY, USA).

For the NF-κB (p65) nuclear translocation assay, DENV-2 infected cells with or without *C. nutans* treatment were harvested at 24 h. The cells were fixed and permeabilized as previously described before adding NF-κB (p65) (D14E12) XP Rabbit antibody (CST#8242, 1:100) (Cell Signaling Technology Inc, Boston, MA, USA). The plate was incubated for 1 h before adding donkey anti-rabbit IgG conjugated with Alexa Fluor 555 (dilution 1:2000) (Invitrogen, Carlsbad, CA, USA) and Hoechst® 33342. The localization of NF-κB p65 in intracellular was detected under a fluorescence microscope.

4.8. Real-Time PCR

The real-time PCR was conducted to determine the effect of *C. nutans* extract on the virus internalization step. Briefly, Huh7 cells were pre-chilled and DENV-2 at an MOI of 0.05 was infected to the cells. The treatment was incubated at 4 °C for 30 min. Then, *C. nutans* extract (62.5 and 125 μg/mL) was added for 2 h at 37 °C. After 24 h of incubation, the infected cells were collected, and the amount of DENV-2 virus was quantified using qRT-PCR technique with specific dengue primers (Table S3). Furthermore, the number of DENV-infected cells and the reduction in intracellular E protein in infected cells were measured to confirm the inhibitory activity by IFA and cell-based ELISA as previously described.

To evaluate the effect of *C. nutans* extract on cytokine gene expression, Huh7 cells were infected with DENV-2 (MOI 0.1) and treated with *C. nutans* extract at 62.5 and 125 μg/mL for 24, 48, and 72 h. At the harvest times, the RNA was isolated using TRIzol reagent (Invitrogen, Carlsbad, CA, USA) and used as the template for reverse transcription using cDNA Tetro cDNA Synthesis Kit (Bioline Reagents Ltd., London, UK). Real-time PCR was performed using a real-time RT-PCR kit according to the manufacturer's instructions (Luna® Universal qPCR Master Mix, New England Biolabs, Ipswich, MA, USA) and specific primers (Table S3) to determine expressions of cytokine, chemokine, and inflammatory enzyme encoding genes were used. The expression was normalized with those of GAPDH and compared to the non-treatment control (set as 1.0). The data are represented in relative normalized expression values.

4.9. Immunoblotting Assay

The immunoblotting assay was performed to measure the effect of C. *nutans* extract on NF-κB expression and signaling inhibition. Huh7 cells (1×10^5 cell/well) were cultured as a monolayer. Then, DENV-2 (MOI 0.1) was used to infect the cells as described above followed by 125 μg/mL of C. *nutans* treatment. Dexamethasone (DEX) (Sigma-Aldrich, MO, USA) at a concentration of 200 μM was used as a positive control of NF-κB signaling inhibition. The virus-infected cells were harvested at 24 h and lysed using the RIPA buffer. The NF-κB protein and its phosphorylated form were determined by specific primary antibodies: phospho-NF-κB (p65) (Ser536) (93H1) (CST#3033, 1:100), NF-κB (p65) (D14E12) XP (CST#8242, 1:100) (Cell Signaling Technology Inc, Boston, MA, USA). The protein expression was detected with a secondary goat anti-rabbit antibody conjugated with horseradish peroxidase (Abclonal, Woburn, MA, USA). The GAPDH protein was used as an internal control. The protein intensity was quantified by ImageJ software (version 1.52a) and compared with the non-treatment control.

4.10. Determination of Total Phenolic Content

The total phenolic content of the C. *nutans* extract was determined by using the standard Folin-Ciocalteu assay. After incubating 50% Folin–Ciocalteu's reagent (Merck, Kenilworth, NJ, USA) with various concentrations of C. *nutans* extract for 1 h, the absorbance was measured at a wavelength of 725 nm. The total phenolic content was calculated against the gallic acid standard curve and expressed as mg of gallic acid equivalents per gram extract.

4.11. Determination of the Phytochemical Composition of C. nutans Extract by HPLC

HPLC was performed to study the phytochemical profile of C. *nutans* extract using an HPLC instrument (Agilent Technologies, Santa Clara, CA, USA) with a ZORBAX Eclipse XDB-C18 column (4.6 × 150 mm, 5 μm (Agilent Technologies)). The linear gradient system of the mobile phase (deionized water: methanol in a ratio of 70:30 (*v*/*v*)) at a flow rate of 1 mL/min for 30 min was used. The absorbance was collected at a UV photodiode array detector (267 nm) for the evaluation of the gallic acid peak.

4.12. Molecular Docking of Dengue 2 Virus Envelope Proteins and NS3 Protease with Gallic Acid

Crystal structures of two dengue 2 virus envelope proteins were obtained from the Protein Data Bank [23,24] along with a structure of dengue NS3 protease [25] (https://www.rcsb.org/ accessed on 5 November 2023). Molecular docking was performed using the GOLD Protein–ligand docking software (GOLD version 5.3.0). Upon molecular docking, hydrogens and waters were removed from the structures. The structure of gallic acid as a target ligand was obtained from MolView (https://molview.org/ accessed on 6 November 2023). The binding sites were referenced with those of ligands extracted from the reported structures. The fitness score for each protein–ligand interaction was obtained from ChemPLP. The resulting docking poses and detailed interactions of the amino acid residues and the ligand were visualized using Discovery Studio Visualizer.

4.13. Statistical Analysis

GraphPad Prism 9 Software (GraphPad Software, Inc., La Jolla, CA, USA, version 9.5.1) was used to analyze statistics. All data were tested at least in three independent experiments. A bar graph was used represent the mean ± standard error of mean (SEM). Significant differences in means were compared between the treated group and the non-treatment control using one-way ANOVA, followed by Tukey's multiple comparisons tests. Statistical significance was considered when the p-value was less than 0.05.

5. Conclusions

This study demonstrated the dual antiviral and anti-inflammatory activity of *C. nutans* extract on DENV-infected Huh7 cells. *C. nutans* extract inhibited DENV at the early step of infection. The treatment also remarkably lowered the pro-inflammatory cytokine expression and inflammation induced by DENV through suppression of the NF-κB signaling cascade. The result from our study showed the benefit of *C. nutans* extract treatment during DENV infection.

Supplementary Materials: The following supporting information can be downloaded at: https://www.mdpi.com/article/10.3390/antibiotics13080705/s1, Figure S1: The cytotoxicity of *C. nutans* extracts in Huh7 cell; Figure S2: Effect of *C. nutans* on lowering p-NF-κB p65 and NF-κB p65 protein expression; Table S1: The effect of *C. nutans* extract to inhibit inflammatory cytokine/chemokine of Huh7 infected cells; Table S2: Anti-dengue activity and anti-inflammation activity of bioactive compounds found in *C. nutans* extract; Table S3: Sequences of primers for the qRT-PCR analysis. References [22,51–60] are cited in the Supplementary Materials.

Author Contributions: Conceptualization, A.P. and K.J.; methodology, A.P., K.J. and S.P.; software, P.S. and K.C; validation, A.P. and K.C.; formal analysis, A.P. and K.J; investigation, K.J., S.P., R.S. and S.S.; resources, A.P.; data curation, A.P.; writing—original draft preparation, K.J. and P.S.; writing—review and editing, A.P. and K.C.; visualization, A.P.; supervision, A.P.; project administration, A.P.; funding acquisition, A.P. All authors have read and agreed to the published version of the manuscript.

Funding: This research was funded by Fundamental Fund 2024, Chiang Mai University. Kanyaluck Jantakee was supported by the CMU Presidential Scholarship (Post-Doctoral Fellowship, 2022) via the Office of Research Administration, Chiang Mai University.

Institutional Review Board Statement: Not applicable.

Informed Consent Statement: Not applicable.

Data Availability Statement: The data used to support the findings of this study are available from the corresponding author upon reasonable request.

Acknowledgments: The authors would like to thank Pa-thai Yenchitsomanus, Division of Molecular Medicine, Faculty of Medicine Siriraj Hospital, Mahidol University, Thailand, for providing the DENV used in this study and Kasetsart University for providing the GOLD docking program.

Conflicts of Interest: The authors declare no conflicts of interest.

References

1. WHO. Dengue and Severe Dengue. 2022. Available online: https://www.who.int/news-room/fact-sheets/detail/dengue-and-severe-dengue (accessed on 5 December 2022).
2. Bhatt, S.; Gething, P.W.; Brady, O.J.; Messina, J.P.; Farlow, A.W.; Moyes, C.L.; Drake, J.M.; Brownstein, J.S.; Hoen, A.G.; Sankoh, O.; et al. The global distribution and burden of dengue. *Nature* **2013**, *496*, 504–507. [CrossRef]
3. Li, Q.; Kang, C. Structures and dynamics of dengue virus nonstructural membrane proteins. *Membranes* **2022**, *12*, 231. [CrossRef]
4. Zerfu, B.; Kassa, T.; Legesse, M. Epidemiology, biology, pathogenesis, clinical manifestations, and diagnosis of dengue virus infection, and its trend in Ethiopia: A comprehensive literature review. *Trop. Med. Health* **2023**, *51*, 11. [CrossRef]
5. Ahmed, I.; Ahamed, R.; Nahar, S.; Bari, L.F.; Dewan, S.M. Immunization against dengue virus infection is coercive: A timely call. *Health Sci. Rep.* **2024**, *7*, e2170. [CrossRef]
6. Acosta, E.G.; Castilla, V.; Damonte, E.B. Alternative infectious entry pathways for dengue virus serotypes into mammalian cells. *Cell Microbiol.* **2009**, *11*, 1533–1549. [CrossRef]
7. Hou, J.; Baker, L.A.; Zhou, L.; Klein, R.S. Viral interactions with the blood-brain barrier: Old dog, new tricks. *Tissue Barriers* **2016**, *4*, e1142492. [CrossRef]
8. Miller, J.L.; de Wet, B.J.; Martinez-Pomares, L.; Radcliffe, C.M.; Dwek, R.A.; Rudd, P.M.; Gordon, S. The mannose receptor mediates dengue virus infection of macrophages. *PLoS Pathog.* **2008**, *4*, e17. [CrossRef]
9. Huerre, M.R.; Lan, N.T.; Marianneau, P.; Hue, N.B.; Khun, H.; Hung, N.T.; Khen, N.T.; Drouet, M.T.; Huong, V.T.; Ha, D.Q.; et al. Liver histopathology and biological correlates in five cases of fatal dengue fever in Vietnamese children. *Virchows Arch.* **2001**, *438*, 107–115. [CrossRef]
10. Povoa, T.F.; Oliveira, E.R.A.; Basilio-de-Oliveira, C.A.; Nuovo, G.J.; Chagas, V.L.A.; Salomao, N.G.; Alves, A.M.B.; Mota, E.M.; Paes, M.V. Correction: Peripheral Organs of Dengue Fatal Cases Present Strong Pro-Inflammatory Response with Participation of IFN-Gamma-, TNF-Alpha- and RANTES-Producing Cells. *PLoS ONE* **2018**, *13*, e0195140. [CrossRef]

11. Conceicao, T.M.; El-Bacha, T.; Villas-Boas, C.S.; Coello, G.; Ramirez, J.; Montero-Lomeli, M.; Da Poian, A.T. Gene expression analysis during dengue virus infection in HepG2 cells reveals virus control of innate immune response. *J. Infect.* **2010**, *60*, 65–75. [CrossRef]
12. Ferreira, R.A.; de Oliveira, S.A.; Gandini, M.; Ferreira Lda, C.; Correa, G.; Abiraude, F.M.; Reid, M.M.; Cruz, O.G.; Kubelka, C.F. Circulating cytokines and chemokines associated with plasma leakage and hepatic dysfunction in Brazilian children with dengue fever. *Acta Trop.* **2015**, *149*, 138–147. [CrossRef] [PubMed]
13. Alcalá, A.C.; Maravillas, J.L.; Meza, D.; Ramirez, O.T.; Ludert, J.E.; Palomares, L.A. Dengue virus NS1 uses scavenger receptor B1 as a cell receptor in cultured cells. *J. Virol.* **2022**, *96*, e01664-21. [CrossRef] [PubMed]
14. Bourgeois, N.M.; Wei, L.; Ho, N.N.; Neal, M.L.; Seferos, D.; Tongogara, T.; Mast, F.D.; Aitchison, J.D.; Kaushansky, A. Multiple receptor tyrosine kinases regulate dengue infection of hepatocytes. *Front. Cell. Infect. Microbiol.* **2024**, *14*, 1264525. [CrossRef] [PubMed]
15. National Drug Information. 2023. Available online: http://ndi.fda.moph.go.th/drug_national (accessed on 15 January 2023).
16. Chia, T.Y.; Gan, C.Y.; Murugaiyah, V.; Hashmi, S.F.; Fatima, T.; Ibrahim, L.; Abdulla, M.H.; Alswailmi, F.K.; Johns, E.J.; Ahmad, A. A Narrative Review on the Phytochemistry, Pharmacology and Therapeutic Potentials of *Clinacanthus nutans* (Burm. f.) Lindau Leaves as an Alternative Source of Future Medicine. *Molecules* **2021**, *27*, 139. [CrossRef] [PubMed]
17. Zulkipli, I.N.; Rajabalaya, R.; Idris, A.; Sulaiman, N.A.; David, S.R. *Clinacanthus nutans*: A review on ethnomedicinal uses, chemical constituents and pharmacological properties. *Pharm. Biol.* **2017**, *55*, 1093–1113. [CrossRef] [PubMed]
18. Jiamton, S.; Chanyachailert, P.; Nanchaipruek, Y.; Jantanapornchai, N.; Patthamalai, P.; Limphoka, P.; Nokdhes, Y.-N.; Jantaravinid, J. Efficacy and Safety of *Clinacanthus nutans* Lindau Cream vs. Podophyllin for the Treatment of Adults with Condyloma Acuminata. *Evid. Based Complement. Altern. Med.* **2022**, *2022*, 1577716. [CrossRef]
19. Chan, S.M.; Khoo, K.S.; Sekaran, S.D.; Sit, N.W. Mode-Dependent Antiviral Activity of Medicinal Plant Extracts against the Mosquito-Borne Chikungunya Virus. *Plants* **2021**, *10*, 1658. [CrossRef]
20. Ismail, N.Z.; Adebayo, I.A.; Mohamad Zain, N.N.; Arsad, H. Molecular docking of compounds from *Clinacanthus nutans* extract detected by GC-MS analysis with the SARS-CoV-2 main protease and ACE2 protein. *Nat. Prod. Res.* **2022**, *36*, 2848–2852. [CrossRef] [PubMed]
21. Lai, Y.C.; Chao, C.H.; Yeh, T.M. Roles of Macrophage Migration Inhibitory Factor in Dengue Pathogenesis: From Pathogenic Factor to Therapeutic Target. *Microorganisms* **2020**, *8*, 891. [CrossRef]
22. Panya, A.; Jantakee, K.; Punwong, S.; Thongyim, S.; Kaewkod, T.; Yenchitsomanus, P.T.; Tragoolpua, Y.; Pandith, H. Triphala in Traditional Ayurvedic Medicine Inhibits Dengue Virus Infection in Huh7 Hepatoma Cells. *Pharmaceuticals* **2021**, *14*, 1236. [CrossRef] [PubMed]
23. Modis, Y.; Ogata, S.; Clements, D.; Harrison, S.C. Crystal Structure of the Dengue 2 Virus Envelope Protein. Available online: https://www.wwpdb.org/pdb?id=pdb_00001oan (accessed on 7 July 2024).
24. Modis, Y.; Ogata, S.; Clements, D.; Harrison, S.C. Crystal Structure of the Dengue 2 Virus Envelope Protein in Complex with n-Octyl-beta-D-glucoside. Available online: https://www.wwpdb.org/pdb?id=pdb_00001oke (accessed on 7 July 2024).
25. Gibbs, A.; Tounge, B.; Steele, R. NMR Structure of an Inhibitor Bound Dengue NS3 Protease. 2014. Available online: https://www.wwpdb.org/pdb?id=pdb_00002m9p (accessed on 4 November 2023).
26. Artpradit, C.; Robinson, L.N.; Gavrilov, B.K.; Rurak, T.T.; Ruchirawat, M.; Sasisekharan, R. Recognition of heparan sulfate by clinical strains of dengue virus serotype 1 using recombinant subviral particles. *Virus Res.* **2013**, *176*, 69–77. [CrossRef] [PubMed]
27. Lee, E.; Wright, P.J.; Davidson, A.; Lobigs, M. Virulence attenuation of Dengue virus due to augmented glycosaminoglycan-binding affinity and restriction in extraneural dissemination. *J. Gen. Virol.* **2006**, *87*, 2791–2801. [CrossRef]
28. Ang, F.; Wong, A.P.; Ng, M.M.; Chu, J.J. Small interference RNA profiling reveals the essential role of human membrane trafficking genes in mediating the infectious entry of dengue virus. *Virol. J.* **2010**, *7*, 24. [CrossRef] [PubMed]
29. Che, P.; Tang, H.; Li, Q. The interaction between claudin-1 and dengue viral prM/M protein for its entry. *Virology* **2013**, *446*, 303–313. [CrossRef]
30. Panya, A.; Bangphoomi, K.; Choowongkomon, K.; Yenchitsomanus, P.T. Peptide inhibitors against dengue virus infection. *Chem. Biol. Drug Des.* **2014**, *84*, 148–157. [CrossRef] [PubMed]
31. Panya, A.; Sawasdee, N.; Junking, M.; Srisawat, C.; Choowongkomon, K.; Yenchitsomanus, P.T. A peptide inhibitor derived from the conserved ectodomain region of DENV membrane (M) protein with activity against dengue virus infection. *Chem. Biol. Drug Des.* **2015**, *86*, 1093–1104. [CrossRef] [PubMed]
32. Panya, A.; Yongpitakwattana, P.; Budchart, P.; Sawasdee, N.; Krobthong, S.; Paemanee, A.; Roytrakul, S.; Rattanabunyong, S.; Choowongkomon, K.; Yenchitsomanus, P.T. Novel bioactive peptides demonstrating anti-dengue virus activity isolated from the Asian medicinal plant *Acacia catechu*. *Chem. Biol. Drug Des.* **2019**, *93*, 100–109. [CrossRef] [PubMed]
33. Saokaew, N.; Poungpair, O.; Panya, A.; Tarasuk, M.; Sawasdee, N.; Limjindaporn, T.; Chaicumpa, W.; Yenchitsomanus, P. Human monoclonal single-chain antibodies specific to dengue virus envelope protein. *Lett. Appl. Microbiol.* **2014**, *58*, 270–277. [CrossRef] [PubMed]
34. Pongmuangmul, S.; Phumiamorn, S.; Sanguansermsri, P.; Wongkattiya, N.; Fraser, I.H.; Sanguansermsri, D. Anti-herpes simplex virus activities of monogalactosyl diglyceride and digalactosyl diglyceride from *Clinacanthus nutans*, a traditional Thai herbal medicine. *Asian Pac. J. Trop. Biomed.* **2016**, *6*, 192–197. [CrossRef]

35. Sookmai, W.E.T.; Pientong, C.; Sakdarat, S.; Kongyingyoes, B. The anti-papillomavirus infectivity of *Clinacanthus nutans* compounds. *Srinagarind Med. J.* **2011**, *26*, 260–263.
36. Rodenhuis-Zybert, I.A.; Wilschut, J.; Smit, J.M. Dengue virus life cycle: Viral and host factors modulating infectivity. *Cell Mol. Life Sci.* **2010**, *67*, 2773–2786. [CrossRef] [PubMed]
37. Khanam, A.; Gutierrez-Barbosa, H.; Lyke, K.E.; Chua, J.V. Immune-Mediated Pathogenesis in Dengue Virus Infection. *Viruses* **2022**, *14*, 2575. [CrossRef] [PubMed]
38. Imad, H.A.; Phumratanaprapin, W.; Phonrat, B.; Chotivanich, K.; Charunwatthana, P.; Muangnoicharoen, S.; Khusmith, S.; Tantawichien, T.; Phadungsombat, J.; Nakayama, E.; et al. Cytokine Expression in Dengue Fever and Dengue Hemorrhagic Fever Patients with Bleeding and Severe Hepatitis. *Am. J. Trop. Med. Hyg.* **2020**, *102*, 943–950. [CrossRef] [PubMed]
39. Leowattana, W.; Leowattana, T. Dengue hemorrhagic fever and the liver. *World J. Hepatol.* **2021**, *13*, 1968–1976. [CrossRef] [PubMed]
40. Tarasuk, M.; Songprakhon, P.; Chieochansin, T.; Choomee, K.; Na-Bangchang, K.; Yenchitsomanus, P.T. Alpha-mangostin inhibits viral replication and suppresses nuclear factor kappa B (NF-kappaB)-mediated inflammation in dengue virus infection. *Sci. Rep.* **2022**, *12*, 16088. [CrossRef] [PubMed]
41. Chelyn, J.L.; Omar, M.H.; Mohd Yousof, N.S.; Ranggasamy, R.; Wasiman, M.I.; Ismail, Z. Analysis of flavone C-glycosides in the leaves of *Clinacanthus nutans* (Burm. f.) Lindau by HPTLC and HPLC-UV/DAD. *Sci. World J.* **2014**, *2014*, 724267. [CrossRef] [PubMed]
42. De Melo, G.O.; Muzitano, M.F.; Legora-Machado, A.; Almeida, T.A.; De Oliveira, D.B.; Kaiser, C.R.; Koatz, V.L.; Costa, S.S. C-glycosylflavones from the aerial parts of Eleusine indica inhibit LPS-induced mouse lung inflammation. *Planta Med.* **2005**, *71*, 362–363. [CrossRef]
43. Ong, W.Y.; Herr, D.R.; Sun, G.Y.; Lin, T.N. Anti-Inflammatory Effects of Phytochemical Components of *Clinacanthus nutans*. *Molecules* **2022**, *27*, 3607. [CrossRef] [PubMed]
44. Cao, H.; Wang, X.; Zhang, B.; Ren, M. The protective effect of vitexinin septic encephalopathy by reducing leukocyte-endothelial adhesion and inflammatory response. *Ann. Palliat. Med.* **2020**, *9*, 2079–2089. [CrossRef]
45. Duan, S.; Du, X.; Chen, S.; Liang, J.; Huang, S.; Hou, S.; Gao, J.; Ding, P. Effect of vitexin on alleviating liver inflammation in a dextran sulfate sodium (DSS)-induced colitis model. *Biomed. Pharmacother.* **2020**, *121*, 109683. [CrossRef]
46. Saha, S.; Buttari, B.; Panieri, E.; Profumo, E.; Saso, L. An Overview of Nrf2 Signaling Pathway and Its Role in Inflammation. *Molecules* **2020**, *25*, 5474. [CrossRef]
47. Wang, W.; Cheng, H.; Gu, X.; Yin, X. The natural flavonoid glycoside vitexin displays preclinical antitumor activity by suppressing NF-kappaB signaling in nasopharyngeal carcinoma. *Onco Targets Ther.* **2019**, *12*, 4461–4468. [CrossRef] [PubMed]
48. Deng, J.; Ma, Y.; He, Y.; Yang, H.; Chen, Y.; Wang, L.; Huang, D.; Qiu, S.; Tao, X.; Chen, W. A Network Pharmacology-Based Investigation to the Pharmacodynamic Material Basis and Mechanisms of the Anti-Inflammatory and Anti-Viral Effect of Isatis indigotica. *Drug Des. Dev. Ther.* **2021**, *15*, 3193–3206. [CrossRef] [PubMed]
49. Thangaraj, K.; Vaiyapuri, M. Orientin, a C-glycosyl dietary flavone, suppresses colonic cell proliferation and mitigates NF-kappaB mediated inflammatory response in 1,2-dimethylhydrazine induced colorectal carcinogenesis. *Biomed. Pharmacother.* **2017**, *96*, 1253–1266. [CrossRef] [PubMed]
50. Khoo, L.W.; Mediani, A.; Zolkeflee, N.K.Z.; Leong, S.W.; Ismail, I.S.; Khatib, A. Phytochemical diversity of *Clinacanthus nutans* extracts and their bioactivity correlations elucidated by NMR based metabolomics. *Phytochem. Lett.* **2015**, *14*, 123–133. [CrossRef]
51. Lin, C.-M.; Chen, H.-H.; Lung, C.-W.; Chen, H.-J. Antiviral and Immunomodulatory Activities of *Clinacanthus nutans* (Burm. f.) Lindau. *Int. J. Mol. Sci.* **2023**, *24*, 10789. [CrossRef] [PubMed]
52. Khoo, L.W.; Kow, S.A.; Lee, M.T.; Tan, C.P.; Shaari, K.; Tham, C.L.; Abas, F. A Comprehensive Review on Phytochemistry and Pharmacological Activities of *Clinacanthus nutans* (Burm.f.) Lindau. *Evid. Based Complement. Altern. Med.* **2018**, *2018*, 1–39. [CrossRef]
53. Lam, K.Y.; Ling, A.P.K.; Koh, R.Y.; Wong, Y.P.; Say, Y.H. A Review on Medicinal Properties of Orientin. *Adv. Pharmacol. Sci.* **2016**, *2016*, 4104595. [CrossRef]
54. Ninfali, P.; Antonelli, A.; Magnani, M.; Scarpa, E.S. Antiviral Properties of Flavonoids and Delivery Strategies. *Nutrients* **2020**, *12*, 2534. [CrossRef]
55. Nikfarjam, B.A.; Hajiali, F.; Adineh, M.; Nassiri-Asl, M. Anti-inflammatory Effects of Quercetin and Vitexin on Activated Human Peripheral Blood Neutrophils:-The effects of quercetin and vitexin on human neutro-phils. *J. Pharmacopunct.* **2017**, *20*, 127–131.
56. Da Fonseca, J.M.; Reis, A.C.C.; Pereira, G.R.; de Moura, H.M.M.; Filho, J.D.S.; Silva, B.d.M.; Brandão, G.C. Chromatographic profile of xanthones and flavonoids in the anti-dengue extracts of *Fridericia samydoides* (Cham.) L.G. Lohmann (Bignoniaceae). *Braz. J. Pharm. Sci.* **2022**, *58*. [CrossRef]
57. Anilkumar, K.; Reddy, G.V.; Azad, R.; Yarla, N.S.; Dharmapuri, G.; Srivastava, A.; Kamal, M.A.; Pallu, R. Evaluation of Anti-Inflammatory Properties of Isoorientin Isolated from Tubers of *Pueraria tuberosa*. *Oxidative Med. Cell. Longev.* **2017**, *2017*, 5498054. [CrossRef] [PubMed]
58. Lv, H.; Yu, Z.; Zheng, Y.; Wang, L.; Qin, X.; Cheng, G.; Ci, X. Isovitexin exerts anti-inflammatory and an-ti-oxidant activities on lipopolysaccharide-induced acute lung injury by inhibiting MAPK and NF-κB and activating HO-1/Nrf2 pathways. *Int. J. Biol. Sci.* **2016**, *12*, 72–86. [CrossRef] [PubMed]

59. Huang, D.; Guo, W.; Gao, J.; Chen, J.; Olatunji, J.O. *Clinacanthus nutans* (Burm. f.) Lindau Ethanol Extract Inhibits Hepatoma in Mice through Upregulation of the Immune Response. *Molecules* **2015**, *20*, 17405–17428. [CrossRef]
60. BenSaad, L.A.; Kim, K.H.; Quah, C.C.; Kim, W.R.; Shahimi, M. Anti-inflammatory potential of ellagic acid, gallic acid and punicalagin A&B isolated from *Punica granatum*. *BMC Complement. Altern. Med.* **2017**, *17*, 47. [CrossRef]

antibiotics

MDPI

Article

Screening Tests for the Interaction of *Rubus idaeus* and *Rubus occidentalis* Extracts with Antibiotics against Gram-Positive and Gram-Negative Human Pathogens

Rafał Hałasa [1,*], Urszula Mizerska [2], Marta Kula [3] and Mirosława Krauze-Baranowska [4]

1 Department of Pharmaceutical Microbiology, Medical University of Gdansk, Al. Gen. J. Hallera 107, 80-416 Gdansk, Poland
2 Department of Polymeric Nanomaterials, Centre of Molecular and Macromolecular Studies, Polish Academy of Sciences, ul. Sienkiewicza 112, 90-363 Lodz, Poland; urszula.mizerska@cbmm.lodz.pl
3 Moderna Poland sp.zoo, Rondo Ignacego Daszyńskiego 1, 00-843 Warszawa, Poland; marta.kula@modernatx.com
4 Department of Pharmacognosy, Medical University of Gdansk, Al. Gen. J. Hallera 107, 80-416 Gdansk, Poland; miroslawa.krauze-baranowska@gumed.edu.pl
* Correspondence: rafal.halasa@gumed.edu.pl

Abstract: WHO (World Health Organization) reports from recent years warn about the growing number of antibiotic-resistant bacterial strains. Therefore, there is an urgent need to constantly search for new substances effective in the fight against microorganisms. Plants are a rich source of chemical compounds with antibacterial properties. These compounds, classified as secondary metabolites, may act independently or support the action of currently used antibiotics. Due to the large number of metabolites isolated from the plant kingdom and new plant species being studied, there is a need to develop new strategies/techniques or modifications of currently applied methods that can be used to select plant extracts or chemical compounds isolated from them that enter into positive, synergistic interactions with currently used antibiotics. One such method is the dual-disk synergy test (DDST). It involves the diffusion of active compounds in the agar environment and influencing the growth of microorganisms grown on it. The method was used to assess the interaction of extracts from the fruit and shoots of some cultivated varieties of *Rubus idaeus* and *Rubus occidentalis* with selected antibiotics. The research was conducted on strains of bacteria pathogenic to humans, including *Staphylococcus aureus*, *Corynebacterium diphtheriae*, *Escherichia coli*, *Pseudomonas aeruginosa*, *Helicobacter pylori*, and *Candida albicans*, showing synergy, antagonism, or lack of interaction of the tested substances—plant extract and antibiotic. As a result, it was found that the diffusion method is useful in screening tests to assess the impact of antibiotic–herbal substance interactions on Gram-positive and Gram-negative microorganisms.

Keywords: additivity; antagonisms; antibiotics; double-disk synergy test; extracts; *Rubus* sp.; synergism

Citation: Hałasa, R.; Mizerska, U.; Kula, M.; Krauze-Baranowska, M. Screening Tests for the Interaction of *Rubus idaeus* and *Rubus occidentalis* Extracts with Antibiotics against Gram-Positive and Gram-Negative Human Pathogens. *Antibiotics* **2024**, *13*, 653. https://doi.org/10.3390/antibiotics13070653

Academic Editors: Anna Duda-Madej, Joanna Kozłowska, Katarzyna Wińska and Marc Maresca

Received: 31 May 2024
Revised: 4 July 2024
Accepted: 13 July 2024
Published: 15 July 2024

1. Introduction

The natural environment has huge plant resources, and they vary depending on the climatic zones. Herbal substances contain multi-component mixtures of various biologically active compounds, such as phenolic compounds, terpenoids, essential oils, alkaloids, lectins, and many others [1–3]. Many teams of researchers are looking for new chemical compounds with high biological activity and virtually no side effects. Currently, scientists' research interests focus mainly on plants, and especially on secondary metabolites with potential anticancer and/or antimicrobial effects [1].

For most compounds isolated from plants, biological properties have been determined, among others, to have antioxidant, anti-inflammatory, and antimicrobial properties. The mechanisms of action of plant metabolites on microorganisms differ from those of currently used antibiotics, which may reduce the risk of cross-resistance and is due to differences in their chemical structures.

Some of them, e.g., phenolic components of essential oils, are known for their strong antimicrobial activity; others, in combination with currently used antibiotics, may become substances that modify bacterial resistance [3–5]. In recent years, many studies have focused on herbal extract–antibiotic interactions [3,6], especially since an increase in the antibacterial activity of aminoglycosides, quinolones, macrolides, and tetracyclines in the presence of phytochemicals has been demonstrated [3,7]. Multidrug therapies may have additive or synergistic effects, and sometimes antagonistic effects may be observed. A negligible effect is observed when two substances have the same binding site in the cell. An additive effect occurs when it is the sum of the effects of individual substances against a given microorganism. Synergism means that the strength of activity of a combination of substances is greater than the sum of the effects of individual compounds. However, the antagonistic effect leads to a decrease in the activity of the combination of substances compared to their individual activity [3,8].

It is generally believed that extracts obtained using organic solvents, including ethanol and methanol, have greater antibacterial potential than water extracts. This is due to the dissolution of more active herbal ingredients with antimicrobial properties in organic solvents [9,10].

All currently used methods for determining the susceptibility of microorganisms are standardized, controlled, and updated by the European Committee on Antimicrobial Susceptibility Testing (EUCAST) or the Clinical and Laboratory Standards Institute (CLSI). In vitro methods used in clinical laboratories are divided into phenotypic and genotypic. The most popular phenotypic methods are diffusion and well techniques (qualitative methods) and dilution with agar or broth (quantitative methods). These techniques can be used to pre-select plant extracts with antimicrobial activity [11].

In the disk diffusion test, a bacterial suspension with a density of 1×10^8 CFU/mL (McFarland 0.5) is inoculated uniformly onto the surface of a Müller–Hinton (MH) agar plate [12]. A paper disk (6 mm in diameter) containing a standard amount of antimicrobial substance is placed on the agar surface and allowed to diffuse into the medium. After incubation in the growing bacteria, a clear area of "no growth" appears around the disk—a zone of inhibition. Its size depends on the diffusion rate of the tested substance and the degree of sensitivity of the microorganism. Strains resistant to the antimicrobial compound reach the edge of the disk. The diameter of the inhibition zone is analyzed based on the accepted interpretation standards for various antibiotics. The double-disk synergy test is a test commonly recommended by EUCAST to detect extended-spectrum β-lactamase (ESBL) resistance mechanisms in Gram-negative bacilli [13,14]. It is also used to detect synergisms between different antibiotics in combating resistant bacteria. This test is a variation of the disk diffusion method, in which two separate disks containing two substances are placed at a distance no greater than the limits of the growth inhibition zones of the compounds. After incubation, any deformations and growth inhibition are observed and interpreted according to Section 3.4. Another variation of the disk diffusion method involves adding the concentrated herbal extract to a cooled agar medium and pouring it onto Petri dishes. The bacterial suspension is then inoculated and antibiotic disks are placed. After incubation, zones of inhibition are checked and compared to established interpretation standards for various antibiotics. If the zone of inhibition increases compared to the norm, this is interpreted as synergism/additivity; if the zone decreases, this is interpreted as antagonism. On the other hand, when the growth inhibition zone remains unchanged, is referred to as no impact [15,16].

In well diffusion agar, the bacterial inoculum is mixed with a liquid, cooled agar medium; poured onto a Petri dish; and allowed to solidify. Then, using a sterile drill (4–8 mm in diameter), wells are cut and the test substances are introduced into each well. After incubation, the plate is observed for obvious zones of inhibition around the well. A variation of this method involves evenly spreading the inoculum suspension on a solid agar plate and adding an antimicrobial substance to each well [2,15,16].

Other methods useful for testing the antibacterial activity of herbal extracts are bioautography (based on planar chromatography techniques), the broth and agar dilution methods, and the killing time assay and the checkerboard assay. All quantitative methods are precise and enable the determination of active concentrations such as MIC—minimum inhibitory concentration and MBC—minimum bactericidal concentration. However, they are of little use in screening tests due to the high consumption of reagents and their absorbent properties [2,15,16].

Depending on the extraction method and solvents used, the composition of the obtained extracts may vary in terms of quantity. The most commonly used method for differentiating the antimicrobial activity of the tested herbal extracts is the disk diffusion method. This is a simple technique in which sterile disks soaked in a specific concentration of analyzed extract are placed on bacteria seeded on an agar medium. Samples are most often incubated in an aerobic atmosphere at 37 °C for 24 h. Zones of growth inhibition around the disk indicate the presence of active compounds.

The aim of the study was to assess the usefulness of disk diffusion methods in examining the interactions of herbal extracts with antibiotics in terms of their effect on microorganisms. Moreover, taking into account previous literature data on the antimicrobial activity of extracts from the fruits and shoots of some varieties of *R. idaeus* and *R. occidentalis*, it was also interesting to assess the types of interactions between these extracts and antibiotics in terms of their impact on selected microorganisms. At the same time, based on previously published scientific research results, it was decided to explain the role of some chemical compounds present in the tested extracts in the observed interactions with antibiotics.

2. Results

2.1. Extract Preparation and Phytochemical Analysis

Methods of preparation and phytochemical analysis of extracts (*R. occidentalis* "Litacz" fruits, *R. idaeus* "Laszka" fruits, *R. idaeus* "Poranna Rosa" fruits, *R. idaeus* "Willamette" shoot) have been described in our recently published article (Hałasa et al.) [17]. The content of phenolic compounds in dry extracts from the fruits and shoots of *R. idaeus* and *R. occidentalis* varieties is presented in Table 1.

Table 1. Content of phenolic compounds in dry extracts from the fruits and shoots of *R. idaeus* and *R. occidentalis* varieties [mg/100 g d. w.] in Hałasa et al. [17].

Chemical Compound	Fruits *R. id.* "PR"	Fruits *R. id.* "Lasz"	Fruits *R. occ.* "Lit"	Shoots *R. id* "Will"
Ellagic acid	32.1	24.6	29.6	455.6
		Ellagic acid derivatives		
Ellagic acid pentoside	26.8	56.4	23.5	195.1
Ellagic acid methylpentoside [1]	24.5	22.9	27.9	51.6
Ellagic acid acetylpentoside [1]	21	22.7	20.6	174.0
Sum of contents	72.3	102.0	72.0	420.7
		Ellagitannins		
Sanguiin H-2 [1]	19.9	20.0	19.7	44.3
Sanguiin H-6 [1]	140.5	189.0	82.9	339.5
Sanguiin H-10 isomer [1]	132.8	84.3	116.1	183.6
Lambertianin C [1]	–	–	–	31.1
Sum of contents	293.2	293.3	218.7	598.5
		Anthocyanins		
Cyanidin 3-*O*-glucoside	–	56.4	253.1	–
Cyanidin 3-*O*-rutinoside [2]	–	14.0	1195.0	–
Cyanidin 3-*O*-sophoroside [2]	–	122.1	–	–

Table 1. *Cont.*

Chemical Compound	Fruits *R. id.* "PR"	Fruits *R. id.* "Lasz"	Fruits *R. occ.* "Lit"	Shoots *R. id* "Will"
Cyanidin 3-*O*-(2^G-glucosylrutinoside) [2]	–	16.1	–	–
Cyanidin 3-*O*-(2^G-xylosylrutinoside) [2] Cyanidin 3-*O*-sambubioside [2]	–	–	1312.0 *	–
Pelargonidin 3-*O*-rutinoside [2]	–	–	37.4	–
Sum of contents	–	208.6	2797.5	–
Others—phenolic acids, flavonoids				
Gallic acid	–	–	–	20.7
Caffeic acid	–	–	–	74.9
Protocatechic acid	–	–	15.4	20.6
Epicatechin	65.5	–	–	1538.2
Quercetin 3-*O*-glucuronide	15.7	10.1	74.4	166.9

Abbreviations: *R. id.* "PR"—*R. idaeus* "Poranna Rosa," *R. id.* "Lasz"—*R. idaeus* "Laszka," *R. occ.* "Lit"—*R. occidentalis* "Litacz," *R. id.* "Will"—*R. idaeus* "Willamette"; [1] calculated on ellagic acid; [2] calculated on cyanidin 3-*O*-glucoside; * given as a sum due to separation as a single peak.

2.2. Determination of Susceptibility to Antibiotics by Agar Disk Diffusion Method

The zones of bacterial growth inhibition by the antibiotics and extracts used in the tests are included in Tables 2 and 3.

Table 2. Zones of growth inhibition of Gram-positive microorganisms for antibiotics and extracts obtained by the disk diffusion method.

	β-hemolytic *Streptococcus* Group A PCM465	β-hemolytic *Streptococcus* Group B	*Staphylococcus aureus* ATCC6538	*Staphylococcus epidermidis* ATCC14990	*Enterococcus hirae* ATCC10541	*Enterococcus faecalis* ATCC51299	*Corynebacterium dyphtheriae*	*Cadida albicans* ATCC10248
Oxacillin	N	N	40 ± 0.1	40 ± 0.2	N	N	N	N
Penicillin	58 ± 0.05	12 ± 0.1	26 ± 0.2	10 ± 0.2	6 ± 0.00	6 ± 0.00	30 ± 0.2	N
Amoxicillin	60 ± 0.05	28 ± 0.05	N	N	N	N	N	N
Ampicillin	N	N	N	N	24 ± 0.1	22 ± 0.2	N	N
Cefotaxime	60 ± 0.13	6 ± 0.00	N	N	N	N	48 ± 0.2	N
Ceftriaxone	25 ± 0.21	6 ± 0.00	N	N	N	N	48 ± 0.25	N
Cefepime	20 ± 0.11	6 ± 0.00	N	N	N	N	N	N
Imipenem	N	N	N	N	N	N	6 ± 0.00	N
Netilmycin	N	N	20 ± 0.2	24 ± 0.2	N	N	N	N
Gentamicin (CN10)	N	N	28 ± 0.25	35 ± 0.2	N	N	50 ± 0.2	N
Gentamicin (CN120)	N	N	N	N	15 ± 0.2	12 ± 0.2	N	N
Streptomycin (S300)	N	N	N	N	11 ± 0.1	10 ± 0.1	N	N
Kanamycin	N	N	23 ± 0.22	23 ± 0.2	N	N	N	N
Amikacin	N	N	20 ± 0.1	25 ± 0.52	N	N	N	N
Tobramycin	N	N	23 ± 0.21	25 ± 0.3	N	N	N	N
Erythromycin	20 ± 0.2	15 ± 0.25	32 ± 0.2	22 ± 0.1	N	N	66 ± 0.34	N

Table 2. *Cont.*

	β-hemolytic *Streptococcus* Group A PCM465	β-hemolytic *Streptococcus* Group B	*Staphylococcus aureus* ATCC6538	*Staphylococcus epidermidis* ATCC14990	*Enterococcus hirae* ATCC10541	*Enterococcus faecalis* ATCC51299	*Corynebacterium dyphtheriae*	*Cadida albicans* ATCC10248
Telithromycin	N	N	N	N	N	N	68 ± 0.35	N
Daptomycin	38 ± 0.21	18 ± 0.2	N	N	N	N	N	N
Clindamycin	38 ± 0.22	6 ± 0.00	25 ± 0.2	40 ± 0.23	N	N	50 ± 0.45	N
Linezolid	42 ± 0.1	32 ± 0.2	42 ± 0.1	50 ± 0.2	30 ± 0.2	26 ± 0.26	70 ± 0.55	N
Ofloxacillin	28 ± 0.25	12 ± 034	32 ± 0.21	20 ± 0.25	N	N	N	N
Ciprofloxacin	N	N	40 ± 0.3	45 ± 0.3	N	N	N	N
Levofloxacin	28 ± 0.31	22 ± 0.25	39 ± 0.25	25 ± 0.25	N	N	N	N
Vancomycin	N	N	25 ± 0.2	26 ± 0.31	20 ± 0.36	14 ± 0.15	40	N
Teicoplanin	28 ± 0.1	16 ± 0.2	20 ± 0.2	25 ± 0.1	20 ± 0.2	15 ± 0.2	N	N
Fusidic acid	N	N	25 ± 0.2	40 ± 0.1	N	N	N	N
Rifampicin	N	N	40 ± 0.13	50 ± 0.22	29 ± 0.21	20 ± 0.26	50 ± 0.2	N
Cotrimoxazole	N	N	35 ± 0.2	30 ± 0.22	N	N	29 ± 0.21	N
Mupirocin	N	N	30 ± 0.35	35 ± 0.22	N	N	N	N
Tetracycline	35 ± 0.2	10 ± 0.21	25 ± 0.2	10 ± 0.11	25 ± 0.15	20 ± 0.32	40 ± 0.1	N
Tigecycline	32 ± 0.21	18 ± 0.21	N	N	N	N	N	N
Chloramphenicol	N	N	35 ± 0.1	19 ± 0.25	29 ± 0.2	9 ± 0.11	N	N
Miconazole	N	N	N	N	N	N	N	14 ± 0.1
Caspafungin	N	N	N	N	N	N	N	15 ± 0.1
Nystatin	N	N	N	N	N	N	N	22 ± 0.1
Econazole	N	N	N	N	N	N	N	15 ± 0.1
Extract of *Rubus occidentalis* "Litacz" fruits	9 ± 0.5 *	9 ± 0.5 *	19 ± 0.1 *	23 ± 0.1 *	6 ± 0.1	6 ± 0.1	20 ± 0.1 *	6 ± 0.1
Extract of *Rubus idaeus* "Laszka" fruits	10 ± 0.5 *	9 ± 0.8 *	20 ± 0.1 *	24 ± 0.1 *	6 ± 0.1	6 ± 0.1	19 ± 0.1 *	6 ± 0.1
Extract of *Rubus idaeus*" Poranna Rosa" fruits	10 ± 0.5 *	11 ± 0.3 *	24 ± 0.1 *	25 ± 0.1 *	6 ± 0.1	6 ± 0.1	22 ± 0.1 *	6 ± 0.1
Extract of *Rubus idaeus* "Willamette" shoot	13 ± 0.1 *	10 ± 0.1 *	23 ± 0.1 *	25 ± 0.1 *	6 ± 0.1	6 ± 0.1	25 ± 0.1 *	6 ± 0.1

N—not tested. The results are presented as mean values ± standard deviation (±SD) from 3 independent experiments. * $p < 0.05$.

Table 3. Zones of inhibition of growth of Gram-negative microorganisms for antibiotics and extracts obtained by the disk diffusion method.

	***Escherichia coli* ATCC8739**	***Klebsiella pneumoniae* ATCC13883**	***Proteus vulgaris* ATCC13883**	***Pseudomonas aeruginosa* ATCC9027**	***Acinetobacter baumanii* 2021**	***Stenotrophomonas maltophilia*12755**	***Helicobacter pylori* ATCC10231**
Amoxicillin	30 ± 0.21	7 ± 0.1	25 ± 0.21	N	N	N	24 ± 0.2
Amoxicillin/clavulanic acid	25 ± 0.12	11 ± 0.2	24 ± 0.12	N	N	N	N
Ampicillin	20 ± 0.2	6 ± 0.00	24	N	N	N	N
Ampicillin/sulbactam	N	N	N	6 ± 0.00	6 ± 0.00	N	N
Cefuroxime	26 ± 0.12	20 ± 0.23	25 ± 0.23	N	N	N	N
Cefotaxime	25 ± 0.2	28 ± 0.22	25 ± 0.1	N	N	N	N
Ceftriaxone	24 ± 0.25	25 ± 0.21	26 ± 0.21	N	N	N	N
Cefazolin	30 ± 0.24	25 ± 0.1	20 ± 0.00	N	N	N	N
Ceftazidime	N	N	N	31 ± 0.23	N	30 ± 0.5	N
Cefepime	40 ± 0.2	20 ± 0.23	21 ± 0.2	20 ± 0.05	16 ± 0.2	N	N
Ticarcillin	25 ± 0.12	12 ± 0.12	28 ± 0.21	25 ± 0.1	16 ± 0.2	N	N
Ticarcillin/clavulanic acid	25 ± 0.2	24 ± 0.2	35 ± 0.2	30 ± 0.21	N	N	N
Piperacillin	35 ± 0.21	28 ± 0.2	28 ± 0.21	N	N	N	N
Piperacillin/tazobactam	32 ± 0.2	29 ± 0.2	35 ± 0.13	37 ± 0.2	30	N	N
Imipenem	25 ± 0.12	7 ± 0.05	19 ± 0.12	30 ± 0.2	6 ± 0.00	N	N
Meropenem	33 ± 0.1	20 ± 0.2	18 ± 0.13	30 ± 0.2	6 ± 0.00	N	N
Ertapenem	33 ± 0.21	17 ± 0.2	18 ± 0.21	N	N	N	N
Doripenem	31 ± 0.22	6 ± 0.00	20 ± 0.21	N	N	N	N
Aztreonam	25 ± 0.2	6 ± 0.00	24 ± 0.1	35 ± 0.2	N	N	N
Clarithromycin	N	N	N	N	N	N	8 ± 0.12
Netilmicin	17 ± 0.25	N	25 ± 0.1	22 ± 0.11	20 ± 0.1	N	N
Gentamicin	19 ± 0.2	N	25 ± 0.23	18 ± 0.12	N	N	N
Amikacin	18 ± 0.1	26 ± 0.2	26 ± 0.21	25 ± 0.2	N	N	N
Tobramycin	19 ± 0.1	10 ± 0.1	29 ± 0.12	N	N	N	N
Ciprofloxacin	35 ± 0.26	N	40 ± 0.35	40 ± 0.35	N	N	N
Levofloxacin	N	N	N	36 ± 0.56	N	35 ± 0.6	30 ± 0.35
Tetracycline	28 ± 0.2	28 ± 0.2	18 ± 0.34	N	N	N	35 ± 0.1
Tigecycline	20 ± 0.1	N	20 ± 0.1	N	N	N	N
Chloramphenicol	31 ± 0.2	30 ± 0.1	28 ± 0.1	18 ± 0.2	15 ± 0.2	N	N
Colistin	N	N	N	17 ± 0.1	14 ± 0.1	18 ± 0.1	N
Metronidazole	N	N	N	N	N	N	6 ± 0.00
Cotrimoxazole	20 ± 0.2	35 ± 0.35	25 ± 0.25	N	N	15 ± 0.25	N

Table 3. *Cont.*

	Escherichia coli ATCC8739	*Klebsiella pneumoniae* ATCC13883	*Proteus vulgaris* ATCC13883	*Pseudomonas aeruginosa* ATCC9027	*Acinetobacter baumanii* 2021	*Stenotrophomonas maltophilia*12755	*Helicobacter pylori* ATCC10231
Extract of *Rubus occidentalis* "Litacz" fruits	6 ± 0.1	6 ± 0.1	6 ± 0.1	6 ± 0.1	6 ± 0.1	6 ± 0.1	6 ± 0.1
Extract of *Rubus idaeus* "Laszka" fruits	6 ± 0.1	6 ± 0.1	6 ± 0.1	6 ± 0.1	6 ± 0.1	6 ± 0.1	6 ± 0.1
Extract of *Rubus idaeus* "Poranna Rosa" fruits	6 ± 0.1	6 ± 0.1	6 ± 0.1	6 ± 0.1	6 ± 0.1	6 ± 0.1	6 ± 0.1
Extract of *Rubus idaeus* "Willamette" shoot	6 ± 0.1	6 ± 0.1	6 ± 0.1	6 ± 0.1	6 ± 0.1	6 ± 0.1	6 ± 0.1

N—not tested. The results are presented as mean values ± standard deviation (±SD) from 3 independent experiments.

For diffusion tests, sets of disks recommended by EUCAST and CLSI were used. Different sensitivity to the antibiotics used was observed, depending on the strains. Generally, strains from the ATCC collection appeared to be sensitive to most antibiotics, whereas clinical strains of β-hemolytic *Streptococcus* groups B, *Acinetobacter baumannii* 2021, and *Stenotrophomonas maltophilia* 12755 were predominantly resistant to the antibiotics tested.

2.3. Screening for Antibiotic–Extract Interactions Using the Double-Disk Synergy Test (DDST)

Examples of antibiotic–extract interaction using the double-disk synergy test are shown in Figure 1. The results of the DDST test for individual bacteria are shown in the subsequent figures in this chapter.

Among streptococci (Figures 2 and 3), synergy of activity between the tested extracts and antibiotics was observed only against β-hemolytic *Streptococcus. R. idaeus* "Willamette" shoot extract showed a synergistic effect with penicillin, amoxicillin, cefotaxime, ceftriaxone, and cefepime (antibiotics from the β-lactam group), as well as erythromycin, lincomycin, linezolid, tetracycline, and tigecycline against β-hemolytic *Streptococcus* group A, with cefotaxime and tigecycline against β-hemolytic *Streptococcus* group B. *R. idaeus* "Laszka" fruit extract presented a synergistic effect with amoxicillin, ceftriaxone, tetracycline, and tigecycline against β-hemolytic *Streptococcus* group A, and amoxicillin and tigecycline against β-hemolytic *Streptococcus* of group B. *R. occidentalis* "Litacz" fruit extract showed a synergistic effect with amoxicillin and tetracycline against β-hemolytic *Streptococcus* group A. *R. idaeus* "Laszka" fruit extract and *R. occidentalis* "Litacz" fruit extract presented synergism with amoxicillin and tigecycline against β-hemolytic *Streptococcus* group B. *R. idaeus* "Poranna Rosa" fruit extract presented a synergistic effect with amoxicillin, cefotaxime, and tetracycline against β-hemolytic *Streptococcus* group A, and against β-hemolytic *Streptococcus* group B with amoxicillin, linezolid, and tigecycline.

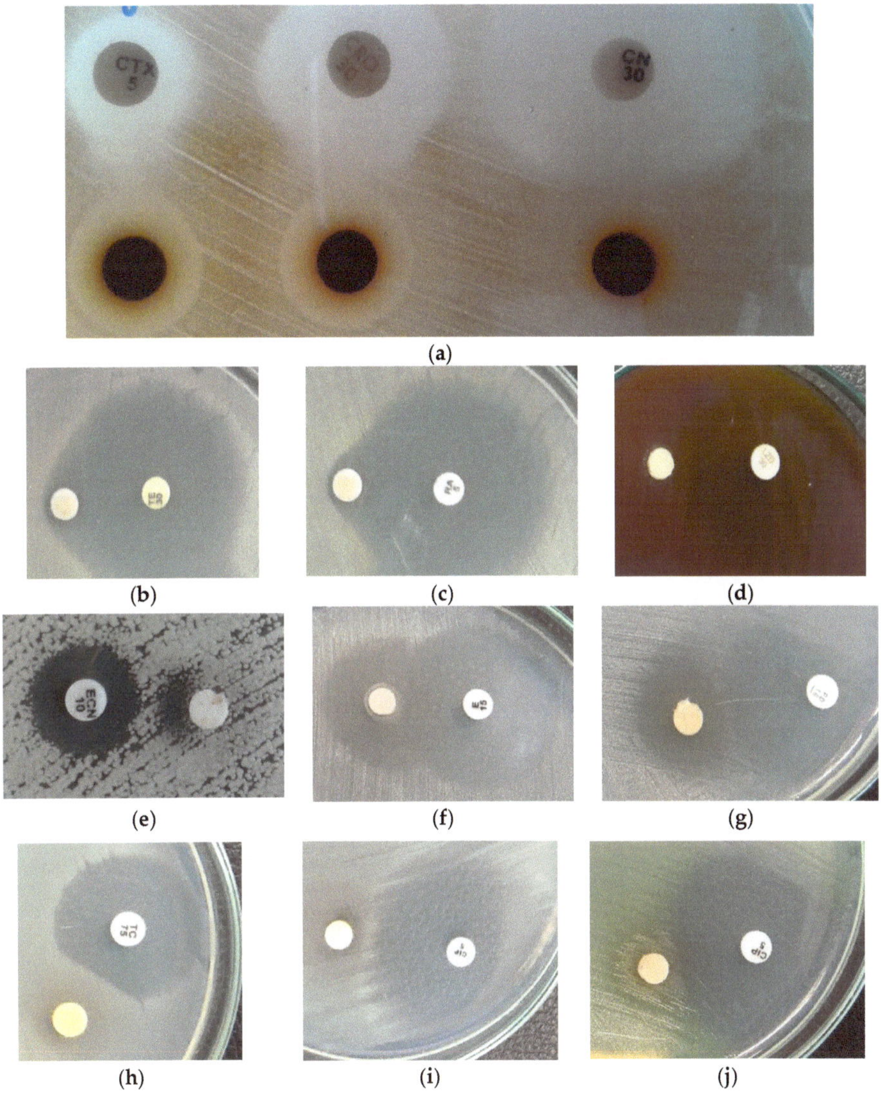

Figure 1. Examples of antibiotic–extract interactions using the double-disk synergy test. (**a**) *S. aureus*—extract of *R. idaeus* "Willamette" shoot—cefazolin (CTX5) and ceftriaxone (CRO30) synergism; extract of *R. idaeus* "Willamette" shoot—gentamicin (GN30)—antagonism; (**b**) *S. aureus*—extract of *R. idaeus* "Poranna Rosa" fruits—tetracycline (Te30)—synergism; (**c**) *S. aureus*—extract of *R. idaeus* "Poranna Rosa" fruits—rifampicin (RA5)—synergism; (**d**) β-hemolytic *Streptococcus* group B—extract of *R. idaeus* "Poranna Rosa" fruits—linezolid (LZD30)—synergism; (**e**) *C. albicans*—extract of *R. idaeus* "Poranna Rosa" fruits—econazole (ECN10)—synergism; (**f**) *S. epidermidis*—extract of *R. idaeus* "Willamette" shoot—erythromycin (E15)—antagonism; (**g**) *S. epidermidis*—extract of *R. idaeus* "Willamette" shoot—fusidic acid (FD10)—antagonism; (**h**) *E.coli* extract of *R. occidentalis* "Litacz" fruits—ticarcillin (TC75)—antagonism; (**i**) *P. vulgaris*—extract of *R. idaeus* "Poranna Rosa" fruits—ciprofloxacin (CIP5); (**j**) *P. aeruginosa*—extract of *R. idaeus* "Willamette" shoot—ciprofloxacin (CIP5).

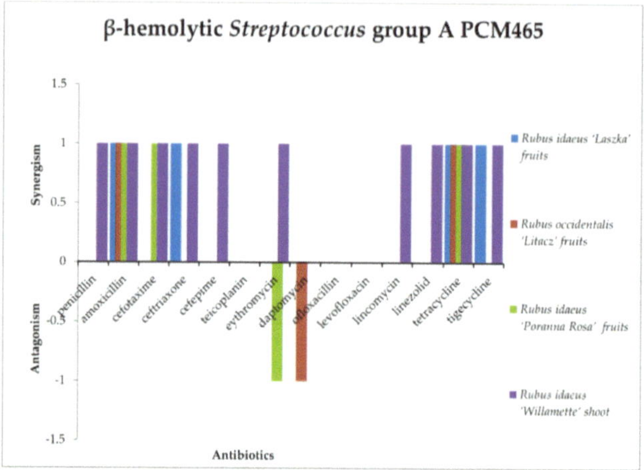

Figure 2. Interactions of *R. idaeus* and *R. occidentalis* extracts with antibiotics against β-hemolytic *Streptococcus* group A PCM465.

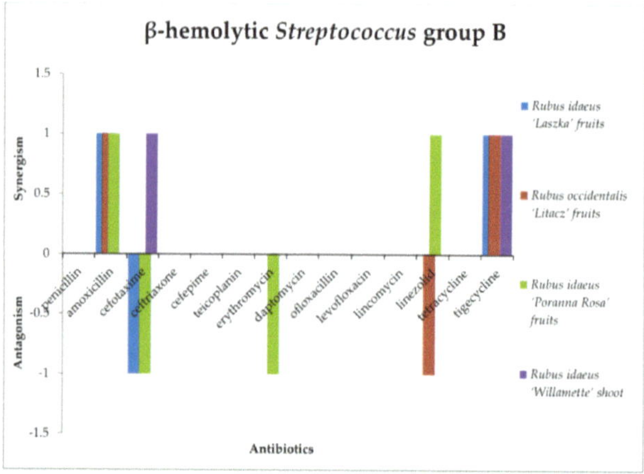

Figure 3. Interactions of *R. idaeus* and *R. occidentalis* extracts with antibiotics against β-hemolytic *Streptococcus* group B.

The following compounds were antagonistic to some antibiotics against β-hemolytic *Streptococcus* group A. Only *R. idaeus* "Laszka" fruit extract and *R. idaeus* "Poranna Rosa" fruit extract in combination with erythromycin and daptomycin, respectively, produced an antagonistic effect, whereas, against *Streptococcus* β-hemolytic group B, the antagonistic effect occurred in the case of *R. idaeus* "Laszka" fruit extract (in combination with cefotaxime), *R. idaeus* "Poranna Rosa" fruit extract (in combination with cefotaxime and erythromycin), and Rubus occidentalis "Litacz" fruit extract (in combination with linezolid).

Against β-hemolytic *Streptococcus* group A PCM465, no interaction was observed between *Rubus idaeus* "Laszka" fruit, *Rubus occidentalis* "Litacz" fruit, *Rubus idaeus* "Poranna Rosa" fruit, and *Rubus idaeus* "Willamette" shoot extracts and teicoplanin, ofloxacillin, or levofloxation. No interaction was observed between *Rubus idaeus* "Laszka" fruit, *Rubus occidentalis* "Litacz" fruit, and *Rubus idaeus* "Poranna Rosa" fruit extracts and penicillin, cefotaxime, cefepime lincomycin, or linezolid. No interaction was observed between *Rubus idaeus* "Laszka" fruit, *Rubus idaeus* "Poranna Rosa" fruit, and *Rubus idaeus* "Willamette" shoot extracts and daptomycin. No interaction was observed between *Rubus idaeus* "Laszka" fruit and *Rubus occidentalis* "Litacz" fruit extracts and erythromycin. No interaction was observed between *Rubus occidentalis* "Litacz" fruit and *Rubus idaeus* "Poranna Rosa" fruit extracts and tigecycline or between *Rubus occidentalis* "Litacz" fruit and *Rubus idaeus* "Poranna Rosa" fruit extracts and ceftriaxone. In studies with β-hemolytic *Streptococcus* group B, no interaction was observed between *Rubus idaeus* "Laszka" fruit, *Rubus occidentalis* "Litacz" fruit, *Rubus idaeus* "Poranna Rosa" fruit, and *Rubus idaeus* "Willamette" shoot extracts and penicillin, ceftriaxone, cefepime teicoplanin, erythromycin, daptomycin, ofloxacillin, levofloxation, lincomycin, linezolid, or tetracycline, or between *Rubus idaeus* "Poranna Rosa" fruit extract with cefotaxime

A synergistic effect of the tested extracts in combination with antibiotics against staphylococci (Figures 4 and 5) was noted. The mixtures of *R. idaeus* "Willamette" shoot extract with penicillin, ofloxacin, and fusidic acid were active against *Staphylococcus aureus* ATCC6538. *R. idaeus* "Laszka" fruit extract was effective with penicillin, linezolid, tetracycline, mupirocin, and fusidic acid against *S. aureus* and with tetracycline and mupirocin against *Staphylococcus epidermidis* ATCC14990. The combination of *R. occidentalis* "Litacz" fruit extract with linezolid, chloramphenicol, tetracycline, mupirocin, fusidic acid, and rifampicin against *S. aureus* yielded the same effect as mupirocin against *S. epidermidis*. In turn, *R. idaeus* "Poranna Rosa" fruit extract in combination with penicillin, oxacillin, tetracycline, mupirocin, fusidic acid, and rifampicin was effective against *S. aureus*. However, the same extract for *S. epidermidis* produced a synergistic effect with oxacillin, tetracycline, and mupirocin.

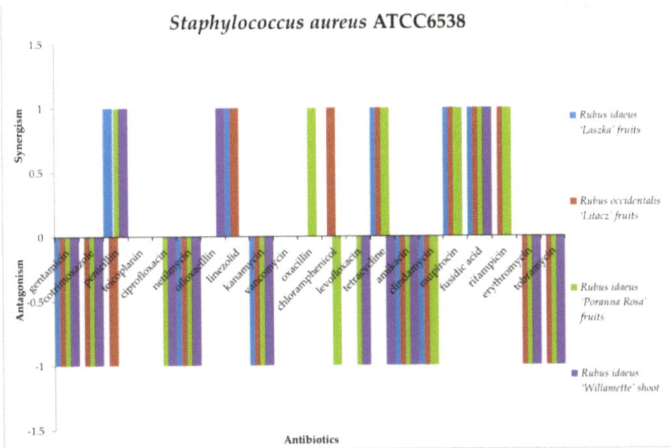

Figure 4. Interactions of *R. idaeus* and *R. occidentalis* extracts with antibiotics against *Staphylococcus aureus* ATCC6538.

All extracts in combination with the antibiotics from the aminoglycoside group (gentamycin, kanamycin, amikacin, tobramycin, netilmicin) produced an antagonistic effect against *S. aureus*. Additionally, the same effect was shown by *R. occidentalis* "Litacz" fruit, *R. idaeus* "Poranna Rosa" fruit, and *R. idaeus* "Willamette" shoot extract with co-trimoxazole, erythromycin, and clindamycin. *R. idaeus* "Poranna Rosa" fruit extract and *R. idaeus* "Willamette" shoot extract acted antagonistically with fluoroquinolones (ciprofloxacin, levofloxacin). Antagonism was demonstrated by *R. idaeus* "Poranna Rosa" fruit with chloramphenicol extract, *R. idaeus* "Willamette" shoot extract with tetracycline, R. *idaeus* "Laszka" fruit extract with clindamycin, and *R. occidentalis* "Litacz" fruit extract with penicillin. Against *S. epidermidis*, most of the interactions between the antibiotics used and the extracts turned out to be antagonistic. The exception was penicillin and teicoplanin, where there were no interactions with the extracts. *R. idaeus* "Willamette" shoot extract showed only antagonistic interactions or no interactions with the following antibiotics: gentamycin, co-trimoxazole, penicillin, teicoplanin, ciprofloxacin, mupirocin, vancomycin, oxacillin, mupirocin, and rifampicin. In the case of rifampicin, antagonism was shown only with *R. idaeus* "Laszka" fruit extract.

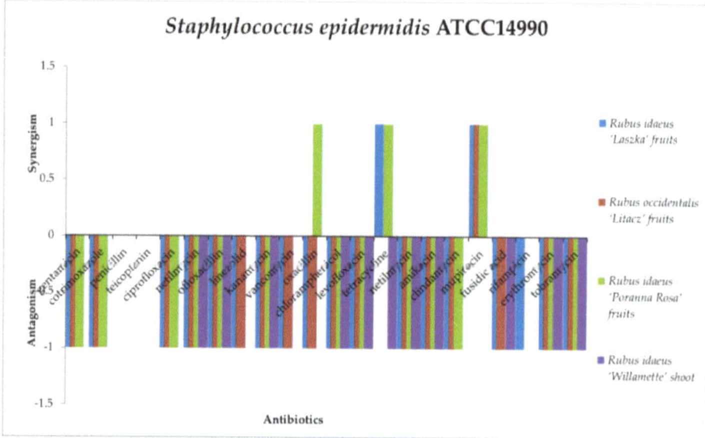

Figure 5. Interactions of *R. idaeus* and *R. occidentalis* extracts with antibiotics against *Staphylococcus epidermidis* ATCC14990.

In the case of *S. aureus* ATCC6538, no interaction was observed between *Rubus idaeus* "Laszka" fruit, *Rubus occidentalis* "Litacz" fruit, *Rubus idaeus* "Poranna Rosa" fruit, and *Rubus idaeus* "Willamette" shoot extracts and teicoplanin, ofloxacillin, or vancomycin. No interaction was observed between *Rubus occidentalis* "Litacz" fruit, *Rubus idaeus* "Laszka" fruit, and *Rubus idaeus* "Willamette" shoot extracts and ciprofloxacin or oxacillin. No interaction was observed between *Rubus idaeus* "Laszka" fruit and *Rubus idaeus* "Laszka" fruit extracts and levofloxacin. No interaction was observed between *Rubus idaeus* "Laszka" fruit and *Rubus idaeus* "Willamette" shoot extracts and chloramphenicol. No interaction was observed between *Rubus idaeus* "Willamette" shoot gentamicin extract and linezolid, clindamycin, mupirocin, rifampicin, azithromycin, or tigecycline. No interaction was observed between *Rubus idaeus* "Poranna Rosa" fruit extract and linezolid or between *Rubus idaeus* "Laszka" fruit and cotrimoxazole. In *S. epidermidis* studies, no interactions were observed in the case of *Rubus idaeus* "Laszka" fruit, Rubus occidentalis "Litacz" fruit, *Rubus idaeus* "Poranna Rosa" fruit and *Rubus idaeus* "Willamette" shoot extracts and penicillin or teicoplanin. No interaction was observed between Rubus occidentalis "Litacz" fruit and *Rubus idaeus* "Willamette" shoot extracts and rifampicin. No interaction was observed between *Rubus idaeus* "Poranna Rosa" fruit and *Rubus idaeus* "Willamette" shoot extracts and linezolid or vancomycin. No interaction was observed between Rubus occidentalis "Litacz" fruit extract with tetracycline or between *Rubus idaeus* "Willamette" shoot extract and gentamicin, cotrimoxazole, ciprofloxacin, oxacillin, clindamycin, mupirocin, ampicillin, or daptomycin.

The *R. idaeus* "Willamette" shoot extract did not show any effects with the antibiotics used against *Enterococcus hirae* ATCC10541 (Figure 6) and *Enterococcus faecalis* ATCC51299 (Figure 7). Only *R. idaeus* "Laszka" fruit extract with ampicillin and *R. occidentalis* "Litacz" fruit extract with penicillin and rifampicin had an antagonistic effect against *E. hirae*. However, none of the tested extracts showed an antagonistic effect in combination with the antibiotics used against *E. faecalis*.

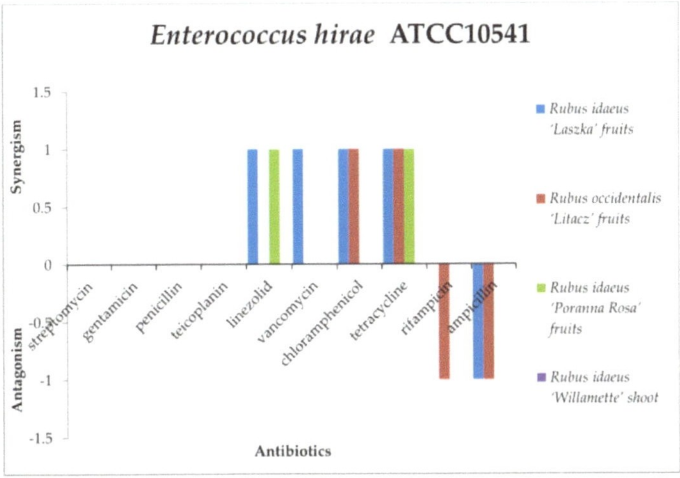

Figure 6. Interactions of *R. idaeus* and *R. occidentalis* extracts with antibiotics against *Enterococcus hirae* ATCC10541.

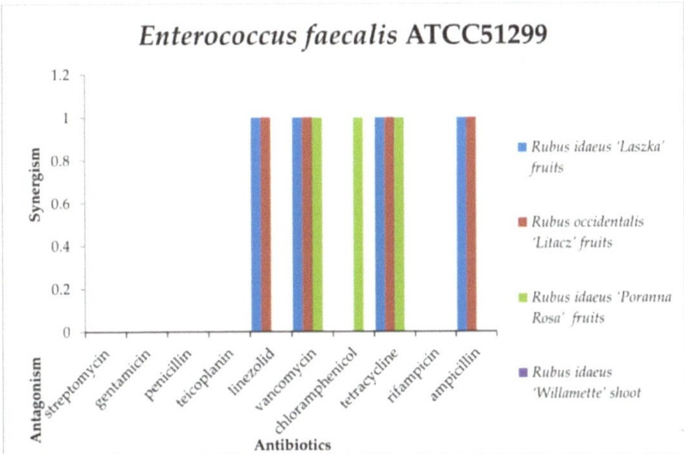

Figure 7. Interactions of *R. idaeus* and *R. occidentalis* extracts with antibiotics against *Enterococcus faecalis* ATCC51299.

It was observed that *R. idaeus* "Laszka" fruit extract in combination with linezolid, vancomycin, and tetracycline produced a synergistic effect against *E. hirae* and *E. faecalis*. Additionally, the extract had the same effect with chloramphenicol against *E. hirae* and with ampicillin against *E. faecalis*. The same phenomenon was observed for *R. occidentalis* "Litacz" fruit extract and tetracycline against *E. hirae* and *E. faecalis*; chloramphenicol with *R. occidentalis* "Litacz" fruit extract against *E. hirae*; linezolid, vancomycin, ampicillin against *E. faecalis*; *R. idaeus* "Poranna Rosa" fruit extract with tetracycline against *E. hirae* and *E. faecalis*; *R. idaeus* "Poranna Rosa" fruit extract with linezolid against *E. hirae*; and with vancomycin and chloramphenicol against *E. faecalis*.

In the case of E. hirae, *Rubus idaeus* "Laszka" fruit, *Rubus occidentalis* "Litacz" fruit, and *Rubus idaeus* "Poranna Rosa" fruit extracts did not show any interaction with streptomycin, gentamicin, penicillin, teicoplanin, or vancomycin. Similarly, *Rubus idaeus* "Laszka" fruit and *Rubus idaeus* "Poranna Rosa" fruit extracts did not show any interaction with rifampicin, nor did *Rubus occidentalis* "Litacz" fruit extract with linezolid. *Rubus idaeus* "Poranna Rosa" fruit extract did not show any interaction with chloramphenicol or ampicillin. *Rubus idaeus* "Willamette" shoot extract did not interact with any of the tested antibiotics. In studies, *E. faecalis Rubus idaeus* "Laszka" fruit and *Rubus occidentalis* "Litacz" fruit extracts did not interact with streptomycin, gentamicin, penicillin, teicoplanin, or rifampicin. *Rubus idaeus* "Poranna Rosa" fruit extract did not interact with linezolid or ampicillin. Similarly, *Rubus idaeus* "Laszka" fruit and *Rubus occidentalis* "Litacz" fruit extracts did not interact with chloramphenicol. *Rubus idaeus* "Willamette" shoot extract did not show any interaction with any of the tested antibiotics.

Only *R. occidentalis* "Litacz" fruit extract in combination with clindamycin had an antagonistic effect against *Corynebacterium diphtheria* (Figure 8). However, several synergistic interactions between the tested substances were noted. This effect was observed for all the tested extracts with tetracycline and doxycycline. Moreover, the phenomenon of synergism was noticed for the following: *R. idaeus* "Willamette" shoot extract, *R. occidentalis* "Litacz" fruit, *R. idaeus* "Poranna Rosa" fruit with cefotaxime; *R. occidentalis* "Litacz" fruit and *R. idaeus* "Poranna Rosa" fruit extract with ceftriaxone; *R. idaeus* "Willamette" shoot and *R. occidentalis* "Litacz" fruit extract with erythromycin and telithromycin; *R. idaeus* "Laszka" fruit and *R. occidentalis* "Litacz" fruit extract with penicillin and gentamicin; *R. idaeus* "Willamette" shoot; *R. idaeus* "Poranna Rosa" fruit extract with vancomycin; *R. occidentalis* "Litacz" fruit and *R. idaeus* "Poranna Rosa" fruit extract with rifampicin; *R. idaeus* "Poranna Rosa" fruit extract with oxacillin, mupirocin, and fusidic acid; *R. idaeus* "Willamette" shoot extract with linezolid; and *R. occidentalis* "Litacz" fruit extract with cotrimoxazole.

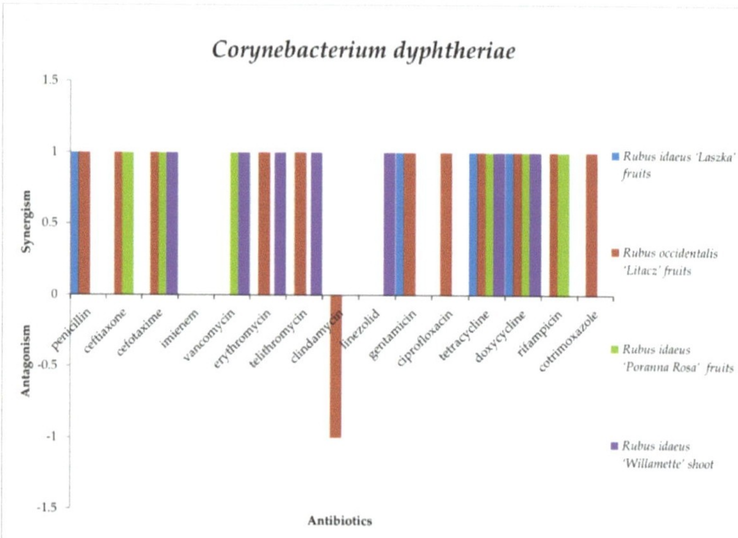

Figure 8. Interactions of *R. idaeus* and *R. occidentalis* extracts with antibiotics against *Corynebacterium diphtheriae*.

Against C. diphtheria, *Rubus idaeus* "Laszka" fruit, *Rubus occidentalis* "Litacz" fruit, *Rubus idaeus* "Poranna Rosa" fruit, and *Rubus idaeus* "Willamette" shoot extracts did not show any significant interactions with imipenem. *Rubus idaeus* "Laszka" fruit, *Rubus occidentalis* "Litacz" fruit, and *Rubus idaeus* "Poranna Rosa" fruit extracts did not interact with linezolid. *Rubus idaeus* "Laszka" fruit, *Rubus idaeus* "Poranna Rosa" fruit, and *Rubus idaeus* "Willamette" shoot extracts did not work with clindamycin. *Rubus idaeus* "Laszka" fruit and *Rubus occidentalis* "Litacz" fruit extracts did not interact with vancomycin. Similarly, *Rubus idaeus* "Laszka" fruit and *Rubus idaeus* "Poranna Rosa" fruit extracts did not interact with erythromycin or telithromycin. *Rubus idaeus* "Poranna Rosa" fruit and *Rubus idaeus* "Willamette" shoot extracts did not work with penicillin, gentamicin, ciprofloxacin, or cotrimoxazole. *Rubus idaeus* "Laszka" fruit and *Rubus idaeus* "Willamette" shoot extracts did not interact with cefotaxime, and *Rubus idaeus* "Willamette" shoot extract did not interact with rifampicin.

In the case of *Escherichia coli* ATCC8739 (Figure 9), the antagonistic effect was exerted by the combination of all extracts with the following antibiotics: ticarcillin, ticarcillin/clavulanic acid, gentamicin, tobramycin, amikacin, and netilmicin (aminoglycoside group), as well as ciprofloxacin and cotrimoxazole. The same effect was observed for the following: *R. occidentalis* "Litacz" fruit, *R. idaeus* "Poranna Rosa" fruit extract, and *R. idaeus* "Willamette" shoot with cefotaxime and doripenem; *R. idaeus* "Poranna Rosa" fruit and *R. idaeus* "Litacz" fruit extract with ertapenem; *R. occidentalis* "Litacz" fruit and *R. idaeus* "Willamette" shoot extract with cefepime; *R. idaeus* "Poranna Rosa" fruit and *R. idaeus* "Willamette" shoot extract with piperacillin, piperacillin/tazobactam, and aztreonam; and *R. idaeus* "Laszka" fruit cefuroxime extract. *R. idaeus* "Willamette" shoot extract showed antagonism with amoxicillin, cefuroxime, ceftriaxone, tetracycline. tigecycline, and chloramphenicol.

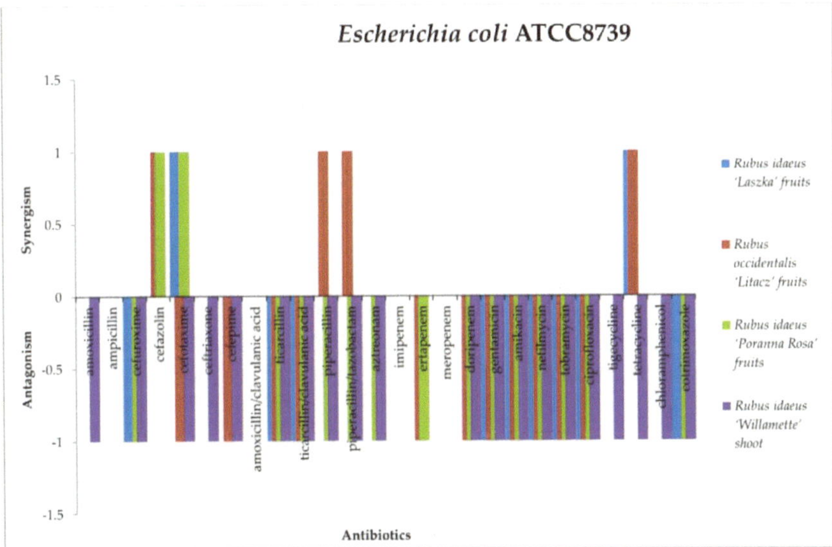

Figure 9. Interactions of *R. idaeus* and *R. occidentalis* extracts with antibiotics against *Escherichia coli* ATCC8739.

A synergistic effect of extracts with antibiotics against *E. coli* was also noted: *R. idaeus* "Laszka" fruit and *R. idaeus* "Poranna Rosa" fruit extract with cefotaxime, *R. idaeus* "Laszka" fruits and *R. occidentalis* "Litacz" fruit extracts with tetracycline, and *R. occidentalis* "Litacz" fruit extract with cefazolin, piperacillin, and piperacillin/tazobactam.

No interactions were demonstrated by *Rubus idaeus* "Laszka" fruits, *Rubus occidentalis* "Litacz" fruits, *Rubus idaeus* "Poranna Rosa" fruits, or *Rubus idaeus* "Willamette" shoot extracts with ampicillin, amoxicillin/clavulanic acid, imipenem, or meropenem. No interactions were demonstrated by *Rubus idaeus* "Laszka" fruits, *Rubus occidentalis* "Litacz" fruits, and *Rubus idaeus* "Poranna Rosa" fruit extracts with amoxicillin, ceftriaxone, tigecycline, or chloramphenicol. No interactions were demonstrated by *Rubus idaeus* "Laszka" fruit and *Rubus idaeus* "Willamette" shoot extracts with cefazolin or ertapenem. No interactions were demonstrated by *Rubus occidentalis* "Litacz" fruit and *Rubus idaeus* "Poranna Rosa" fruit extracts with cotrimoxazole. No interactions were demonstrated by *Rubus idaeus*

"Laszka" fruit extract with cefepime, piperacillin, piperacillin/tazobactam, aztreonam, or doripenem. No interactions were demonstrated by *Rubus occidentalis* "Litacz" fruit extract with cefuroxime or aztreonam, and no interactions were demonstrated by *Rubus idaeus* "Poranna Rosa" fruit extract with cefepime and tetracycline.

An antagonistic effect against *Klebsiella pneumoniae* ATCC13883 (Figure 10) was observed for the following combinations: *R. occidentalis* "Litacz" fruit, *R. idaeus* "Laszka" fruit, *R. idaeus* "Poranna Rosa" fruit, and *R. idaeus* "Willamette" shoot extract with gentamicin, amikacin, tobramycin, netilmicin (aminoglycoside group), ciprofloxacin, tigecycline, and cotrimoxazole; *R. occidentalis* "Litacz" fruit and *R. idaeus* "Willamette" shoot extract against cefotaxime, piperacillin/tazobactam, and chloramphenicol; *R. idaeus* "Poranna Rosa" fruit and *R. idaeus* "Willamette" shoot extract with cefepime, piperacillin, and doripenem; *R. idaeus* "Laszka" fruit extract with amoxicillin/clavulanic acid; *R. idaeus* "Poranna Rosa" fruit extract with ertapenem; and *R. idaeus* "Willamette" shoot extract with amoxicillin, cefuroxime, cefotaxime, ticarcillin, ticarcillin/clavulanic acid, aztreonam, and tetracycline.A synergistic effect against *K. pneumoniae* was noted for *R. idaeus* "Laszka" fruit extract with cefotaxime, doripenem, and tetracycline; *R. idaeus* "Poranna Rosa" fruit extract with cefuroxime, cefazolin, aztreonam, ciprofloxacin, and chloramphenicol; and *R. occidentalis* "Litacz" fruit extract with doripenem.

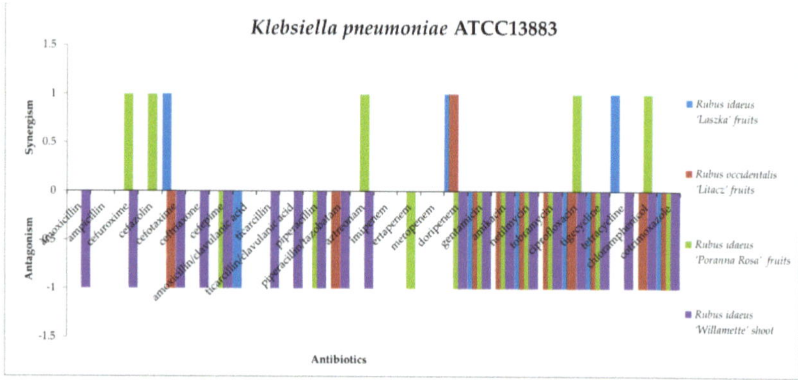

Figure 10. Interactions of *R. idaeus* and *R. occidentalis* extracts with antibiotics against *Klebsiella pneumoniae* ATCC13883.

Rubus idaeus "Laszka" fruit, *Rubus occidentalis* "Litacz" fruit, *Rubus idaeus* "Poranna Rosa" fruit, and *Rubus idaeus* "Willamette" shoot extracts with amoxicillin and ampicillin showed no interaction. *Rubus idaeus* "Laszka" fruit, *Rubus occidentalis* "Litacz" fruit, and *Rubus idaeus* "Willamette" shoot extracts with cefuroxime, cefazolin, ticarcillin, and piperacillin showed no interaction. Similarly, *Rubus occidentalis* "Litacz" fruit, *Rubus idaeus* "Poranna Rosa" fruit, and *Rubus idaeus* "Willamette" shoot extracts showed no interaction with amoxicillin/clavulanic. *Rubus idaeus* "Laszka" fruit and *Rubus occidentalis* "Litacz" fruit extracts showed no interaction with ceftriaxone, cefepime, or aztreonam. *Rubus idaeus* "Poranna Rosa" fruit and *Rubus idaeus* "Willamette" shoot extracts showed no interaction with cefotaxime or tetracycline. *Rubus idaeus* "Laszka" fruit, *Rubus occidentalis* "Litacz" fruit, and *Rubus idaeus* "Poranna Rosa" fruit extracts showed no interaction with ticarcillin/clavulanic acid. *Rubus idaeus* "Laszka" fruit, *Rubus idaeus* "Poranna Rosa" fruit, and *Rubus idaeus* "Willamette" shoot extracts showed no interaction with piperacillin/tazobatam, imipenem, or meropenem. *Rubus idaeus* "Laszka" fruit extract showed no interaction with amikacin or chloramphenicol. *Rubus idaeus* "Laszka" fruit extract showed no interaction with ertapenem or tobramycin. *Rubus idaeus* "Willamette" shoot extract showed no interaction with tetracycline.

Antagonistic activity against *Proteus vulgaris* NTCT4635 (Figure 11) was exerted by the following: *R. occidentalis* "Laszka" fruit, *R. idaeus* "Litacz" fruit, *R. idaeus* "Poranna Rosa" fruit, and *R. idaeus* "Willamette" shoot extracts with cefuroxime, gentamicin, netilmicin, amikacin, and tobramycin; *R. occidentalis* "Litacz" fruit and *R. idaeus* "Willamette" shoot extract in the presence of cefepime; *R. idaeus* "Laszka" fruit extract with tetracycline; *R. occidentalis* "Litacz" fruit extract with ticarcillin and piperacillin/tazobactam; *R. idaeus* "Poranna Rosa" fruit extract with ticarcillin, ertapenem, meropenem, doripenem, and cotrimoxazole; and *R. idaeus* "Willamette" shoot extract with ceftriaxone, amoxicillin/clavulanic acid, piperacillin, aztreonam, and ciprofloxacin.

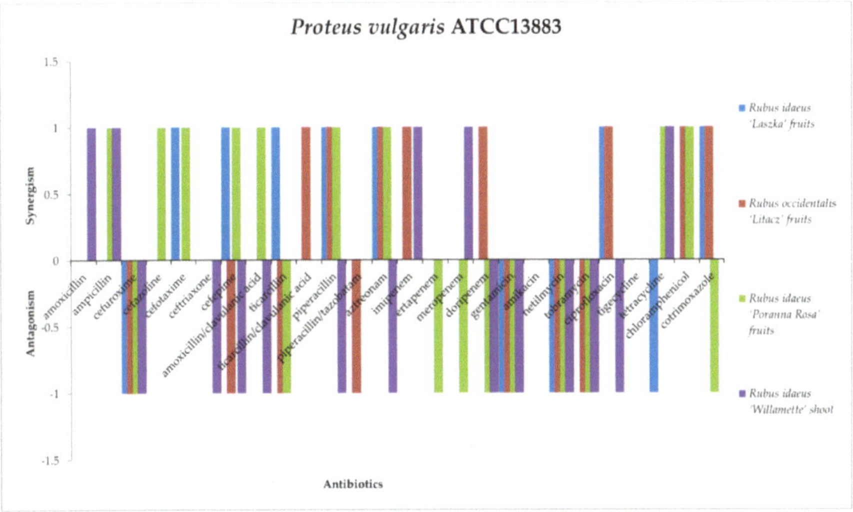

Figure 11. Interactions of *R. idaeus* and *R. occidentalis* extracts with antibiotics against *Proteus vulgaris* NTCT4635.

In turn, a synergistic effect of extract with antibiotics against *P. vulgaris* was observed for the following combinations: *R. idaeus* "Willamette" shoot extract with amoxicillin, ampicillin, imipenem, meropenem, and tetracycline; *R. idaeus* "Laszka" fruit extract with cefotaxime, cefepime, ticarcillin, piperacillin, aztreonam, ciprofloxacin, and cotrimoxazole; *R. occidentalis* "Litacz" fruit extract with ticarcillin/clavulanic acid, piperacillin, aztreonam, imipenem, doripenem, ciprofloxacin, chloramphenicol, and cotrimoxazole; and *R. idaeus* "Poranna Rosa" fruit extract with ampicillin, cefazolin, cefotaxime, cefepime, amoxicillin/clavulanic acid, piperacillin, aztreonam, tetracycline, and chloramphenicol.

There was no interaction between *Rubus occidentalis* "Litacz" fruit, *Rubus idaeus* "Poranna Rosa" fruit, *Rubus idaeus* "Laszka" fruit, and *Rubus idaeus* "Willamette" shoot extracts and tigecycline or amikacin; *Rubus idaeus* "Laszka" fruit, *Rubus occidentalis* "Litacz" fruit, *Rubus idaeus* "Poranna Rosa" fruit extracts and amoxicillin or ceftriaxone; *Rubus idaeus* "Laszka" fruit and *Rubus occidentalis* "Litacz" fruit extracts and ampicillin and meropenem; *Rubus idaeus* "Laszka" fruit, *Rubus occidentalis* "Litacz" fruit, and *Rubus idaeus* "Willamette" shoot extracts and cefazolin, amoxicillin/clavulanic acid, or ertapenem; *Rubus idaeus* "Laszka" fruit, *Rubus idaeus* "Poranna Rosa" fruit, and *Rubus idaeus* "Willamette" shoot extracts and ticarcillin/clavulanic acid or piperacillin/tazobatam; *Rubus occidentalis* "Litacz" fruit and *Rubus idaeus* "Willamette" shoot extracts and cefotaxime; *Rubus idaeus* "Laszka" fruit and *Rubus idaeus* "Poranna Rosa" fruit extracts and imipenem; *Rubus idaeus* "Laszka" fruit and *Rubus idaeus* "Willamette" shoot extracts and chloramphenicol; *Rubus idaeus* "Laszka" fruit extract with doripenem or tobramycin; *Rubus occidentalis* "Litacz" fruit extract and tetracycline; *Rubus idaeus* "Poranna Rosa" fruit extract and ciprofloxacin; or *Rubus idaeus* "Willamette" shoot extract with ticarcillin or cotrimoxazole.

The antagonistic effects against *Pseudomonas aeruginosa* ATCC9027 (Figure 12) were exerted by the following: *R. idaeus* "Laszka" fruit extract with imipenem; *R. occidentalis* "Litacz" fruit extract with meropenem; *R. idaeus* "Poranna Rosa" fruit extract with imipenem and meropenem; *R. occidentalis* "Litacz" fruit and *R. idaeus* "Poranna Rosa" fruit extract with ciprofloxacin; *R. idaeus* "Willamette" shoot extract with ciprofloxacin and levofloxacin; *R. occidentalis* "Litacz" fruit extract with gentamicin; *R. idaeus* "Poranna Rosa" fruit extract with tobramycin; *R. idaeus* "Willamette" shoot extract with gentamicin, netilmicin, and tobramycin; *R. occidentalis* "Litacz" fruit extract with cefepime and ticarcillin; *R. idaeus* "Poranna Rosa" fruit extract with ticarcillin/clavulanic acid and aztreonam; *R. idaeus* "Willamette" shoot extract with cefazolin, ticarcillin, ticarcillin/clavulanic acid, piperacillin/tazobactam, and aztreonam; *R. idaeus* "Poranna Rosa" fruit extract with chloramphenicol; and *R. idaeus* "Willamette" shoot extract with chloramphenicol and colistin.

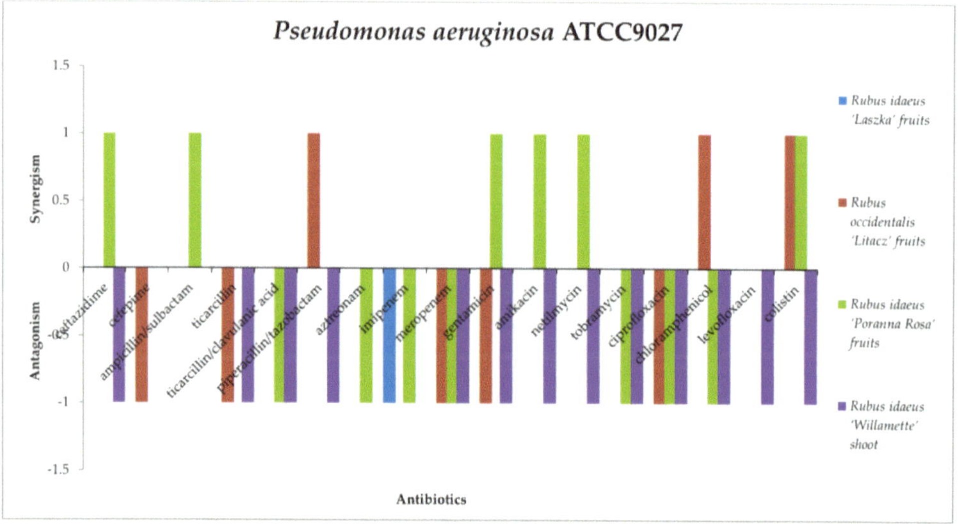

Figure 12. Interactions of *R. idaeus* and *R. occidentalis* extracts with antibiotics against *Pseudomonas aeruginosa* ATCC9027.

Synergism against *P. aeruginosa* was demonstrated by *R. occidentalis* "Litacz" fruit extract with piperacillin/tazobactam, chloramphenicol, and colistin, and by *R. idaeus* "Poranna Rosa" fruit extract with ceftazidime, ampicillin/sulbactam, gentamicin amikacin, netilmicin, and colistin. However, *R. idaeus* "Willamette" shoot and *R. idaeus* "Laszka" fruit extract did not show synergistic interactions with the tested antibiotics.

No interactions were observed between *Rubus idaeus* "Laszka" fruit, *Rubus idaeus* "Poranna Rosa" fruit, *Rubus idaeus* "Willamette" shoot extracts and cefepime; *Rubus idaeus* "Laszka" fruit and *Rubus occidentalis* "Litacz" fruit; extracts and ceftazidime, amikacin, netilmicin, tobramycin, chloramphenicol, or ticarcillin/clavulanic acid; *Rubus idaeus* "Laszka" fruit, *Rubus occidentalis* "Litacz" fruit and *Rubus idaeus* "Willamette" shoot extracts and ampicin/sulbactam or aztreonam; *Rubus idaeus* "Laszka" fruits and *Rubus idaeus* "Poranna Rosa" fruit extracts and ticarcillin, piperacillin/tazobactam, or levofloxacin; *Rubus occidentalis* "Litacz" fruit and *Rubus idaeus* "Willamette" shoot extracts and imipenem; *Rubus idaeus* "Laszka" fruit extract and meropenem, gentamicin, ciprofloxacin, or colistin; or *Rubus occidentalis* "Litacz" fruit extract and imipenem.

R. idaeus "Poranna Rosa" fruit extract did not show antagonism with the antibiotics used against *Acinetobacter baumannii* 2021 (Figure 13). However, the antagonistic effects were exerted by *R. idaeus* "Laszka" fruit extract with ticarcillin, meropenem, and ciprofloxacin; *R. occidentalis* "Litacz" fruit extract with piperacillin/clavulanic acid and meropenem; and *R. idaeus* "Willamette" shoot extract with ceftazidime, ticarcillin/clavulanic acid, piperacillin/tazobactam, gentamicin, amikacin, netilmicin, tobramycin, and colistin.

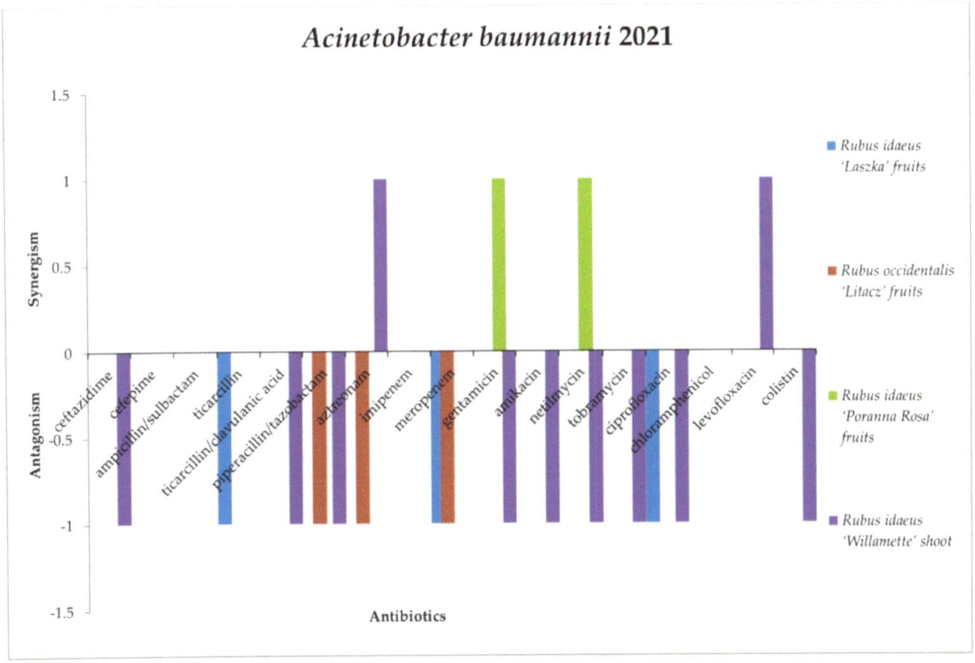

Figure 13. Interactions of *R. idaeus* and *R. occidentalis* extracts with antibiotics against *Acinetobacter baumannii* 2021.

R. idaeus "Willamette" shoot extract and *R. idaeus* "Laszka" fruit extract did not show synergism with any antibiotics used against *Acinetobacter baumannii* 2021. However, synergism of extracts with antibiotics was recorded for *R. occidentalis* "Litacz" fruit extract with piperacillin/tazobactam, chloramphenicol, and colistin, and *R. idaeus* "Poranna Rosa" fruit extract and ceftazidime, ampicillin/sulbactam, gentamicin amikacin, netilmycin, and colistin.

No effects were shown by *Rubus idaeus* "Laszka" fruit, *Rubus occidentalis* "Litacz" fruit, *Rubus idaeus* "Poranna Rosa" fruit, and *Rubus idaeus* "Willamette" shoot extracts with cefepime, ampicin/sulbactam, imipenem, or chloramphenicol; *Rubus idaeus* "Laszka" fruit, *Rubus occidentalis* "Litacz" fruit, and *Rubus idaeus* "Poranna Rosa" fruit extracts with ceftazidim or ticarcillin/clavulanic acid; *Rubus occidentalis* "Litacz" fruit, *Rubus idaeus* "Poranna Rosa" fruit, and *Rubus idaeus* "Willamette" shoot extracts and ticarcillin; *Rubus idaeus* "Laszka" fruit and *Rubus idaeus* "Poranna Rosa" fruit extracts with piperacillin/tazobactam; *Rubus idaeus* "Laszka" fruit, *Rubus occidentalis* "Litacz" fruit, and *Rubus idaeus* "Poranna Rosa" fruit extracts from amikacin, tobramycin, levofloxacin, or colistin; *Rubus idaeus* "Laszka" fruit *and Rubus occidentalis* "Litacz" fruit extracts with gentamicin and netilmycin; *Rubus occidentalis* "Litacz" fruit and *Rubus idaeus* "Poranna Rosa" fruit extracts from ciprofloxacin; *Rubus idaeus* "Laszka" fruit and *Rubus idaeus* "Poranna Rosa" fruit extracts with aztreonam; or *Rubus idaeus* "Willamette" shoot and *Rubus idaeus* "Poranna Rosa" fruit extracts with meropenem.

Against *Stenotrophomonas maltophilia* 12755, antagonistic activity was exerted only by *R. idaeus* "Willamette" shoot extract with ceftazidime, levofloxacin, cotrimoxazole, and colistin.

No effects were shown by *Rubus idaeus* "Laszka" fruit, *Rubus occidentalis* "Litacz" fruit, and *Rubus idaeus* "Poranna Rosa" fruit extracts from levofloxacin, cotrimoxazole, colistin, or ceftazidim.

In turn, against *Helicobacter pylori* ATCC 10231, synergistic effects were exerted only by *R. idaeus* "Laszka" fruit extract, *R. occidentalis* "Litacz" fruit extract, and *R. idaeus* "Poranna Rosa" fruit extract with tetracycline.

Against *Candida albicans* ATCC10248 (Figure 14), a synergistic effect was detected with antimycotics *R. idaeus* "Laszka" fruit and *R. idaeus* "Poranna Rosa" fruit extract with miconazole, and with *R. idaeus* "Willamette" shoot extract with miconazole, econazole, ketoconazole, and fluconazole. No effects were shown by *Rubus idaeus* "Laszka" fruit, *Rubus occidentalis* "Litacz" fruit, *Rubus idaeus* "Poranna Rosa" fruit, and *Rubus idaeus* "Willamette" shoot extracts with caspafungin, nystatin, amphotericin B, or flucytosine; *Rubus idaeus* "Laszka" fruit, *Rubus occidentalis* "Litacz" fruit, and *Rubus idaeus* "Poranna Rosa" fruit extracts with econazole and ketoconazole; or *Rubus occidentalis* "Litacz" fruit miconazole extract.

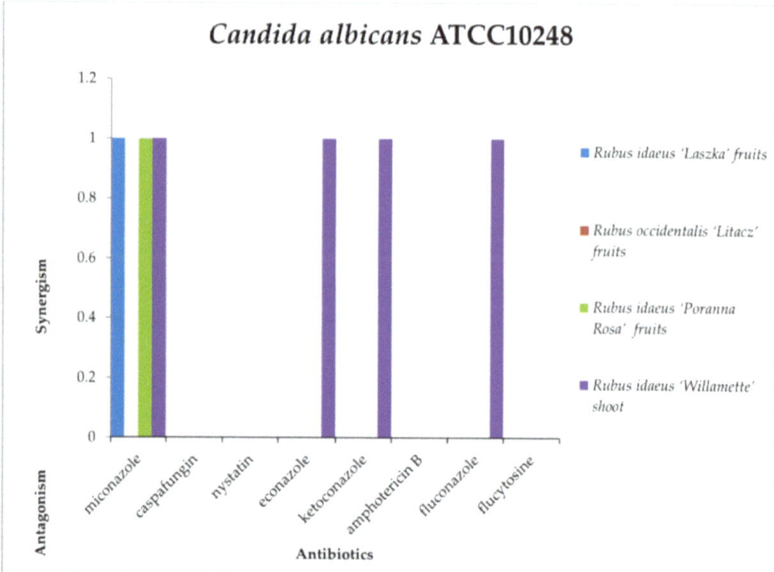

Figure 14. Interactions of *R. idaeus* and *R. occidentalis* extracts with antimycotics against *Candida albicans* ATCC10248.

3. Materials and Methods

3.1. Strains, Growth Medium, Antibiotics

Gram-positive bacteria: β-hemolytic *Streptococcus* group A PCM465, β-hemolytic *Streptococcus* groups B (clinical isolates, collection of the Department of Pharmaceutical Microbiology, Medical University of Gdańsk), *Corynebacterium diphtheria* (collection of the Department of Pharmaceutical Microbiology, Medical University of Gdańsk), *Enterococcus hirae* ATCC10541, *Enterococcus faecalis* ATCC51299, *Staphylococcus aureus* ATCC6538, *Staphylococcus epidermidis* ATCC14990, *Candida albicans* ATCC10231. Gram-negative bacteria: *Escherichia coli* ATCC8739, *Klebsiella pneumoniae* ATCC13883, *Proteus vulgaris* NTCT4635, *Pseudomonas aeruginosa* ATCC9027, *Acinetobacter baumannii* 2021 (collection of the Department of Pharmaceutical Microbiology, Medical University of Gdańsk), *Stenotrophomonas maltophilia* 12755 (clinical isolates, collection of the Department of Pharmaceutical Microbiology, Medical University of Gdańsk), and *Helicobacter pylori* ATCC10231. Columbia agar with 5% sheep blood (Becton Dickinson, Franklin Lakes, NJ, USA) for β-hemolytic *Strpetococcus* groups A and B, (strains growth in 5% CO_2 atmosphere at 37 °C for 48 h), *Corynebacterium diphtheria*, (growth in an aerobic atmosphere for 48 h at 37 °C). *Enterococcus hirae, Enterococcus faecalis, Staphylococcus aureus, Staphylococcus epidermidis, Escherichia coli, Klebsiella pneumoniae, Proteus vulgaris, Pseudomonas aeruginosa, Acinetobacter baumanii*, grown in Mueller–Hinton agar (cation-adjusted MH, Becton Dickinson) in an aerobic atmosphere at 37 °C for 48 h. *Helicobacter pylori* grown in tryptic soy broth (TSA, Becton Dickinson) supplemented with 5% sheep blood in microaerophilic conditions (GENbag microaer, BioMerieux, Lyon, France) at 37 °C for 72–96 h (materials from ATCC). *Candida albicans* grown in Sabourauda agar (Biomaxima, Lublin, Poland) at 37 °C for 48 h.

In our research on antibiotic paper disks (BioMaxima S.A., Lublin, Poland), amikacin 30 µg/mL (AK30), amoxicillin 10 µg/mL (AX10), amoxicillin 25 µg/mL (AX25), amoxacillin/clavulanic acid 3 µg/mL (AMC3), amoxicillin/clavulanic acid 20/10 µg/mL (AMC30), ampicillin 10 µg/mL (AM10), ampicin/sulbactam 20/10 µg/mL (SAM30), azitomycin 15 µg/mL (AZM15), aztreonam 30 µg/mL (ATM30), caspafungi (Kf), cefazolin 5 µg/mL (CTX5), cefepime 30 µg/mL (FEP30), cefotaxime 30 µg/mL (CTX30), cefoxitin 30 µg/mL (FOX30), cefpodoxime 10 µg/mL (CPD10), ceftazidime 30 µg/mL (CAZ30), ceftriaxone 30 µg/mL (CRO30), cefuroxime 30 µg/mL (CXM30), chloramphenicol 30 µg/mL (C30), ciprofloxacin 5 µg/mL (CIP5), clindamycin 2 µg/mL (DA2), colistin 50 µg/mL (CT50), cotrimoxazole 25 µg/mL (SXT25), daptomycin 30 µg/mL (DAP30), doripenem 10 µg/mL (DOR10), doxycycline 30 µg/mL (DO30), econazole 10 µg/mL (ECN10), ertapenem 10 µg/mL (ETP10), erythromycin 15 µg/mL (E15), fusidic acid 10 µg/mL (FA10), gentamicin 10 µg/mL (CN10), gentamicin 120 µg/mL (CN120), imienem 10 µg/mL (IMP10), kanamycin 30 µg/mL (K30), ketoconazole 10 µg/mL (KCA10), levofloxacin 5 µg/mL (LEV5), linezolid 30 µg/mL (LZN30), meropenem 10 µg/mL (MEM10), miconazole 10 µg/mL (MCL10), moxifloxacin10 µg/mL (MUP10), mupirocin 200 µg/mL (MUP200), netilmycin 10 µg/mL (NET10), nystatin 100 units/mL (NS100), ofloxacillin 5 µg/mL (OFX5), optochin (OP), oxacillin 1 µg/mL (OX1), oxacillin 5 µg/mL (OX5), penicillin 10 units/mL (P10), piperacillin 30 µg/mL (PRL30), piperacillin/tazobactam 100/10 µg/mL (TZP110), rifampicna 5 µg/mL (RA5), streptomycin 300 µg/mL (S300), teicoplanin 30 µg/mL (TEC30), telithomycin 15 µg/mL (TELl15), tetracycline 30 µg/mL (Te30), ticarcillin 75 µg/mL (TC75), ticarcillin/clavulanic acid 75 µg/mL (TIM75), tigecycline 15 µg/mL (TGC15), tobramycin 10 µg/mL (Tob10), and vancomycin 5 µg/mL (Va5) were used.

3.2. Preparation of Extracts

The methods of preparation and phytochemical analysis (HPLC-DAD-ES-IMS) of extracts (*R. occidentalis* "Litacz" fruit, *R. idaeus* "Laszka" fruit, *R. idaeus* "Poranna Rosa" fruit, *R. idaeus* "Willamette" shoot) have been described in our recently published article (Hałasa et al.) [17].

3.3. Determination of Susceptibility to Antibiotics by the Agar Disk Diffusion Method

Disk diffusion for antibiotics was performed according to EUCAST standards. Antibiotic susceptibility was interpreted according to the EUCAST clinical breakpoints (v11.0) (2020). Columbia agar with 5% sheep blood, Muller–Hinton agar, TSA supplemented with 5% sheep blood or Sabourauda agar (Section 3.1), and antibiotic paper disks (Section 3.1) were used. Freeze-dried methanolic extracts were dissolved in sterile deionized water. Sterile filter paper disks for extracts were 6 mm in diameter.

3.4. Screening for Extract–Antibiotic Interactions Using a Double-Disk Synergy Test (DDST)

The double–disk synergy test (DDST) is a test commonly recommended by the EUCST to detect extended-spectrum β-lactamase (ESBL) resistance mechanisms in Gram-negative bacilli [13,14]. It is also used to detect synergisms between different antibiotics in combating resistant bacteria.

The double-disk synergy test (DDST) was performed with some modifications. An appropriate medium was used depending on the species of microorganism being tested, as well as sterile paper disks (6 mm in diameter) for the extracts and commercially available antibiotic disks. The disks were placed on a plate with a medium inoculated with the tested strain of the microorganism. The distance between the arranged disks should be equal to or greater than the sum of the radii of the growth inhibition zones of the individual tested components. After pre-incubation, the plates were subjected to further incubation according to the requirements for a given microorganism. After incubation, the results of the analyzed extract–antibiotic interaction were interpreted on the basis of the obtained growth inhibition zones (Figure 15). In the case of additive or indifferent results, two independent zones of growth inhibition were observed. In the presence of a synergistic interaction, enhancement or bridging was observed at or near the interface of two growth inhibition zones. In the case of antagonism, a cutoff near the junction of two inhibition zones or a separation of the inhibition zones was observed. In the absence of growth inhibition zones (lack of activity) for both components separately, only the presence of both at the same time made it possible to inhibit the growth of microorganisms.

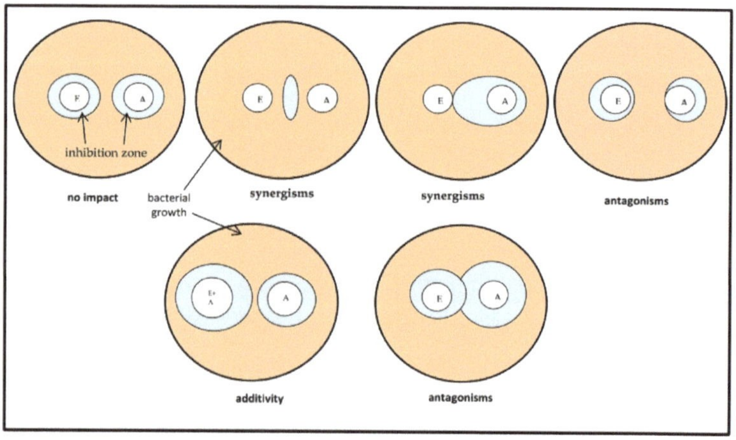

Figure 15. Diagram showing extract (E) –antibiotic (A) interactions against bacteria using the double-disk synergy test (DDST).

In order to check the additivity of the activity of extracts and antibiotics, a study was carried out in which disks with an antibiotic were placed on bacteria and the extract in a volume of 15 µL/mL was added. After incubation, the size of the growth inhibition zones was checked. The control unit consisted of disks containing only the antibiotic.

Since the zones of inhibition of antibiotics against different bacteria are different, and the interactions of synergism and antagonism change the shape of the zone of growth inhibition (mainly antibiotic), we decided to standardize the results we obtained: synergism and additivity were assigned "+1," antagonism "−1," and no interactions "0."

3.5. Statistical Analysis

All experiments were performed at least 3 times. Standard deviation was calculated using Microsoft Excel 2010. Inhibition zone data were analyzed in STATISTICA v. 13.3 using the independent samples t-test. For Gram-positive bacteria, lisnezolid zones were used as a reference, and for C albicans, nystatin zones were used. For Gram-negative bacteria, meropenem growth inhibition zones were used as a reference, and for *S. maltophilia* and *H.*

pylori, levofloxacin was used. The *p*-value was considered statistically significant when it was less than 0.05.

4. Discussion

The overuse and/or inappropriate use of antibiotics as antimicrobial agents by medical professionals in recent decades, and the widespread use of antibiotics in industry, including animal husbandry and agriculture, has led to the spread of antibiotic resistance among microorganisms. Currently, the most resistant nosocomial pathogens of clinical importance form a group called "ESKAPE." This group includes species such as *Enterococcus faecium*, *Staphylococcus aureus*, *Klebsiella pneumoniae*, *Acinetobacter baumannii*, *Pseudomonas aeruginosa*, and *Enterobacter* spp., abbreviated as "ESKAPE" [18,19]. The rate of increase in resistance among microorganisms is negatively correlated with the number of effective antibiotics and the discovery of new active compounds. In the face of such data, the WHO has recognized the growing level of antibiotic resistance, accompanied by an increasing number of multidrug-resistant microorganisms, as a global problem, a return to the "pre-antibiotic era," and a serious threat to public health [20,21]. The Center for Disease Control (CDC) reported that 2 million illnesses in the United States each year are caused solely by antibiotic-resistant microorganisms; 23,000 of them are fatal [22]. Moreover, it is estimated that by 2050, 10 million people may die due to infections caused by antibiotic-resistant bacteria [23], and the costs of treating infections with resistant strains will be in the order of millions per year.

As part of the implementation of new strategies for searching for new compounds with antibacterial activity, plant extracts seem to be a potential source of antimicrobial molecules. Biologically active compounds present in plants are divided into different classes depending on their chemical structure and properties: constituents of essential oils, phenols, alkaloids, saponins, and peptides [24]. Phenols constitute the largest group of secondary plant metabolites. Most phenols present in plants with potential antimicrobial activity are classified into six groups: flavonoids, phenolic acids, tannins, stilbenoids, quinones, and coumarins.

The presence of many chemical compounds in extracts obtained from plants and essential oils raises the possibility that more than one mechanism is responsible for the antimicrobial activity. Therefore, it is believed that the herbal extracts act through several antimicrobial mechanisms at the cellular level, e.g., distribution of the cytoplasmic membrane, disruption and inactivation of the external activities of the membrane of Gram-negative bacteria, shedding lipopolysaccharides, coagulation of the internal content of the cell, and prevention of enzyme production [9,10,25].

The popularity of the disk diffusion method in the initial selection of preparations obtained from plants (herbal extracts) in terms of their antibacterial activity is due to its simplicity, low cost, and repeatability. However, many factors influence obtaining a correct and reliable result, such as the selection of the appropriate medium, its pH, the depth of the agar layer, as well as the incubation conditions, including the humidity of the process, which is partly related to the evaporation of water, which is a constituent of culture medium. The solvent used for the extraction of plant material also affects the diameter of the bacterial inhibition zone achieved and the scope of diffusion of active compounds. The possibility of precipitation of water-insoluble compounds on the disk will also block the process of diffusion of the tested substances in the agar and ultimately affect the growth of the used microorganisms [2,26,27]. The molecular size of the compounds contained in the extract will determine the rate of diffusion between the polymerized agar molecules. As a consequence, there may be results of questionable quality. In the case of polar compounds that diffuse more slowly in the agar, the lack of an inhibition zone does not necessarily mean a lack of antimicrobial activity. Some authors emphasize that the main disadvantage of the disk diffusion test is that it is qualitative rather than quantitative and does not distinguish between bactericidal and bacteriostatic effects. It is believed that disk diffusion can be used to screen low-molecular-weight antimicrobial compounds, while well diffusion is preferred for high-molecular-weight compounds [28].

These negative aspects may also affect the application of disk diffusion to the double-disk synergy test. It should also be remembered that when using the double-disk synergy test to assess interactions, attention should be paid to important details such as the size of the growth inhibition zones of both tested components (herbal extract and antibiotic), which translates into the distance of the disks from each other and the area of joint impact of both components on growing microorganisms.

Despite certain limitations of disk diffusion, presented above, its use in the double-disk synergy test made it possible to determine the interactions between the tested herbal extracts from fruits and stems of some taxa of the *Rubus* genus and selected antibiotics aimed at enhancing the antimicrobial effect. Moreover, after taking into account some differences and similarities in the chemical composition of the tested plant extracts, it made it possible to clarify the type of certain interactions based on the presence of specific biologically active compounds, primarily in terms of their antimicrobial activity (previously described in the literature). It should also be remembered that when using the double-disk synergy test to assess interactions, attention should be paid to important details such as the size of the growth inhibition zones of both tested components (herbal extract and antibiotic), which translates into the distance of the disks from each other and the area of joint impact of both components on growing microorganisms.

This research used antibiotics, bacterial strains, and one yeast. Interactions between herbal extract and a given antibiotic were observed in relation to extracts from the fruits of two cultivar varieties of red raspberry, namely, *Rubus idaeus* "Laszka" and "Poranna Rosa," and the fruit of one cultivar of black raspberry, *Rubus occidentalis* "Litacz," as well as an extract from the shoots of the "Willamette" variety of *R. idaeus*. The tested extracts were analyzed in terms of chemical composition—compounds from the group of polyphenols and simple phenols—using the HPLC-DAD-ESI-MS method, and their contents were determined [17]. All tested extracts contained high levels of polyphenols comprising the groups of ellagitannins, flavonoids, and phenolic acids, and in two of them anthocyanins were also present (*R. idaeus* "Laszka," *R. occidentalis* "Litacz"). The extract from shoots of *R. idaeus* "Willamette" was characterized particularly by the highest levels of ellagitannins and free and conjugated ellagic acid (4–6-fold and 2-fold more than fruit extracts, respectively). Ellagitannins are classified as natural antibiotics, and their mechanism of antibacterial action includes, among others, iron chelation, inhibition of cell wall synthesis, disruption of the cell membrane, and inhibition of fatty acid biosynthetic pathways [29].

Anthocyanins were present in only two fruit extracts, and the "Litacz" fruit extract contained four times more anthocyanins. Anthocyanins are also classified as polyphenols with antibacterial activity, but the potency of this action is much weaker compared to tannins. Synergistic interactions with antibiotics were revealed, with 33 for the "Willamette" shoot extract, 31 for the "Laszka" fruit extract, 58 for the "Poranna Rosa" fruit extract (no anthocyanins), and 44 for the fruit of the "Litacz" variety.

Our studies showed that all tested extracts presented synergistic effects most often in combination with tetracycline, tigecycline, penicillin, amoxicillin, and cefotaxime against all tested bacteria. Synergy with tetracyclines has been described for gallic acid as a constituent of herbal extracts. Moreover, studies on the effect of gallic acid on MDR (multidrug-resistant) *E. coli* strains have shown a synergistic effect with ampicillin, oxytetracycline, cefoxitin, and cefotaxime and norfloxacin, ofloxacin, amikacin, and tobramycin [4,30]. However, in the case of *P. aeruginosa*, it influenced the integrity of the cell wall, and a synergistic effect was observed with tetracycline, sulfamethoxazole, and streptomycin [31,32]. Synergy of gallic acid with tetracycline effectively inhibits the in vitro growth of Gram-negative and -positive pathogens [33,34]. *S. aureus* treated with gallic acid lost tolerance to low osmotic pressure. In the case of *E. coli* and *P. aeruginosa* cells, gallic acid caused the outflow of intracellular K^+ ions, which was related to the permeability of the cytoplasmic membrane [32].

However, in the tested extracts from the genus *Rubus*, gallic acid was identified by the HPLC-DAD-ESI-MS method only in the extract from the shoots of *Rubus idaeus* "Willamette." The phenols contained in the berries caused a decrease in the permeability of the outer membrane, similarly to EDTA *(ethylenediaminetetraacetic acid)*, by chelating divalent cations, which changed the membrane stabilization by intercalation into the membrane structure [32], as shown in the example of *Salmonella enterica* serovar Typhimurium and *S. enterica* serovar Infantis.

Martinia et al. [35] concluded that polyphenols contained in *R. ulmifolius* extracts act on *H. pylori* ion pumps, i.e., enzymes that regulate the flow of copper and other metal cations through membranes. Additionally, quercetin is a chelator of divalent ions [36]. The mentioned ions (Mg^{2+}, Ca^{2+}) inactivate tetracyclines [18]. A simultaneous destabilization of the cell membrane was observed to be caused by the synergistic effect of quercetin [33,37], epicatechin [38], ellagic acid derivatives, and caffeic and gallic acid. Literature data indicate that the main mode of action is an increase in cytoplasmic membrane permeability (quercetin) [39,40] intercalated into the cytoplasmic membrane (caffeic acid) [39,41]. Siriwong et al. [42] tested the combination of quercetin and ceftazidime against *Streptococcus pyogenes* and observed, in addition to increased permeability of the cytoplasmic membrane, changes in cell morphology, damage to peptidoglycan and the cell membrane, and decreased levels of nucleic acid. However, increased levels of proteins in bacterial cells were also observed. Epicatechin gallate affected the integrity of the cytoplasmic membrane and inhibited bacterial gene expression [38]. Together with quercetin, it influenced the activity of β-lactamase, reduced the content of fatty acids, and increased the permeability of the cytoplasmic membrane [39,40].

The synergy of phytochemicals with penicillin, amoxicillin, and other beta-lactams has previously been reported. A synergistic effect of quercetin with ampicillin, cefradine, ceftriaxone, imipenem, or methicillin has been demonstrated [36]. The synergy mechanism of quercetin with amoxicillin against MDR *Staphylococcus epidermidis* is as follows: inhibition of peptidoglycan synthesis, activation of β-lactamase, reduction of fatty acid content, and increase in cytoplasmic membrane permeability [39,40]. Epicatechin revealed a synergistic effect against clinical strains of MRSA (methicillin-resistant *Staphylococcus aureus*), with beta-lactam antibiotics: ampicillin, ampicillin/sulbactam, cefazolin, cefepime, and imipenem/cilastatin [37]. (−)-Epigallocatechin gallate influences the staphylococcal virulence factors, has a direct effect on the bacterial cell wall of MRSA and MSSA clinical and standard strains, and shows synergism with oxacillin, ampicillin/sulbactam, penicillin, imipenem, banipenem, meropenem tetracycline, and oxytetracycline [43–48].

In our research, the action of epicatechin and quercetin is shown in the interactions of all extracts with ampicillin and amoxicillin against Gram-positive bacteria: *Streptococcus* sp., *Staphylococcus* sp., *Enterococcus* sp., and *C. diphtheriae*. The phenomenon of synergism was observed against Gram-negative bacteria (*E. coli, K. pneumoniae, P. vulgaris*) for "Poranna Rosa" and "Laszka" fruit extracts in combination with the following antibiotics: cefotaxime, cefepime, cefuroxime, cefazoline, and amoxicillin/clavulanic acid. In the case of *P. aeruginosa* and *A. baumanii*, the "Poranna Rosa" fruit extract produced synergism with ceftazidime, amoxicillin/sulbactam, gentamicin, amikacin, netilmicin, and colistin.

Literature data show that caffeic acid produced synergy with cefotaxime against clinical and reference strains of MRSA and MSSA (methicillin-susceptible *Staphylococcus aureus*) [38,39]. In turn, gallic acid presented the same effect with several beta-lactam antibiotics (ampicillin, cefoxitin, cefotaxime). In the extracts we tested, unfortunately, epicatechin was present only in extracts from "Poranna Rosa" and "Willamete," whereas caffeic acid and gallic acid were detected only in shoot extract, and the amount of quercetin in all the extracts we tested was relatively small. However, it should be emphasized that the tested extracts may contain other ingredients that act synergistically with beta-lactams.

Caffeic acid intercalated into the cytoplasmic membrane, and the compounds showed synergy with clindamycin and erythromycin against clinical and reference strains of MRSA and MSSA [38,39]. Protocatechuic acid ethyl ester acted synergistically with clindamycin against clinical and standard strains of MRSA and MSSA [49]. The presence of both compounds in the extract we tested was visible in the synergism with erythromycin towards β-hemolytic *Streptococcus* group A and *C. diphtheriae*.

In turn, the synergism of rifampicin with the "Poranna Rosa" and "Litacz" extracts in relation to *S. aureus* and *C. diphtheriae* may indicate the effect of quercetin, which, in combination with rifampicin and ciprofloxacin, acted synergistically against MRSA strains [49,50].

Quercetin and its derivatives (rutin [quercetin-3-O-rhamnoglucoside], quercetin-3-glucoside [isoquercetin], quercetin-3-O-rhamnoside–quercitrin) have several other identified mechanisms of action on bacterial cells, including (i) action on fatty acid synthesis pathway II (FASII type II fatty acid synthase). In *P. aeruginosa*, it inhibits the Fab Z enzyme—dehydratase of the beta-hydroxyacyl-acyl carrier protein (ACP), and in *Mycobacterium* it inhibits FASI (type I FAS fatty acid synthase), which interferes with the synthesis of mycolic acids. In turn, in *H. pylori*, it inhibits 3-hydroxyacyl-ACP dehydrase in FAS II [36,51]. The fluidity and stability of the cell membrane can be affected. (ii) Damage to the cell wall by inhibition of D-alanine-D-alanine ligase, which is responsible for the production of the final peptidoglycan precursor, UDPMurNAc-pentapeptide [36,52]. (iii) Activation of β-lactamase [39,40]. Perhaps these mechanisms are also involved in the revealed synergism with beta-lactam antibiotics.

It should be noted that there are synergisms that we cannot explain and that have not been reported so far. Fruit extracts showed synergism with vancomycin and linezolid against *Enterococcus* spp. and with linezolid, fusidic acid, and mupirocin against *Staphylococcus* spp. However, mupirocin, vancomycin, and fusidic acid in combination with the "Poranna Rosa" extract acted synergistically against *C. diphtheriae*. Based on the results obtained, it should be added that *C. diphtheriae* turned out to be a very sensitive bacterium to the combinations of herbal extracts and antibiotics used. Another bacterium with unusual synergisms in contrast to other Gram-negative bacilli turned out to be *P. vulgaris*. Only against this bacterium did the extract from the shoots of *R. idaeus* "Willamette" show synergism with imipenem, meropenem, amoxicillin, and tetracycline.

Many studies have shown that antibiotics such as beta-lactams, aminoglycosides, and fluoroquinolones, regardless of their specific targets, induce oxidative stress, and thus participate in bacterial killing by antibiotic ROS (reactive oxygen species) [36].

The literature usually contains information about the synergistic interaction of phytochemicals and their mechanisms of action on bacteria, but there is little research on antagonistic interactions.

The fruit and shoot extracts we tested showed antagonism with aminoglycoside antibiotics (gentamicin, tobramycin, amikacin, netilmicin) against *S. aureus*. *E. coli*, *K. pneumoniae*, and *P. vulgaris*. In the case of *P. aeruginosa* and *A. baumannii*, the antagonism phenomenon was dependent on the type of extract. The extract from the shoots of the "Willamette" variety had an antagonistic effect with all tested antibiotics against both bacteria. In, turn, only "Poranna Rosa" fruit extract (with tobramycin) and "Litacz" fruit extract (with gentamicin) showed antagonism towards *P. aeruginosa*. "Litacz" fruit extract had no effect with aminoglycosides against *P. aeruginosa* or *A. baumannii*.

Studies have shown that isoquercetin presents antagonistic effects with aminoglycoside antibiotics, e.g., neomycin, kanamycin, gentamicin, and amikacin against *E. coli* strains. Moreover, neither isoquercetin nor quercetin affected the activity of aminoglycosides against a multi-resistant strain of *S. aureus* [34,53]. Researchers believe that the mechanism of action of quercetin on *E. coli* and *S. aureus* [38] involves damage to the bacterial cell wall and membrane. Moreover, the activity of extracellular alkaline phosphatase and β-galactosidase, as well as the concentration of soluble proteins in *E. coli* and *S. aureus*, increased significantly. On the other hand, while ATP (adenosine triphosphate) synthase activity increased significantly in *S. aureus*, it remained unchanged in *E. coli* [33]. However, there is a different mechanism at work here; it has been shown that quercetin (and its derivatives) and epicatechin [36] inhibit ATP synthase of the F1F0 type, which is a key enzyme in the interconversion of cellular energy [54]. During ATP synthesis, this large protein complex uses the proton gradient and associated membrane potential to synthesize ATP. It can also reverse and hydrolyze ATP, creating a proton gradient. The proton gradient is necessary for the entry of aminoglycosides (gentamicin, amikacin, neomycin, streptomycin, spectinomycin, tobramycin) into the bacterial cell.

The results of phytochemical studies of the tested extracts, presented in Table 1, combined with the mechanisms of action quoted above, suggest that the shoot extracts containing quercetin and epicatechin have an antagonistic effect against the entire group of aminoglycosides, while the "Poranna Rosa" fruit extract (containing quercin and epicatechin in smaller amounts than in the shoot extract) should have lower activity. However, the lowest level of antagonism or its absence should be produced by "Litacz" fruit extracts and "Laszka" fruit extracts (both extracts contained only quercetin). Our results show that antagonism towards aminoglycosides was present in the action of all extracts, regardless of their composition. However, the strength of the antagonistic effect in relation to the quantitative composition of the extracts could be determined using quantitative methods.

Quercetin and its derivative have other mechanisms of action, i.e., inhibition of the activity of bacterial topoisomerase in MRSA (synergism with ciprofloxacin) [4,38,55]. Ohemeng et al. [56] described the inhibition of *E. coli* DNA gyrase by combining with the B subunit [35]. Studies were also conducted in which the effect on gyrase in *Mycobacterium smegmatis* and *M. tuberculosis* was present [38,57].

Topoisomerase is also a target for a group of antibiotics used in therapy, i.e., quinolones (fluoroquinolones), which also have an additional site of attachment in bacterial cells, i.e., DNA gyrase [18]. Quinolones interact with gyrase in Gram-negative bacterial cells and with topoisomerase in Gram-positive bacterial cells. At this point, an antagonistic effect may occur between the ingredients of the extracts, i.e., quercetin and antibiotics. Antagonism between ciprofloxacin and/or levofloxacin and the tested extract from "Wilamette" shoots (against *S. aureus*, *S. epidermidis*, *E. coli*, *K. pneuminiae*, *P. vulgaris*, *P. aeruginosa*) was shown in our studies. A similar profile of plant extract–antibiotic interactions was recorded for fruit extracts from "Poranna Rosa" and "Litacz." The extract from *R. idaeus* "Laszka" fruit had a synergistic effect only against *E. coli* and *A. baumannii*.

Rubus idaeus "Willamette" shoot extract showed antagonism against Gram-negative bacilli mainly with ticarcillin/clavulanic acid, piperacillin/tazobactam, and aminoglycosides. Moreover, the "Willamette" shoot extract with colistin had an antagonistic effect against *P. aeruginosa*, *S. maltophilia*, and *A. baumannii*. The fruit extract of *R. idaeus* "Laszka" showed antagonism with imipenem, as did meropenem against *P. aeruginosa* and *A. baumannii*. The antagonism of *R. idaeus* "Poranna Rosa" fruit extract against *S. aureus* has been demonstrated. The fruit extract of *R. occidentalis* "Litacz" showed antagonism with cefotaxime against *E. coli* and *K. pneumoniae*, and with ticarcillin and piperacillin against *E. coli*, *K. pneumoniae*, *P. vulgaris*, *P. aeruginosa*, and *A. baumannii*. However, "Poranna Rosa" fruit extract presented antagonism with doripenem, ertapenem, meropenem, piperacillin, and ticarcillin against *E. coli*, *K. pneumoniae*, *P. vulgaris*, and *P. aeruginosa*. In turn, the "Litacz" fruit extract showed antagonism with doripenem, imipenem, ertapenem, and meropenem against *E. coli*, *P. aeruginosa*, and *A. baumannii*. The same extract showed antagonism with cefepime against *E. coli*, *P. vulgaris*, and *P. aeruginosa*, and with cefotaxime against *E. coli* and *K. pneumoniae*. Cotrimoxazole was an antagonist of the shoot extract of *R. idaeus* "Willamette" against *S. aureus*, and with *R. idaeus* "Laszka" fruit extract against *E. coli*, *K. pneumoniae*, and *M. catarrhalis*. The phenomenon of antagonism was observed also in the case of *Rubus idaeus* "Poranna Rosa" fruit extract with cotrimoxazole against *S. aureus*, *E. coli*, *K. pneumoniae* and *P. aeruginosa*. The fruit extract of *R. occidentalis* "Litacz" showed antagonism with cotrimoxazole against *S. aureus*, *E. coli*, and *K. pneumoniae*. Next, the "Litacz" fruit extract showed antagonism with penicillin against *S. aureus* and *E. hirae*, and with clindamycin against *S. aureus* and *Corynebacterium*. It is important to explain the mechanism of this relationship. It is currently unclear which component of the extracts causes the synergistic interactions.

It should be noted that all beta-lactam antibiotics, which include penicillins [18,58], cephalosporins, carbapenems, and monobactams, bind and block specific penicillin-binding proteins (PBPs), thereby inhibiting peptidoglycan synthesis. PBPs exhibit the activity of D,D-transpeptidases and D,D-carboxypeptidases, which are involved in cross-linking-adjacent pentapeptide chains of peptidoglycan [18,59]. Peptidoglycan (PGN) is a component of the bacterial cell wall, and its metabolism plays a major role in determining the structure and shape of bacterial cells, antibiotic resistance, and host–bacteria interactions. Bacteria produce qualitatively and quantitatively different sets of BPB proteins. This results in different activity and range of action of individual beta-lactams. A hypothesis can be formulated that the antagonism of the tested *Rubus* extracts with some beta-lactam antibiotics results from blocking the target site for this antibiotic—the specific BPB protein. At the same time, this may explain the synergism of the tested *Rubus* extracts with another beta-lactam antibiotic that binds to another PBP protein.

It should be added that the antimicrobial effect of phenols depends on their concentration: At low concentrations, phenols inhibit the activity of microbial enzymes, while at high concentrations they induce protein denaturation [60]. The antimicrobial activity of plant metabolites is generally correlated with antioxidant properties, e.g., phenolic acids (caffeic acid); flavonoids, including their glycoside derivatives; some terpenoids; and others [61–63].

Diffusion methods are common due to their simplicity, low cost, and speed, and unambiguous results can be achieved at the same time for many test samples. However, like all agar-based methods, they are laborious and time-consuming. The lack of growth inhibition zones does not always mean a lack of activity of the extract or compound contained in it because their degree of diffusion in the solidified agar medium is different and limited by the degree of their polarity, solubility, and particle size [15]. Klančnik et al. [64] found that the broth dilution method is more suitable for assessing the antimicrobial activity of plant extracts than the agar diffusion method. Moreover, their experiments allowed them to conclude that both methods (dilution and diffusion) gave comparable results only in the case of Gram-positive bacteria. Similar results were obtained by us.

To sum up, our goal was to demonstrate the use of the modified double-disk synergy test as a screening method for examining the interaction of antibiotics with herbal extracts on the example of and using extracts from fruit and shoots of certain species and varieties of raspberries rich in phenolic compounds (polyphenols and simple phenols) with confirmed antimicrobial activity in numerous studies. We are aware that, based on diffusion methods, determining the real level of antimicrobial activity of plant extracts may not always be possible, because not all antimicrobial active substances contained in the tested extract dissolved well and penetrated the agar layer. However, the use of disk diffusion in the double-disk synergy test enabled the demonstration of various types of interactions between herbal extracts and antibiotics, the most important of which was revealing the enhancement of the antimicrobial effect of the antibiotic when combined with an herbal extract rich in phenolic compounds.

Moreover, based on literature data and the results of chromatographic analyses regarding the chemical composition of the tested herbal extracts, it was possible to partially explain the mechanism of the observed synergistic effects between a given antibiotic and the tested herbal extract.

We believe that the best and most accurate method for testing the interactions between extracts and antibiotics is the checkerboard dilution method and FICI (Fractional Inhibitory Concentration Index) determination. However, based on our own experience with *H. pylori* (results published in Hałasa et al. [17]), the disk diffusion method allowed for the demonstration of the synergism of the tested extracts with tetracycline (which probably requires the presence of antibacterially active quercetin in the extract). In turn, checkerboard tests allowed for the determination of the exact concentration and showed synergistic interactions of the extracts with doxycycline and levofloxacin, as well as additive interactions with amoxicillin.

We believe that our proposed solution, which involves the use of the disk diffusion method to screen for the determination of the interaction of an herbal extract with an antibiotic against bacteria, is effective, characterized by simplicity of implementation and repeatability. This is confirmed by the interactions between antibiotics and herbal extracts revealed in our research—extracts from fruits of some varieties of *R. idaeus* and *R. occidentalis* and an extract from the shoots of one variety of *R. idaeus*.

5. Conclusions

The results of our research show that the proposed method—disk diffusion in the double-disk synergy test (DDST)—allows for the assessment of the interactions between the tested herbal extracts and antibiotics and their impact on the growth of some microorganisms. Using this test, a number of synergistic interactions were revealed between extracts from the fruit and shoots of some cultivars of the *Rubus* genus. *Rubus* extracts, having shown antimicrobial activity in previous studies, also revealed antagonistic effects with some antibiotics. The number of observed antagonistic interactions was greater than the number of observed synergistic interactions. However, extracts from the *Rubus* genus that do not contain anthocyanins but do contain high concentrations of ellagitannins and ellagic acid were characterized by a high number of synergistic interactions and the highest number of antagonistic interactions. Other compounds occurring in lower concentrations in the tested *Rubus* genus extracts may be involved in the mechanisms of the observed interactions, namely, flavonoids, including quercetin and its derivatives and phenolic acids, including caffeic and gallic acid. A full explanation of the mechanism of these interactions requires further research.

Author Contributions: Conceptualization, R.H.; methodology, R.H., U.M. and M.K.-B.; software, R.H.; validation, R.H.; formal analysis, R.H.; investigation, R.H. and U.M.; resources, R.H.; data curation, R.H.; writing—original draft preparation, R.H.; writing—review and editing R.H., U.M. and M.K.-B.; visualization, R.H.; supervision, M.K., U.M. and M.K.-B. All authors have read and agreed to the published version of the manuscript.

Funding: This research received no external funding.

Institutional Review Board Statement: Not applicable.

Informed Consent Statement: Not applicable.

Data Availability Statement: Data are contained within the article.

Acknowledgments: The authors would like to give many thanks to Katarzyna Turecka.

Conflicts of Interest: Author Marta Kula was employed by the company Moderna Poland sp.zoo. The remaining authors declare that the research was conducted in the absence of any commercial or financial relationships that could be construed as a potential conflict of interest.

References

1. Vaou, N.; Stavropoulou, E.; Voidarou, C.; Tsakris, Z.; Rozos, G.; Tsigalou, C.; Bezirtzoglou, E. Interactions between Medical Plant-Derived Bioactive Compounds: Focus on Antimicrobial Combination Effects. *Antibiotics* **2022**, *11*, 1014. [CrossRef] [PubMed]
2. Bubonja-Šonje, M.; Knežević, S.; Abram, M. Challenges to antimicrobial susceptibility testing of plant-derived polyphenolic compounds. *Arh. Hig. Rada. Toksikol.* **2020**, *71*, 300–311. [CrossRef] [PubMed]
3. Cowan, M.M. Plant products as antimicrobial agents. *Clin. Microbiol. Rev.* **1999**, *12*, 564–582. [CrossRef] [PubMed]
4. Barbieri, R.; Coppo, E.; Marchese, A.; Daglia, M.; Sobarzo-Sánchez, E.; Nabavi, S.F.; Nabavi, S.M. Phytochemicals for human disease: An update on plant-derived compounds antibacterial activity. *Microbiol. Res.* **2017**, *196*, 44–68. [CrossRef] [PubMed]
5. Cheesman, M.J.; Ilanko, A.; Blonk, B.; Cock, I.E. Developing new antimicrobial therapies: Are synergistic combinations of plant extracts/compounds with conventional antibiotics the solution? *Pharmacogn. Rev.* **2017**, *11*, 57–72. [CrossRef] [PubMed]
6. Ertürk, Ö. Antibacterial and antifungal activity of ethanolic extracts from eleven spice plants. *Biologia* **2006**, *61*, 275–278. [CrossRef]
7. Matuschek, E.; Brown, D.F.; Kahlmeter, G. Development of the EUCAST disk diffusion antimicrobial susceptibility testing method and its implementation in routine microbiology laboratories. *Clin. Microbiol. Infect.* **2014**, *20*, 255–266. [CrossRef] [PubMed]
8. Assam, J.P.; Dzoyem, J.P.; Pieme, C.A.; Penlap, V.B. In vitro antibacterial activity and acute toxicity studies of aqueous-methanol extract of *Sida rhombifolia* Linn. (Malvaceae). *BMC Complement. Altern. Med.* **2010**, *10*, 40. [CrossRef] [PubMed]

9. Bilal, M.; Rasheed, T.; Iqbal, H.M.N.; Hu, H.; Wang, W.; Zhang, X. Macromolecular agents with antimicrobial potentialities: A drive to combat antimicrobial resistance. *Int. J. Biol. Macroml.* **2017**, *103*, 554–574. [CrossRef] [PubMed]
10. Salem, M.Z.; Ali, H.M.; El-Shanhorey, N.A.; Abdel-Megeed, A. Evaluation ofextracts and essential oil from Callistemon viminalis leaves: Antibacterialand antioxidant activities, total phenolic. *Asian Pac. J. Trop. Med.* **2013**, *6*, 785–791. [CrossRef]
11. AlSheikh, H.M.A.; Sultan, I.; Kumar, V.; Rather, I.A.; Al-Sheikh, H.; Jan, A.T.; Haq, Q.M.R. Plant-Based Phytochemicals as Possible Alternative to Antibiotics in Combating Bacterial Drug Resistance. *Antibiotics* **2020**, *9*, 480. [CrossRef] [PubMed]
12. EUCAST. Clinical Breakpoints-Breakpoints and Guidance. Available online: https://www.eucast.org/clinical_breakpoints (accessed on 2 January 2020).
13. EUCAST. Guidelines for Detection of Resistance Mechanisms and Specific Resistances of Clinical and/or Epidemiological Importance. Available online: https://www.eucast.org/resistance_mechanisms (accessed on 1 July 2017).
14. Verschelden, G.; Noeparast, M.; Stoefs, A.; Van Honacker, E.; Vandoorslaer, K.; Vandervore, L.; Olbrecht, M.; Van Damme, K.; Demuyser, T.; Piérard, D.; et al. Aztreonam-avibactam synergy, a validation and comparison of diagnostic tools. *Front.Microbiol.* **2023**, *14*, 1322180. [CrossRef] [PubMed]
15. Klančnik, A.; Piskernik, S.; Jeršek, B.; Smole Možina, S. Evaluation of diffusion and dilution methods to determine the antibacterial activity of plant extracts. *J. Microbiol. Methods* **2010**, *81*, 121–126. [CrossRef] [PubMed]
16. Brennan-Krohn, T.; Kirby, J.E. When One Drug Is Not Enough: Context, Methodology, and Future Prospects in Antibacterial Synergy Testing. *Clin. Lab. Med.* **2019**, *39*, 345–358. [CrossRef] [PubMed]
17. Hałasa, R.; Turecka, K.; Mizerska, U.; Krauze-Baranowska, M. Anti-Helicobacter pylori Biofilm Extracts from Rubus idaeus and Rubus occidentalis. *Pharmaceutics* **2024**, *16*, 501. [CrossRef] [PubMed]
18. Baran, A.; Kwiatkowska, A.; Potocki, L. Antibiotics and Bacterial Resistance—A Short Story of an Endless Arms Race. *Int. J. Mol. Sci.* **2023**, *24*, 5777. [CrossRef]
19. Gajdács, M.; Albericio, F. Antibiotic Resistance: From the Bench to Patients. *Antibiotics* **2019**, *8*, 129. [CrossRef] [PubMed]
20. World Health Organization. *Antimicrobial Resistance: Global Report on Surveillance*; World Health Organization: Geneva, Switzerland, 2014. Available online: https://apps.who.int/iris/handle/10665/112642 (accessed on 19 December 2021).
21. Lack of New Antibiotics Threatens Global Efforts to Contain Drug-Resistant Infections. Available online: https://www.who.int/news/item/17-01-2020-lack-of-new-antibiotics-threatens-global-efforts-to-contain-drug-resistant-infections (accessed on 19 December 2021).
22. CDC's Antibiotic Resistance Threats in the United States (2019 AR Threats Report). 2019. Available online: https://www.cdc.gov/drugresistance/biggest-threats.html (accessed on 19 December 2021).
23. Antimicrobial Resistance: Tackling a Crisis for the Health and Wealth of Nations/the Review on Antimicrobial Resistance chairedby Jim O'Neill. | Wellcome Collection. Available online: https://wellcomecollection.org/works/rdpck35v (accessed on 19 December 2021).
24. Ferdes, M. Antimicrobial compounds from plants. In *Fighting Antimicrobial Resistance*; Budimir, A., Ed.; IAPC-OBP: Zagreb, Croatia, 2018; pp. 243–271.
25. Djilani, A.; Dicko, A. The therapeutic benefits of essential oils. In *Nutrition, Well-Being and Health*; Bouayed, J., Bohn, T., Eds.; Well-Being Health 7; IntechOpen: Rijeka, Croatia, 2012; pp. 155–179.
26. Bubonja-Sonje, M.; Giacometti, J.; Abram, M. Antioxidant and antilisterial activity of olive oil, cocoa and rosemary extract polyphenols. *Food Chem.* **2011**, *127*, 821–827. [CrossRef]
27. King, T.; Dykes, G.; Kristianti, R. Comparative evaluation of methods commonly used to determine antimicrobial susceptibility to plant extracts and phenolic compounds. *JAOAC Int.* **2008**, *91*, 1423–1429. [CrossRef]
28. Valgas, C.; de Souza, S.M.; Smânia, E.; Smânia, A. Screening methods to determine antibacterial activity of natural products. *Braz. J. Microbiol.* **2007**, *38*, 369–380. [CrossRef]
29. Farha, A.K.; Yang, Q.-Q.; Kim, G.; Li, H.-B.; Zhu, F.; Liu, H.-Y.; Gan, R.-Y.; Corke, H.; Kim, G.; Li, H.-B.; et al. Tannins as an alternative to antibiotics. *Food Biosience* **2020**, *38*, 100751. [CrossRef]
30. Dwivedi, G.R.; Tiwari, N.; Singh, A.; Kumar, A.; Roy, S.; Negi, A.S.; Pal, A.; Chanda, D.; Sharma, A.; Darokar, M.P. Gallic acid-based indanone derivative interacts synergistically with tetracycline by inhibiting efflux pump in multidrug resistant . *Ecoli. Appl. Microbiol. Biotechnol.* **2016**, *100*, 2311–2325. [CrossRef] [PubMed]
31. Alibi, S.; Crespo, D.; Navas, J. Plant-Derivatives Small Molecules with Antibacterial Activity. *Antibiotics* **2021**, *10*, 231. [CrossRef] [PubMed]
32. Oulahal, N.; Degraeve, P. Phenolic-Rich Plant Extracts With Antimicrobial Activity: An Alternative to Food Preservatives and Biocides? *Front. Microbiol.* **2022**, *12*, 753518. [CrossRef] [PubMed]
33. Shin, J.; Prabhakaran, V.-S.; Kim, K.-S. The multi-faceted potential of plant-derived metabolites as antimicrobial agents against multidrug-resistant pathogens. *Microb. Pathog.* **2018**, *116*, 209–214. [CrossRef]
34. Saavedra, M.J.; Borges, A.; Dias, C.; Aires, A.; Bennett, R.N.; Rosa, E.S.; Simões, M. Antimicrobial activity of phenolics and glucosinolate hydrolysis products and their synergy with streptomycin against pathogenic bacteria. *Med. Chem.* **2010**, *6*, 174–183. [CrossRef] [PubMed]
35. Martini, S.; D'Addariob, C.; Colacevichc, A.; Focardi, S.; Borghinic, F.; Santuccid, A.; Figurae, N.; Rossi, C. Antimicrobial activity against *Helicobacter pylori* strains and antioxidant properties of blackberry leaves (*Rubus ulmifolius*) and isolated compounds. *Int. J. Antimicrob. Agents* **2009**, *34*, 50–59. [CrossRef] [PubMed]

36. Górniak, I.; Bartoszewski, R.; Króliczewski, J. Comprehensive review of antimicrobial activities of plant flavonoids. *Phytochem. Rev.* **2019**, *18*, 241–272. [CrossRef]
37. Wang, S.; Yao, J.; Zhou, B.; Yang, J.; Chaudry, M.T.; Wang, M.; Fenglin, X.; Li, Y.; Yin, W. Bacteriostatic effect of quercetin as an antibiotic alternative in vivo and its antibacterial mechanism in vitro. *J. Food Prot.* **2018**, *81*, 68–78. [CrossRef] [PubMed]
38. Miklasińska-Majdanik, M.; Kępa, M.; Wojtyczka, R.D.; Idzik, D.; Wąsik, T.J. Phenolic Compounds Diminish Antibiotic Resistance of *Staphylococcus aureus* Clinical Strains. *Int. J. Environ. Res. Public Health* **2018**, *15*, 2321. [CrossRef]
39. Álvarez-Martínez, F.J.; Barraj'on-Catal'an, E.; Herranz-L'opez, M.; Micol, V. Antibacterial plant compounds, extracts and essential oils: An updated review on their effects and putative mechanisms of action. *Phytomedicine* **2021**, *90*, 153626. [CrossRef] [PubMed]
40. Siriwong, S.; Teethaisong, Y.; Thumanu, K.; Dunkhunthod, B.; Eumkeb, G. The synergy and mode of action of quercetin plus amoxicillin against amoxicillin-resistant *Staphylococcus epidermidis*. *BMC Pharmacol. Toxicol.* **2016**, *17*, 39. [CrossRef] [PubMed]
41. Kepa, M.; Miklasinska-Majdanik, M.; Wojtyczka, R.D.; Idzik, D.; Korzeniowski, K.; Smolen-Dzirba, J.; Wasik, T.J. Antimicrobial Potential of Caffeic Acid against *Staphylococcus aureus* Clinical Strains. *Biomed. Res. Int.* **2018**, *2018*, 7413504. [CrossRef]
42. Siriwong, S.; Thumanu, K.; Hengpratom, T.; Eumkeb, G. Synergy and mode of action of ceftazidime plus quercetin or luteolin on *Streptococcus pyogenes*. *Evid. Based Complement. Alternat. Med.* **2015**, *2015*, 759459. [CrossRef] [PubMed]
43. Zhao, W.H.; Hu, Z.Q.; Okubo, S.; Hara, Y.; Shimamura, T. Mechanism of synergy between epigallocatechin gallate and beta-lactams against methicillin-resistant *Staphylococcus aureus*. *Antimicrob. Agents Chemother.* **2001**, *45*, 1737–1742. [CrossRef] [PubMed]
44. Zhao, W.H.; Hu, Z.Q.; Okubo, S.; Hara, Y.; Shimamura, T. Inhibition of penicillinase by epigallocatechin gallate resulting in restoration of antibacterial activity of penicillin against penicillinase-producing *Staphylococcus aureus*. *Antimicrob. Agents Chemother.* **2002**, *46*, 2266–2268. [CrossRef] [PubMed]
45. Novy, P.; Rondevaldova, J.; Kourimska, L.; Kokoska, L. Synergistic interactions of epigallocatechin gallate and oxytetracycline against various drug resistant *Staphylococcus aureus* strains in vitro. *Phytomedicine* **2013**, *20*, 432–435. [CrossRef] [PubMed]
46. Hu, Z.Q.; Zhao, W.H.; Asano, N.; Yoda, Y.; Hara, Y.; Shimamura, T. Epigallocatechin gallate synergistically enhances the activity of carbapenems against methicillin-resistant. *Antimicrob. Agents Chemother.* **2002**, *46*, 558–560. [CrossRef] [PubMed]
47. Cho, Y.S.; Schiller, N.L.; Oh, K.H. Antibacterial effects of green tea polyphenols on clinical isolates of methicillin-resistant *Staphylococcus aureus*. *Curr. Microbiol.* **2008**, *57*, 542–546. [CrossRef] [PubMed]
48. Sudano Roccaro, A.; Blanco, A.R.; Giuliano, F.; Rusciano, D.; Enea, V. Epigallocatechin-gallate enhances the activity of tetracycline in staphylococci by inhibiting its efflux from bacterial cells. *Antimicrob. Agents Chemother.* **2004**, *48*, 1968–1973. [CrossRef] [PubMed]
49. Miklasińska, M.; Kępa, M.; Wojtyczka, R.D.; Idzik, D.; Zdebik, A.; Orlewska, K.; Wąsik, T.J. Antibacterial activity of protocatechuic acid ethyl ester on *Staphylococcus aureus* clinical strains alone and in combination with antistaphylococcal drugs. *Molecules* **2015**, *20*, 13536–13549. [CrossRef]
50. Lin, R.D.; Chin, Y.P.; Hou, W.C.; Lee, M.H. The effects of antibiotics combined with natural polyphenols against clinical methicillin-resistant *Staphylococcus aureus* (MRSA). *Planta Med.* **2008**, *74*, 840–846. [CrossRef] [PubMed]
51. Zhang, L.; Kong, Y.; Wu, D.; Zhang, H.; Wu, J.; Chen, J.; Ding, J.; Hu, L.; Jiang, H.; Shen, X. Three flavonoids targeting the beta-hydroxyacyl-acyl carrier protein dehydratase from *Helicobacter pylori*: Crystal structure characterization with enzymatic inhibition assay. *Protein Sci.* **2008**, *17*, 1971–1978. [CrossRef] [PubMed]
52. Lu, M.; Li, T.; Wan, J.; Li, X.; Yuan, L.; Sun, S. Antifungal effects of phytocompounds on *Candida* species alone and in combination with fluconazole. *Int. J. Antimicrob. Agents* **2017**, *49*, 125–136. [CrossRef] [PubMed]
53. Veras, H.; Santos, I.; Santos, A.; Matias, E.; Leite, G.; Souza, H.; Costa, J.; Coutinho, H. Comparative evaluation of antibiotic and antibiotic modifying activity of quercetin and isoquercetin in vitro. *Curr. Top. Nutraceut Res.* **2011**, *9*, 25–30.
54. Capaldi, R.A.; Aggeler, R. Mechanism of the F_1F_0-type ATP synthase, a biological rotary motor. *Trends Biochem. Sci.* **2002**, *27*, 154–160. [CrossRef] [PubMed]
55. Bernard, F.X.; Sable, S.; Cameron, B.; Provost, J.; Desnottes, J.F.; Crouzet, J.; Blanche, F. Glycosylated flavones as selective inhibitors of topoisomerase IV. *Antimicrob. Agents. Chemother.* **1997**, *41*, 992–998. [CrossRef] [PubMed]
56. Ohemeng, K.A.; Schwender, C.F.; Fu, K.P.; Barrett, J.F. DNA gyrase inhibitory and antibacterial activity of some flavones. *Bioorg. Med. Chem. Lett.* **1993**, *3*, 25–230. [CrossRef]
57. Suriyanarayanan, B.; Shanmugam, K.; Santhosh, R. Synthetic quercetin inhibits mycobacterial growth possibly by interacting with DNA gyrase. *Rom. Biotechnol. Lett.* **2013**, *18*, 1587–1593.
58. Abraham, E.P. An Enzyme from Bacteria able to Destroy Penicillin. *Nature* **1940**, *146*, 837. [CrossRef]
59. Wise, E.M.; Park, J.T. Penicillin: Its basic site of action as an inhibitor of a peptide cross-linking reaction in cell wall mucopeptide synthesis. *Proc. Natl. Acad. Sci. USA* **1965**, *54*, 75–81. [CrossRef]
60. Hayek, S.A.; Gyawali, R.; Ibrahim, S.A. Antimicrobial natural products. In *Microbial Pathogens and Strategies for Combating Them: Science, Technology and Education*; Mendez-Vilas, A., Ed.; FORMATEX: Badajoz, Spain, 2013; Volume 2, pp. 910–921.
61. Pisoschi, A.M.; Cimpeanu, C.; Predoi, G. Electrochemical methods for total antioxidant capacity and its main contributors determination: A review. *Open Chem.* **2015**, *13*, 824–856. [CrossRef]
62. Nichita, C.; Neagu, G.; Cucu, A.; Vulturescu, V.; Bertesteanu, S.V.G. Antioxidative properties of *Plantago lanceolata* L. extracts evaluated by chemiluminiscence method. *AgroLife Sci. J.* **2016**, *5*, 95–102.

63. Pisoschi, A.M.; Pop, A.; Georgescu, C.; Turcus, V.; Olah, N.K.; Mathem, E. An overview of natural antimicrobials role in food. *Eur. J. Med. Chem.* **2018**, *143*, 922–935. [CrossRef] [PubMed]
64. Mann, C.M.; Markham, J.L. A new method for determining the minimum inhibitory concentration of essential oils. *J. Appl. Microbiol.* **1998**, *84*, 538–544. [CrossRef] [PubMed]

antibiotics

MDPI

Review

The Molecular Mechanisms and Therapeutic Potential of Cranberry, D-Mannose, and Flavonoids against Infectious Diseases: The Example of Urinary Tract Infections

Petros Ioannou * and Stella Baliou

School of Medicine, University of Crete, 71003 Heraklion, Greece
* Correspondence: p.ioannou@uoc.gr

Abstract: The treatment of infectious diseases typically includes the administration of anti-infectives; however, the increasing rates of antimicrobial resistance (AMR) have led to attempts to develop other modalities, such as antimicrobial peptides, nanotechnology, bacteriophages, and natural products. Natural products offer a viable alternative due to their potential affordability, ease of access, and diverse biological activities. Flavonoids, a class of natural polyphenols, demonstrate broad anti-infective properties against viruses, bacteria, fungi, and parasites. Their mechanisms of action include disruption of microbial membranes, inhibition of nucleic acid synthesis, and interference with bacterial enzymes. This review explores the potential of natural compounds, such as flavonoids, as an alternative therapeutic approach to combat infectious diseases. Moreover, it discusses some commonly used natural products, such as cranberry and D-mannose, to manage urinary tract infections (UTIs). Cranberry products and D-mannose both, yet differently, inhibit the adhesion of uropathogenic bacteria to the urothelium, thus reducing the likelihood of UTI occurrence. Some studies, with methodological limitations and small patient samples, provide some encouraging results suggesting the use of these substances in the prevention of recurrent UTIs. While further research is needed to determine optimal dosages, bioavailability, and potential side effects, natural compounds hold promise as a complementary or alternative therapeutic strategy in the fight against infectious diseases.

Keywords: cranberry; D-mannose; infectious diseases; polyphenols; flavonoids

Citation: Ioannou, P.; Baliou, S. The Molecular Mechanisms and Therapeutic Potential of Cranberry, D-Mannose, and Flavonoids against Infectious Diseases: The Example of Urinary Tract Infections. *Antibiotics* **2024**, *13*, 593. https://doi.org/10.3390/antibiotics13070593

Academic Editors: Anna Duda-Madej, Joanna Kozłowska and Katarzyna Wińska

Received: 30 May 2024
Revised: 22 June 2024
Accepted: 25 June 2024
Published: 26 June 2024

1. Introduction

Infectious diseases can be caused by bacteria, viruses, fungi, parasites, or other infectious factors, such as prions [1]. Their treatment classically includes the administration of anti-infectives, such as antibiotics, antivirals, antifungals, or antiparasitic drugs [1]. However, antimicrobial resistance (AMR) is emerging worldwide. It can be a considerable cause of morbidity and mortality in patients suffering from infections from pathogens with significant AMR, such as multi-drug-resistant (MDR) pathogens, extensively-resistant pathogens (XDR), and pan-drug-resistant (PDR) pathogens [2,3].

The development of AMR has led to the need to develop novel methods for treating infections that may not involve anti-infectives. These anti-infective-sparing techniques can involve several methods, such as antimicrobial peptides, bacteriophages, nanotechnology, or other methods [4–6]. In most cases, these methods are at the stage of clinical studies and have not yet been approved. Moreover, these products may be technologically demanding, requiring advanced techniques for design, production, and effective and safe use.

On the other hand, substances from natural products and simple chemical compounds could have anti-infective activity and, in some cases, be used to manage simple infections, such as urinary tract infections [7]. The production of such substances could be less laborious while they could be equally safe as more complex and technologically demanding substances [8]. Such substances, such as cranberry products or D-mannose, have been suggested for the management of recurrent urinary tract infections (UTIs) in the past; however, there is no consensus for supporting their broad use currently [9].

The present study aimed to review the theoretical role of natural compounds, and more specifically, of flavonoids in treating infectious diseases and also to evaluate the role of cranberries and D-mannose in treating UTIs.

2. The Need for and Examples of Alternative Treatments in the Treatment of Infectious Diseases

The increasing rates of AMR among bacterial pathogens are worrisome, and the trends suggest that AMR will be a considerable cause of mortality in the following decades, leaving few anti-infectives available as viable options for treating infections [10,11]. It is estimated that in 2050, if no action is taken, AMR could even be a primary cause of death [12]. The Global Burden of bacterial antimicrobial resistance study that was published in 2022 showed that an estimated 4.95 million deaths in 2019 were associated with bacterial AMR [3]. In terms of geographical distribution, the all-death rate attributable to AMR was estimated to be higher in western sub-Saharan Africa and lowest in Australia [3]. Several pathogens may have significant AMR that is of particular concern from a public health perspective. For example, MDR tuberculosis (infection by *Mycobacterium tuberculosis* with resistance to rifampicin and isoniazid), an infectious disease with high transmissibility, is common in Asia, while one of the most common resistant Gram-positive bacteria, methicillin-resistant *Staphylococcus aureus* (MRSA), is increasingly prevalent in North and South America, South Europe, Asia, and some countries in Africa, with MRSA being associated with higher mortality rates compared to methicillin-susceptible *S. aureus* (MSSA) [3,13,14]. Moreover, *Escherichia coli*, the most common cause of UTIs and one of the most commonly identified Gram-negative bacteria in microbiology laboratories, has increased rates of resistance in the last decades, with very high rates of ampicillin resistance, extended-spectrum beta lactamase production, and even increasing rates of carbapenem resistance among these and other Enterobacteriaceae species [15–18]. To that end, the prevalence of AMR *E. coli* has almost doubled from 2001 to 2010 and keeps rising [19]. Areas like South Asia, Central America, and South Europe have increased rates of *E. coli* strains with resistance to third-generation cephalosporins [3].

The rise of AMR can be associated with huge surcharges, leading to an increased cost of many thousands of US dollars per patient episode, thus causing a substantial burden to the healthcare system and the global economy [20]. Moreover, the rise of AMR can lead to increased consumption of anti-infectives in medicine and other fields, such as animal husbandry, aquaculture, food preservation, and veterinary medicine among others [21]. Indeed, this magnitude of uses of anti-infectives can lead to the selection of resistant species of microorganisms [21]. Moreover, antimicrobial resistance can also spread through mobile genetic elements, as in the case of plasmid transfer through conjugation, transduction with phages, or transformation through uptake of free DNA [22]. Probiotics, on the other hand, are live microorganisms that could protect against infectious diseases, most commonly through a reduction in the likelihood of development of infection through various mechanisms, such as competitive inhibition, immunomodulation, production of mediators such as defensins or other anti-infective compounds, and inhibition of biofilm formation [22]. Even though their role is ambiguous for some conditions, such as *Clostridioides difficile* infection, there are studies suggesting their potential benefit in other instances, even though more research is needed [23]. For example, a recent systematic review showed that probiotic and symbiotic supplements could reduce the likelihood of ventilator-associated pneumonia and sepsis in critically ill patients, reduce the duration of intensive care unit (ICU) and hospital stay, and lower mortality in the ICU [24].

Thus, using methods other than anti-infectives could be associated with benefits such as the lack of development of resistance, the reduced likelihood of adverse reactions, and the possibility of drug delivery precisely to the site of infection. For example, organisms naturally produce antimicrobial peptides to protect them against pathogens, and these peptides could be used to treat infections. Some of them, such as daptomycin and colistin have already been approved for human use, while others are currently used in fields other than medicine, such as food preservatives [5]. Bacteriophages are organisms that exist naturally and target bacteria. This interaction could be exploited to allow their use as anti-infective factors in treating infectious diseases. This could be beneficial since they could have minimal effects on the host's microbiome and have the theoretical advantage of not causing adverse effects. However, although they have several non-medical applications, they have not been approved for human use. At the same time, their production could be technically demanding since they have high selectivity for their target bacteria [4]. Nanotechnology has been mainly exploited for diagnostic purposes in infectious diseases; however, their use in treating infections is highly demanding from a technical perspective, even though they could provide nanobiotics that could deliver the antibiotic substance precisely at the site of infection [6].

3. Literature Search Methodology

To review the role of natural compounds, and, more specifically, of flavonoids in the management of infectious diseases, and also evaluate the role of cranberry products and D-mannose in the treatment of UTIs, a search of the PubMed/Medline database up until 30 April 2024 for eligible articles was conducted. The search term '(flavonoids OR D-mannose OR cranberry) AND infection' was used. Relevant reviews and original studies providing data on the topic were retrieved, evaluated, and added to the synthesis in a liberal way. The two investigators (PI and SB) screened the studies and extracted and synthesized the evidence. The references of the included articles were searched to identify other relevant articles.

4. The Potential of Flavonoids in the Management of Infectious Diseases and Their Underlying Molecular Mechanisms

In many low- and middle-income countries, growing health expenditure is associated with a growing inaccessibility to treatment [25]. On the one hand, increasing AMR and the growing cost associated with it could lead to difficulties in providing treatments [3,20]. On the other hand, natural remedies may be inexpensive and more readily available [26]. In the case of infectious diseases, natural products seem promising because they reduce the burden of pathogens and restore the balance of healthy microbiota. Moreover, they can also suppress or eradicate biofilms [26,27].

Flavonoids are natural polyphenols synthesized by plants, and they have many distinct categories, including isoflavones (daidzein and genistein), flavones (luteolin and apigenin), flavonols (galangin, quercetin, and kaempferol), flavanones (naringerin and hesperetin), flavanonols (taxifolin), and flavan-3-ols (catechin and epicatechin), which are precursors of tannins—catechins [28]. Throughout the plant kingdom, flavonoids are physiologically active phytochemicals used in various herbal remedies, and they constitute the most prevalent and widely distributed phytochemicals in herbs [29]. In foods, flavonoids can be detected in tea, wine, grains, fruits, vegetables, and bark [30]. Accordingly, flavones can be detected in chamomile, red peppers, celery, parsley, mint, and Ginkgo biloba, exerting anticancer, antioxidant, and anti-inflammatory properties. Flavanone-rich foods include grapes, grapefruit, lemons, limes, and oranges [31]. Due to their antioxidant properties, flavones and flavonols are crucial for shielding plants from UV rays [32,33]. Of particular interest, the main flavone in food is luteolin. Cereals and herbs contain luteolin, while vegetables like broccoli and carrots have glycosylated luteolin [34,35]. Luteolin has tumor-suppressive properties, hindering oxidative stress and inflammation and modulating gut microbiota [36,37]. In addition, isoflavonoids from lentils, peas, beans, tofu, and soybeans have been demonstrated to ameliorate vascular complications. Anthocyanins are another category of flavonoids involved in food types (eggplant, red cabbage, blood oranges, cherries, black plums, elderberries, blackberries, strawberries, cranberries, red grapes, and red wine). In this context, the anthocyanins have also proved effective in sustaining cardiovascular health. Accordingly, flavonoid supplements like silymarin, green tea extracts, rutin, curcumin, and quercetin display similar properties [31]. Quercetin, whose chemical name is 2-(3,4-dihydroxyphenyl)-3,5,7-trihydroxychromen-4-one, is a flavonoid studied extensively. Many fruits (apples, berries, etc.) and vegetables (lettuce, broccoli, tomato, etc.) provide daily sources of quercetin in its glycoside type. From a molecular perspective, quercetin has an anticoagulant, antiproliferative, anti-infective, and antioxidant nature that is advantageous for health and longevity [38]. Quercetin has been shown to hinder growth inhibition of the following bacterial species, including *S. sobrinus*, *L. acidophilus*, *S. sanguis*, *A. actinomycetemocomitans*, and *P. intermedia* [39]. Several examples substantiate the action of quercetin to modulate the action of F-type ATPase across the membrane [39].

Flavonoids have anticancer, anti-inflammatory, antioxidant, antibacterial, antiviral, anticancer, antidiabetic, and wound-healing properties [34–36]. Notably, researchers have emphasized the anti-inflammatory effect of flavonoids in managing diseases. Indeed, they seem to reduce inflammation in leukemia, sepsis, asthma, sclerosis, atherosclerosis, psoriasis, allergic rhinitis, ileitis/colitis, and rheumatoid arthritis, attenuating their clinical signs [40,41]. In particular, flavonoids seem to interfere with the production of proinflammatory cytokines, chemokines, plasma proteases, prostaglandins, leukotrienes, interleukins, and nitric oxide [42,43]. For example, Chen et al. have supported that flavonoids suppressed the synthesis of prostaglandin E_2 (PGE_2), nitric oxide (NO), and interleukin-6 (IL-6) in LPS-induced macrophages [42,43]. In another case, it has been documented that quercetin prevents nuclear factor-kappa B (NF-kB) transcriptional translocation in intestinal epithelial cells, thus, hindering acute inflammation [42,43].

A growing body of research has focused on elucidating the antiviral effects of flavonoids against multiple viruses. For example, the suppressive action of flavonoids against the human immunodeficiency virus (HIV) is a significant field of research. For example, Li et al. have also shown that the flavone O-glycoside inhibits HIV-1 entrance into cells expressing CD4 and chemokine co-receptors [44], and it antagonizes HIV-1 reverse transcriptase [44]. In addition to the above, other studies have shown that numerous catechins, including robustaflavone, hinokiflavone, and baicalein, can inhibit HIV-1 reverse transcriptase [45]. Moreover, dietary flavonoids are considered promising for the treatment of coronavirus disease 2019 (COVID-19), and they have been shown to have direct antiviral properties against influenza, Ebola, severe acute respiratory syndrome (SARS), and Middle East respiratory syndrome (MERS) [46–48].

In addition, the antiparasitic effects of flavonoids have also been examined. For example, flavonoids have been proven very promising in the defense against the parasite *Entaemoba histolytica* trophozoites [49]. While further studies are needed to figure out the precise molecular targets, bioavailability, and administration method of flavonoids, they offer a very exciting and practical approach that should be taken into consideration as protection against *E. histolytica* [49].

4.1. The Antibacterial Impact of Flavonoids and Their Underlying Mechanism

The microbicidal activity of flavonoids has received considerable attention. Several examples have substantiated the beneficial effect of flavonoids in managing bacterial infections. Initially, a methanolic extract of the leaves of *B. purpurascens* proved effective in combating some microbial pathogens. This medicinal herb contains a variety of flavonoids, including bergenin, catechin, naringenin, and myricetin. This methanolic extract also showed strong antibacterial effects against *S. aureus* and *Streptococcus* species [31]. In another case, 3,3,4-trihydroxyflavone found in the methanolic fraction of *Justicia wynaadensis* extract had antibacterial efficacy against opportunistic and multi-resistant pathogens. *Enterococcus faecalis*, *Klebsiella pneumoniae*, *Enterobacter aerogenes*, *Escherichia coli*, and *Pseudomonas aeruginosa* are some bacteria which flavones efficiently killed in diabetic wounds [50]. For example, a recent systematic review highlighted the effect of flavonoids on the development of upper respiratory tract infection (URTI). In particular, flavonoid supplementation decreased the onset of URTI by 33% without causing any apparent side effects [51].

From a molecular perspective, the mechanisms underlying the antibacterial activity of flavonoids are the following: modification of membrane permeability, decreased porins in the cell membrane, prevention of bacterial biofilm formation, inhibition of nucleic acid synthesis, inhibition of the synthesis of the cell envelope, reduction in energy metabolism, hindering virulence factors, prevention of extracellular glycans, and changes in bacterial enzymes such as sortase A and proton-translocating F-ATPases [27,28,52]. Other underlying mechanisms that account for the protective effects of flavonoids rely on the modification of bacterial enzymes like proton-translocating F-ATPases and sortase A (SrtA) and on the synthesis of extracellular glucans by glucosyltransferase (GTF), thus reducing the virulence factors [53]. Figure 1 shows the mechanisms of the anti-infective effects of flavonoids.

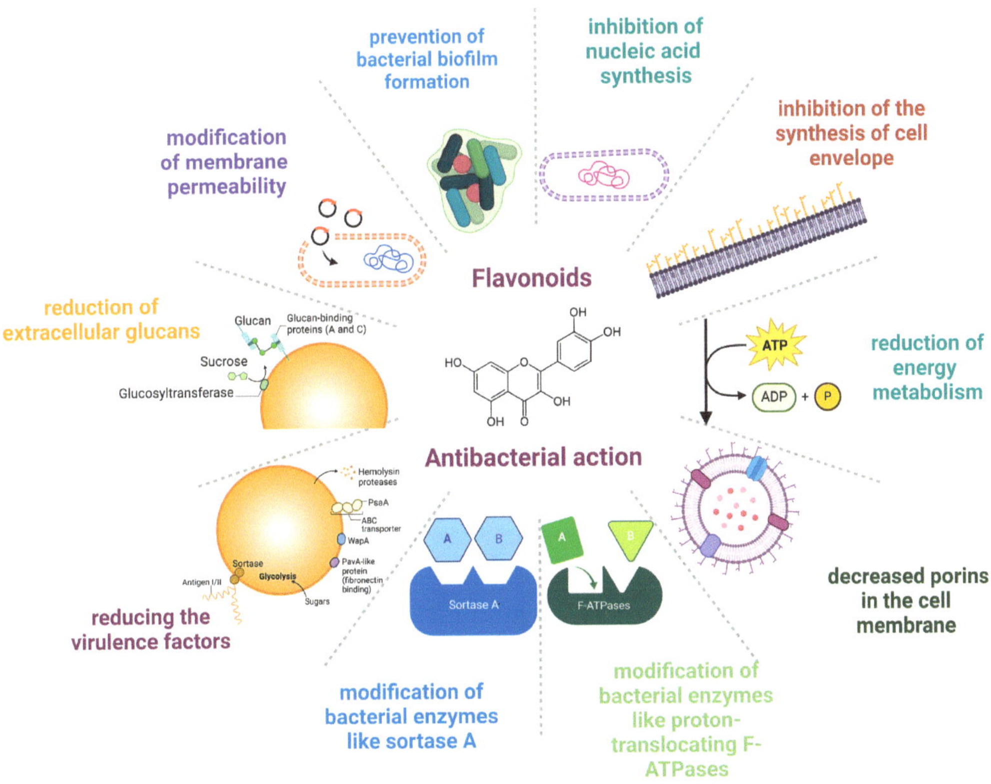

Figure 1. Mechanisms underlying the anti-infective effect of flavonoids. Created with BioRender.com (license date: 29 May 2024).

In more detail, flavonoids have been used in several ways to reduce bacterial growth. For example, flavonoids can interfere with membrane lipid bilayers in two ways, thereby contributing to membrane rupture [54]. The first one involves the more non-polar molecules splitting apart in the hydrophobic membrane's interior, and the second one involves hydrogen bonds forming between lipids' polar head groups and the more hydrophilic flavonoids at the membrane. Several examples of flavonoids have substantiated their capacity to cause membrane rupture. Initially, catechins have been illustrated to cause membrane disruption by hindering the action of membrane enzymes [55]. Other polyphenols, such as the salidroside, rutin (quercetin-3-O-rhamnoglucoside), and quercetin, have been shown to cause membrane damage [56]. Epicatechin (EC), epigallocatechin gallate (EGCG), and a flavonol quercetin are some flavonoid cases with reported membrane damage [57]. Epigallocatechin gallate (EGCG) is the major catechin involved in green tea and is also present in legumes and fruits [58,59]. Epicatechin (EC) can be found in grapes, blackberries, cherries, apples, raspberries, pears, broad beans, cocoa, and tea leaves [60]. The bactericidal properties of EGCG and EC have been attributed to their membrane-disrupting nature [61]. Apart from the membrane disruption, the microbicidal effect of flavonoids can also be attributed to their pro-oxidant properties. According to Fathima and Rao, catechins have been shown to eliminate bacterial growth due to an oxidative burst [57]. In particular, catechins caused the excessive formation of reactive oxygen species (ROS), which in turn modified membrane permeability and induced membrane rupture [57].

Thirdly, the flavonoids' antibacterial effect has been explained by their capacity to prevent biofilm formation. For example, synthetic lipophilic 3-arylideneflavonones (replaced with different phenolic compounds at the C-3 position of the C Ring) are highly effective against *S. aureus*, *Staphylococcus epidermidis*, and *Enterococcus faecalis* due to their capacity to inhibit biofilm formation [62].

Moreover, flavonoids' ability to interfere with nucleic acid synthesis seems to constitute another mechanism underlying their protective action against bacterial pathogens. To support this, some examples of relevant flavonoids have been reported to be myricetin, robinet, and (−)-epigallocatechin, which have been considered to account for the prevention of DNA synthesis. In more detail, the negative impact of flavonoids on DNA and RNA synthesis can be explained by the possibility that the B ring of the flavonoids is involved in intercalation or hydrogen bonding with the stacking of nucleic acid bases [63].

In addition to the above, the antibacterial impact of flavonoids can be the result of their ability to hinder the activity of DNA enzymes. Indeed, polyphenols like quercetin, apigenin, and 3,6,7,3′,4′-pentahydroxyflavone have been proven to hinder bacterial growth of *S. epidermidis*, *S. aureus*, *E. coli*, *S. typhimurium*, and *Stenotrophomonas maltophilia* by limiting the action of DNA gyrase [64].

In addition, flavonoids have been reported to reduce energy metabolism. Some flavonoids have shown suppressive action against the bacterial growth of *S. aureus* and *Micrococcus luteus* but not against *E. coli*. In particular, licochalcones were revealed to hinder oxygen consumption in *M. luteus* and *S. aureus* but not in *E. coli*. Interestingly, licochalcones seemed to interfere with energy metabolism similar to respiratory-inhibiting antibiotics [65]. In this direction, it has been highlighted that flavonoids can downregulate the electron transport chain, hindering ATP synthesis. Various polyphenols have been demonstrated to bind to ATP synthase's unique polyphenol binding site. Since the polyphenol binding pocket is located at the interface between the F1 sector's a, b, and c subunits of ATPase, polyphenols seem to inhibit the action of ATP synthase. In particular, the speculated route of flavonoid inhibitory properties is mediated through binding to ATP synthase's polyphenol binding pocket and preventing the rotation of the ATP synthase c-subunit [66]. Recently, it came to light that flavonoids may interfere with *E. coli* F1FO ATPase [67]. Consistent with the above, Plaper and colleagues have also shown that quercetin can suppress the ATPase activity of *E. coli* DNA gyrase by binding to its GyrB component [68], taking into consideration that quercetin binds to the B subunit of gyrase. Indeed, quercetin blocks the ATP binding pocket by generating hydrogen bonds via 5, 7, and 30 -OH groups to the amino acid residues of DNA gyrase [68]. Accordingly, some flavonoids have been reported to hinder the ATP binding pocket of D-alanine-D-alanine ligase [69].

4.2. The Synergistic Effect of Flavonoids with Antibiotics in the Fight against Antimicrobial Resistance

Flavonoids have received considerable attention recently as the main antibiotic substitute since AMR has rendered increasing numbers of microbial illnesses resistant to treatment [26]. Specific flavonoids can enhance the therapeutic value of the antibiotics' resistance. [27]. Due to the rise of drug-resistant bacteria, novel anti-infective agents have been identified. In this regard, flavonoids are considered great options to combat drug-resistant microbes, given the broad spectrum of their biological activities. Unlike Gram-positive bacteria, Gram-negative bacteria are resistant to a wide spectrum of antibiotics, owing to the limited permeability of their cell membrane. The key mechanism linked to their resistance is MexAB-OprM and AcrAB-TolC efflux pumps, as well as poor porin expression [70]. The increased secretion of antibiotics through efflux pumps is prevalent in MDR bacteria [5]. While flavonoids present in nature have been extensively studied, semi-synthetic or synthetic flavonoids have also been appreciated as promising therapeutics in preventing or even eliminating bacteria at concentrations lower than 1 μg/mL. Synthetic flavonoids feature a substituted pattern that frequently contains hydroxy groups, halogens, or other heteroatomic rings like cations of 1,3-dithiolium, piperidine, or pyridine [71].

Flavonoids in combination with anti-infectives can be used to combat AMR. In this direction, the flavonoid dihydromyricetin (DHM) and the two triterpenoids ursolic acid (UA) and oleanolic acid (OA) were examined for their ability to counteract bacterial growth [72]. Without adversely affecting eukaryotic cells, all of the substances displayed antibacterial capacity against Gram-negative bacteria (*Proteus hauseri*, *E. coli*, and *Campylobacter jejuni*) and Gram-positive bacteria (*S. aureus*, *S. epidermidis*, and *Listeria monocytogenes*). Related to Gram-positive types of bacteria, the triterpenoids demonstrated the most potent antibacterial activity. The same study highlighted that flavonoid dihydromyricetin (DHM) can exert therapeutic benefit against *Staphylococcus* strains when combined with two triterpenoids, ursolic acid (UA) and oleanolic acid (OA) [72]. As a result, DHM's antibacterial activity was strengthened when combined with triterpenoids [72].

Combining Quercetin-3-arabinofuranoside with Myricetin with Procyanidin A2 exerts bactericidal action against *S. mutans* and *S. anginosus*, by hindering the action of glucosyltransferases and F-ATPases [73].

Another combination of the following substances, Luteolin, Morin, Naringin, Quercetin, and Rutin, has been proven helpful in combating *A. naeslundii*, *A. viscosus*, *A. actinomycecomitans*, *E. faecalis*, and *L. casei* [74].

Flavonoids seem beneficial in mitigating candidiasis caused by *Candida albicans*, a fungus frequently inhabiting the oral mucosa. Flavonoids combat antimicrobial-resistant candidiasis by escaping resistance to fluconazole [75]. Apart from fluconazole, other therapeutic options are polyenes, azoles, and echinocandin. Many commercial pharmaceuticals used to treat candidiasis are in the azole class; therefore, the activity of flavonoids will be compared to that of recently synthesized azole compounds and to azole medications currently available as approved medications [76]. In particular, combining certain flavonoids with antifungal azole drugs can exhibit therapeutic benefits against antibiotic-resistant *Candida albicans* [77]. The flavonoids potentiate the therapeutic efficacy of azole drugs, modulating their bioavailability and metabolism, thus contributing to the attenuation of candidiasis [77]. Similarly, quercetin can act synergistically with fluconazole to exhibit antifungal action against *C. albicans* [78]. The mechanisms underlying the therapeutic potential of the aforementioned therapeutics are based on the induction of cell death and DNA damage [78].

From a molecular perspective, flavonoids have been shown to increase antibiotic efficacy by boosting endogenous ROS formation, which might compromise the organism's ability to deal with oxidative stress triggered by the antibiotic. For example, antibiotics like quinolones, b-lactams, and aminoglycosides are known to account for hydroxyl radical generation via the Fenton reaction [79]. Accordingly, the flavonoids served as either hydroxyl radical quenchers or hydroxyl radical quenchers seem to reduce bacterial killing, implying that oxidative burst through hydroxyl radical formation can amplify antibiotic-induced cell death [79]. Table 1 shows some examples of studies that provide data on the activity of different flavonoids against commonly isolated microorganisms, with some of them also suggesting synergy with specific antibiotics.

Table 1. Examples of studies providing data on the activity of flavonoids on common microorganisms.

Type of Flavonoid	Bacterial Species	Synergy with Antibiotic	Reference
Flavones	*Escherichia coli, Staphylococcus aureus*		[64]
Flavonols	*Escherichia coli, Staphylococcus aureus*		[64]
Isoflavones	*Escherichia coli, Staphylococcus aureus*		[64]
Apigenin (flavone)	*Bacillus subtilis, Micrococcus luteus*		[80]
Apigenin (flavone)	*Bacillus subtilis, Escherichia coli, Pseudomonas auruginosa, Proteus vulgaris*		[81]
Apigenin (flavone)	*Escherichia coli, Pseudomonas aeruginosa, Proteus mirabilis, Klebsiella pneumoniae, Salmonella typhimurium, Enterobacter aerogenes, Enterobacter cloacae*		[82,83]
Flavone	*Proteus vulgaris, Proteus mirabilis*		[84]
Flavone	Vancomycin-intermediate *Staphylococcus aureus*	Vancomycin, Oxacillin	[85]
Apigenin (flavone)	MRSA	Ampicillin, Ceftriaxone	[86]
Luteolin (flavone)	MRSA	Ampicillin, Cephradine, Ceftriaxone, Imipenem, Methicillin	[87]
Hesperetin (flavanon)	*Staphylococcus aureus*		[88]
Galangin (flavonol)	MRSA, MSSA *Enterococcus* spp., *Pseudomonas aeruginosa*		[89]
Luteolin (flavone)	*Streptococcus pyogenes*	Ceftazidime	[90]
Quercetin (flavonol)	*Staphylococcus aureus*	Cloxacillin	[91]
Quercetin (flavonol)	MRSA	Ceftriaxone	[87]
Quercetin (flavonol) + Luteolin (flavone)	MRSA Clinical Isolates	Imipenem	[87]
Genistein (isoflavone)	*Staphylococcus aureus*	Norfloxacin	[92]
EGCG—epigallocatechin gallate	MRSA6975, MRSA3202	Tetracycline Oxacillin	[93]

MRSA: methicillin-resistant *Staphylococcus aureus*; MSSA—methicillin-sensitive *Staphylococcus aureus*; the table may not be exhaustive of all studies in the literature.

4.3. Clinical Trials of Flavonoids in Infectious Diseases

Flavonoids have not only been evaluated in vitro. Several clinical studies evaluate their efficacy and safety in managing acute respiratory tract infections. For example, a recent systematic review evaluated 30 randomized controlled trials (RCTs) providing data for more than 5000 patients with respiratory tract infections [94]. The results from the meta-analysis of this systematic review suggested that flavonoids were safe and had adequate clinical efficacy in treating viral acute respiratory tract infections, including COVID-19, influenza, common cold, acute rhinosinusitis, acute non-streptococcal tonsillopharyngitis, acute bronchitis, pneumonia, and infections of the upper respiratory tract. Yet, as the authors of this systematic review also stated, their results should be read cautiously since there were few RCTs per type of acute respiratory tract infection. At the same time, many RCTs had a significant risk of bias and were of low quality. Thus, larger studies should be conducted to draw safe conclusions regarding flavonoids' effect on these infections.

In another study, the effect of luteolin 4′-Neohesperidoside against four antibiotic-resistant microorganisms was evaluated [95]. More specifically, in vitro studies showed that luteolin 4′-Neohesperidoside was effective against MRSA, *K. pneumoniae*, and Shiga toxin-producing *E. coli* (STEC) serogroup O111. Further experiments in vivo revealed a significant decrease in STEC and *K. pneumoniae* colonization and shedding. Those experiments implied that 4′-Neohesperidoside could be used as an antibiotic-sparing agent in patients, even with infections of resistant pathogens, if clinical trials in humans prove it is safe and efficient.

4.4. Limitations and Bioavailability of Flavonoids

The use of flavonoids may have some limitations. Initially, the daily consumption of flavonoids by food sources hinders their bioavailability. In particular, the absorption of flavonoids has been reported to be limited. For example, a study including 97 people revealed that the plasma concentrations of total polyphenolic metabolites were 0–4 μmol/L following 50 mg of aglycone equivalents, and the urinary excretion was 0.3–43% of the ingested dose [96]. Another issue is the purification or bioengineering of flavonoids remains difficult. Last but not least, developing optimal delivery formulations of flavonoids is critical to exert their therapeutic effect. For example, in a review by Thilakarathna et al. addressing issues regarding the bioavailability of flavonoids, an attempt to understand the necessary steps towards improving the bioavailability of flavonoids is made [97]. To that end, improving intestinal absorption using mixtures or delivery vehicles, such as lipid-based systems, could increase absorption for relatively insoluble molecules [98]. Another possibility could be using nanodelivery vehicles that could also lead to increased absorption of insoluble molecules [99]. Additionally, changing the site of absorption inside the gastrointestinal system via enzymatic conversion of the flavonoid could also lead to better absorption [100]. Increasing the metabolic stability of flavonoids through methylation or other modifications could also reduce their elimination after absorption [101]. However, future studies on pharmacokinetics and pharmacodynamics of flavonoids of particular interest in infectious diseases would be needed to allow a better understanding of their absorption, delivery, and elimination, as well as allow for their potential use against infections in the future.

5. Natural Products for the Management of Urinary Tract Infections

Recurrent UTIs, more commonly acute uncomplicated cystitis, can be a cause of considerable discomfort in women, more commonly before menopause. The definition of recurrent UTIs includes at least two infections within six months or at least three within one year. Recurrent simple cystitis is relatively frequent in young, healthy women without immunosuppression or abnormality in their anatomy and physiology of the urinary tract. At least 25% of female college students with an episode of UTI had a recurrence within the six months following the first episode, and isolation of E. coli (the most common cause of UTIs) was associated with a higher rate of recurrence [102,103]. Several risk factors such as sexual intercourse, spermicide use, having a first UTI at a young age, having a mother with a history of recurrent UTIs, urologic factors such as urinary incontinence or post-voiding residual urine, and other biological and genetic factors increase the risk for UTIs [104–113].

Management of patients with recurrent UTIs includes excluding reversible causes that increase the likelihood of infection, such as anatomical obstruction of urine flow, appropriate education about hygiene and practices that could reduce the infection rate, and in case these measures fail, treatment of patients with an agent that could reduce the infection rates [114]. Agents to reduce the recurrence rate of UTIs include using anti-infectives, such as the combination of trimethoprim and sulfamethoxazole, or nitrofurantoin, or non-anti-infective options, like natural products, such as cranberry products, or D-mannose [22]. The use of antibiotics is widely endorsed in the literature and in guidelines; however, adverse events may occur. For example, Clostridioides difficile colitis resulting from changes in the gut microbiome can develop in patients treated with antibiotics [115]. On the other hand, unnecessary use of antibiotics can lead to an increased risk of colonization and infection by pathogens with AMR [116]. Thus, the appropriate selection of the therapeutic agent to be used in patients with UTIs, especially when its symptoms are localized in the lower urinary tract, could be of clinical benefit to the patient and minimize unnecessary potential harm.

5.1. Cranberry Products

Cranberry fruit, which is abundant in polyphenols, exhibits a strong biological defense against *Streptococcus mutans*. In particular, the effects of cranberry extracts containing flavonols (FLAV), anthocyanins (A), and proanthocyanidins (PAC) were examined regarding the virulence parameters related to the formation of *Streptococcus mutans* biofilms and acidogenicity [117]. According to this research, the components of cranberries that effectively impeded *S. mutans* bacterial growth were flavonols and proanthocyanidins [117].

The most common use of cranberry products in infectious diseases involves their use in recurrent UTIs. Cranberry is a plant known as *Vaccinium macrocarpon*, *Vaccinium oxycoccos*, and *Vaccinium erythrocarpum*, in the family Ericaceae. They primarily consist of water and a mixture of fructose, organic acids, flavonoids, ascorbic acid, catechins, proanthocyanidins, anthocyanidins, and triterpenoids [22,118]. The factors considered more clinically relevant for preventing recurrent UTIs in women are anthocyanidins and proanthocyanidins, which are tannins (polyphenols) that function as natural plant defense against pathogens [22,119]. In the last centuries, before the discovery of antibiotics, cranberries had been used for UTIs and various other conditions, such as gastric and liver diseases, cancer, and scurvy [22,120,121]. The protection of cranberries against recurrent UTIs involves several mechanisms [22]. For example, cranberry juice is very acidic, and the associated acidification of urine has been proposed as a potential explanation of the protective effect of cranberry juice against recurrent UTIs [118,120]. A better understanding of the pathophysiology of UTIs and the role of some components of cranberries led to the identification of new potential mechanisms as well. For example, the adherence of bacterial cells on the urothelium, an important first step in the pathophysiology of a UTI, could be inhibited by cranberry products. More specifically, bacterial adhesins enable bacteria such as *E. coli* to bind to carbohydrate receptors on the surface of cells of the urothelium [22,121]. Such adhesins are inhibited by fructose and, maybe more importantly, proanthocyanidins, which are contained in cranberries [122]. Figure 2 shows the pathophysiology of UTIs that is relevant to the way the cranberry products work towards inhibition of UTI development.

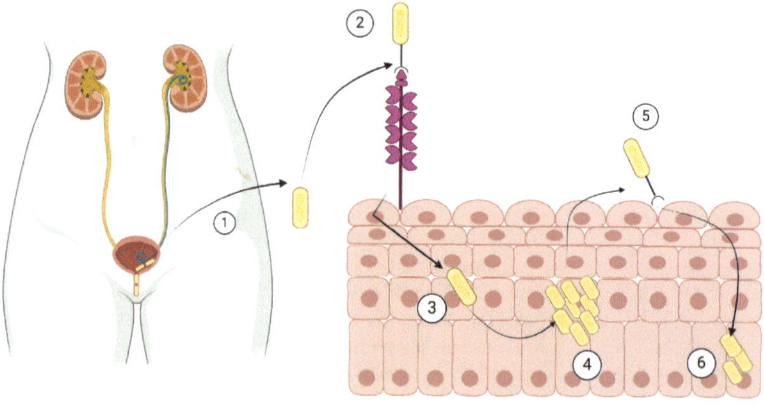

Figure 2. Pathogenesis of urinary tract infection by *Escherichia coli*. (1) *E. coli* ascends the urethra and (2) binds to mannose residues of uroplakins on the urothelium. (3) *E. coli* is then internalized, (4) it multiplies in the urothelial cells, (5) it recolonizes the bladder, and (6) it remains dormant in the urothelial cells. Created with BioRender.com (license date: 28 May 2024).

Proanthocyanidins are polymers and oligomers of flavans, often linked by a single bond, even though, in cranberries, another bond (type A-linkage) can be found and is primarily implicated in the clinical effect of cranberries with respect to the reduction in recurrence of UTIs [123]. The specific association of A-linkage with this clinical effect was shown in human volunteers who consumed products containing different proanthocyanidins, such as grape juice, green tea, apple juice, and dark chocolate. The urine of the volunteers that had consumed cranberry juice had a higher ability to inhibit the adhesion of bacteria, as confirmed in vitro, and this would be attributed to the type A linkage of proanthocyanidins [123]. The exact mechanism that causes the reduced adhesion of bacteria to the urothelium in the presence of proanthocyanidins or other cranberry components is not yet completely defined, even though there are some proposed explanations [22]. For example, cranberry components can competitively inhibit the binding of *E. coli* on the urothelial cells by acting as analogs of those receptors [119]. Additionally, cranberry components could alter the properties of the cell surface of bacteria, thus affecting their ability to adhere to the urothelium. For example, bacterial cells of *E. coli* that have been exposed to cranberry compounds for long periods could present morphological changes, such as changing from rod shaped to spherical [124]. Other compounds of cranberries could also have similar properties. For example, ursolic acid (a pentacyclic triterpenoid) may have a synergistic or complementary role to proanthocyanidins towards the inhibition of bacterial adhesion through altering gene expression and inhibiting the formation of biofilms [125].

Cranberries can be administered in many forms, such as tablets, juice cocktails, or fresh berries. However, the concentration of proanthocyanidins differs between these different forms, and the precise content of the original preparation exists only in whole cranberry fruit [22,126]. It has been estimated that 240 mL of cranberry juice can inhibit the adhesion of 80% of uropathogenic *E. coli* on the urothelium for 10 h after ingestion [127]. A randomized placebo-controlled study in which older women were given 300 mL of cranberry juice containing 36 mg of proanthocyanidins or placebo showed a statistically significant reduction in the possibility of bacteriuria and pyuria of 42% in the group that received the cranberry juice [128]. However, since other studies have suggested that a higher amount of proanthocyanidins are needed for adequate and prolonged inhibition of bacterial adhesion, the optimal concentration and the best formulation remain unknown [22,129–131].

Several studies have evaluated the effect of cranberries in treating and preventing the recurrence of UTIs [131–134]. Importantly, a recent systematic review evaluated and critically appraised the clinical trials assessing the effect of cranberry use in the prevention and treatment of uncomplicated lower UTIs (cystitis) in healthy adult women [7]. In this systematic review, 12 such studies were included for further evaluation. Among them, eleven assessed the effect of different formulations of cranberries on the prevention of recurrent UTIs [135–145]. Ten of those were randomized, the mean or median age in all these studies was up to 55 years, and in most studies, the cranberry product was compared with placebo. In 6 of these 11 studies, the cranberry product was shown to reduce the likelihood of recurrent UTIs. At the same time, adverse events were rare and, in most cases, included gastrointestinal symptoms that, in a small number of women, led to discontinuation of the product [7]. Even though the success rate among these studies was generally inadequate, there were several issues regarding heterogeneity, differences in the proanthocyanidins concentration that was administered, and poor study design, thus limiting the generalizability of the results. Until now, cranberry products are not widely endorsed in guidelines for use in the prevention of UTIs but could be used on a case-by-case basis based on a conditional recommendation [114].

5.2. D-Mannose

D-mannose is a monosaccharide and an isomer of glucose. It is involved in the glycosylation of proteins such as monoclonal antibodies. Following oral administration, D-mannose is rapidly absorbed, can be detected in the plasma in about 30 min, and is excreted in the urine [146]. *E. coli* has virulence factors like pili, which are protein structures that extend from the bacterial cell and allow it to attach to host cells [147]. Type 1 pili, which are composed of Fim proteins, have an adhesion lectin domain that allows them to bind to mannosylated proteins of the host, such as uroplakin 1a, β1 and α3 integrins, and Tamm–Horsfall glycoproteins, which are located on the urothelium [22,148–151]. FimH, the protein that has the adhesion lectin domain responsible for binding to mannosylated proteins of the host, seems to be critical in the pathogenesis of UTIs, as confirmed in mouse models of UTIs by *E. coli*, implying that targeting this protein could be promising in designing potential therapies for UTIs [152]. Due to the similarity between D-mannose and the binding site of urothelial proteins such as uroplakins, an increased concentration of D-mannose in the urine could competitively inhibit the adhesion of the FimH protein on the mannosylated uroplakin receptors, thus preventing the binding of *E. coli* bacterial cells on these receptors [153]. For example, in a rat model of intravesical inoculation with *E. coli*, a 10% D-mannose solution was able to reduce bacteriuria compared to a control of normal saline [153]. Moreover, an in vitro study evaluating the ability of D-mannose to inhibit the adhesion of different *E. coli* strains isolated from urine, the vagina, and the perianal area of women with recurrent UTIs showed that D-mannose was able to inhibit the adhesion of more than 60% of these strains by at least 50%, with complete inhibition shown in the majority of them [154]. In that direction, there are efforts towards developing synthetic mannosides with a much higher affinity for the FimH ligand than D-mannose, such as M4284, which has a 100,000-fold higher binding affinity for FimH. In a mouse model of colonization by an *E. coli* strain isolated from humans with cystitis, oral administration of M4284 led to significantly reduced intestinal colonization, and M4284 was also able to treat UTI in mice when compared to vehicle control [155]. Thus, such synthetic mannosides could be tested in future studies at a preclinical and then at a clinical level to elucidate the role of FimH inhibitors as potential therapeutic agents for UTIs [22,156]. Figure 3 shows the mechanism by which D-mannose can inhibit the binding of *E. coli* on the urothelium.

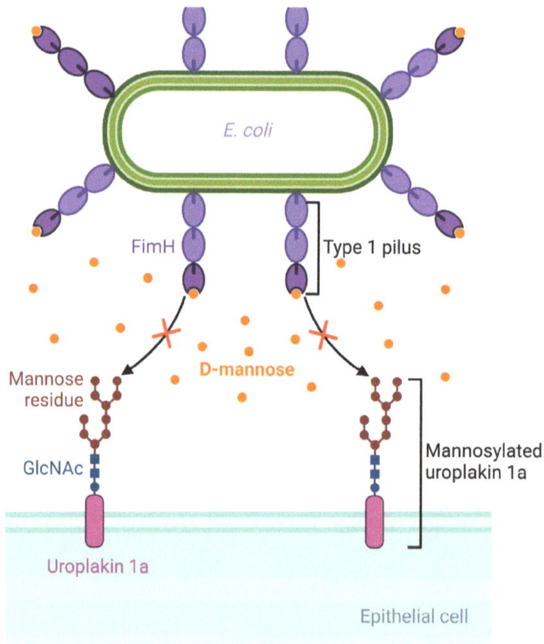

Figure 3. Mechanism of D-mannose inhibition of binding of Escherichia coli to the urothelium. D-mannose has a similar structure to the binding site of the glycoprotein receptors of the urothelium (such as uroplakin). If D-mannose is present in adequate concentrations in the urine, it can saturate FimH adhesin and inhibit bacterial binding to the urothelial receptors. Created with BioRender.com (license date: 28 May 2024).

Clinical evidence regarding the use of D-mannose in managing UTIs is much less than that regarding the use of cranberry products [157–159]. A recent systematic review evaluated and critically appraised the clinical trials assessing the effect of D-mannose use in managing uncomplicated lower UTI (cystitis) in healthy adult women [7]. Two randomized studies were eventually included in the analysis, the mean and median ages were within the range of 42 to 52 years, and the comparison included patients receiving D-mannose or antibiotics (either nitrofurantoin or the combination of trimethoprim and sulfamethoxazole) and, in one study, no treatment in a third group [160,161]. The treatment duration was six months in both studies, and a positive effect was observed in both studies. Adverse effects were uncommon and mainly included diarrhea in about 8% of patients. Importantly, even though in that systematic review the trials with D-mannose were relatively fewer than those with cranberries, the authors conclude that treatment with D-mannose or with a combination of D-mannose and antibiotics could be the most beneficial treatment for reducing recurrences of UTIs [7]. They do, however, conclude that randomized trials with blinding and significantly larger sample sizes are needed to confirm the results shown in their study [7]. Regarding the dose of D-mannose, as summarized in a recent Cochrane systematic review, different doses from 200 mg to 3 g have been used. However, different studies were not comparable due to the different dosing. In most studies, efficacious doses were those from 500 mg to 2 g (or 3 g with a de-escalation to 2 g) [162]. No clear recommendation exists until now regarding their use in recurrent UTIs [114].

6. Natural Compounds as Antibiotic-Sparing Agents in the Era of AMR

Despite antibiotics' activity against infections such as UTIs, the possibility of pathogens developing AMR is a significant drawback. Notably, AMR may develop more frequently in patients having recently received antibiotics and may even develop during antibiotic treatment in some pathogens [163,164]. In particular, UTIs nowadays have increasing resistance [17,165]. Patients on antimicrobial prophylaxis for UTIs may present with breakthrough infections by bacteria with AMR [166]. Thus, using anti-infectives wisely in these patients may be a critical antimicrobial stewardship approach leading to reduced costs, fewer adverse events, and fewer breakthrough infections by resistant pathogens [167]. From that perspective, natural compounds such as cranberry products or D-mannose could be used as antibiotic-sparing agents, thus allowing clinicians to preserve antibiotics for more severe infections [168].

7. Conclusions

Anti-infective treatment of infectious diseases has been long considered the mainstay of treatment of infectious diseases. However, due to the development of AMR, non-anti-infective therapies have emerged as potential alternative options for treating infections. Natural products could be such an example, and they have been used for many years in managing recurrent UTIs. Cranberry products and D-mannose have been periodically evaluated and have provided, at times, some encouraging results, even though more studies with better randomized, controlled, and blind designs should be performed in the future to better establish their safety and efficacy in managing patients with recurrent UTIs. Until then, D-mannose and cranberry products could be used as antibiotic-sparing agents on a case-by-case basis in some patients suffering from recurrent UTIs.

Author Contributions: Conceptualization, P.I. and S.B.; methodology, P.I. and S.B.; software, P.I. and S.B.; validation, P.I. and S.B.; formal analysis, P.I. and S.B.; investigation, P.I. and S.B.; resources, P.I. and S.B.; data curation, P.I. and S.B.; writing—original draft preparation, P.I. and S.B.; writing—review and editing, P.I. and S.B.; visualization, P.I. and S.B.; supervision, P.I.; project administration, P.I. All authors have read and agreed to the published version of the manuscript.

Funding: This research received no external funding.

Institutional Review Board Statement: Not applicable.

Informed Consent Statement: Not applicable.

Data Availability Statement: Not applicable.

Acknowledgments: We would like to thank Miruna Maria Apetroaei for her help in the preparation of the figures.

Conflicts of Interest: The authors declare no conflicts of interest.

References

1. Bennett, J.E.; Dolin, E.; Blaser, M.J. *Mandell, Douglas, and Bennett's Principles and Practice of Infectious Diseases*, 9th ed.; Elsevier: Philadelphia, PA, USA, 2019.
2. Magiorakos, A.-P.; Srinivasan, A.; Carey, R.B.; Carmeli, Y.; Falagas, M.E.; Giske, C.G.; Harbarth, S.; Hindler, J.F.; Kahlmeter, G.; Olsson-Liljequist, B.; et al. Multidrug-Resistant, Extensively Drug-Resistant and Pandrug-Resistant Bacteria: An International Expert Proposal for Interim Standard Definitions for Acquired Resistance. *Clin. Microbiol. Infect. Off. Publ. Eur. Soc. Clin. Microbiol. Infect. Dis.* **2012**, *18*, 268–281. [CrossRef] [PubMed]
3. Antimicrobial Resistance Collaborators. Global Burden of Bacterial Antimicrobial Resistance in 2019: A Systematic Analysis. *Lancet* **2022**, *399*, 629–655. [CrossRef] [PubMed]
4. Ioannou, P.; Baliou, S.; Samonis, G. Bacteriophages in Infectious Diseases and Beyond—A Narrative Review. *Antibiotics* **2023**, *12*, 1012. [CrossRef] [PubMed]
5. Ioannou, P.; Baliou, S.; Kofteridis, D.P. Antimicrobial Peptides in Infectious Diseases and Beyond—A Narrative Review. *Life* **2023**, *13*, 1651. [CrossRef] [PubMed]
6. Ioannou, P.; Baliou, S.; Samonis, G. Nanotechnology in the Diagnosis and Treatment of Antibiotic-Resistant Infections. *Antibiotics* **2024**, *13*, 121. [CrossRef] [PubMed]
7. Konesan, J.; Liu, L.; Mansfield, K.J. The Clinical Trial Outcomes of Cranberry, D-Mannose and NSAIDs in the Prevention or Management of Uncomplicated Urinary Tract Infections in Women: A Systematic Review. *Pathogens* **2022**, *11*, 1471. [CrossRef] [PubMed]
8. Farha, M.A.; Brown, E.D. Strategies for Target Identification of Antimicrobial Natural Products. *Nat. Prod. Rep.* **2016**, *33*, 668–680. [CrossRef]
9. Kwok, M.; McGeorge, S.; Mayer-Coverdale, J.; Graves, B.; Paterson, D.L.; Harris, P.N.A.; Esler, R.; Dowling, C.; Britton, S.; Roberts, M.J. Guideline of Guidelines: Management of Recurrent Urinary Tract Infections in Women. *BJU Int.* **2022**, *130* (Suppl. S3), 11–22. [CrossRef] [PubMed]
10. Cerini, P.; Meduri, F.R.; Tomassetti, F.; Polidori, I.; Brugneti, M.; Nicolai, E.; Bernardini, S.; Pieri, M.; Broccolo, F. Trends in Antibiotic Resistance of Nosocomial and Community-Acquired Infections in Italy. *Antibiotics* **2023**, *12*, 651. [CrossRef]
11. Salam, M.A.; Al-Amin, M.Y.; Salam, M.T.; Pawar, J.S.; Akhter, N.; Rabaan, A.A.; Alqumber, M.A.A. Antimicrobial Resistance: A Growing Serious Threat for Global Public Health. *Healthcare* **2023**, *11*, 1946. [CrossRef]
12. O'Neill, J. Review on Antimicrobial Resistance. Tackling Drug-Resistant Infections Globally. 2016. Available online: https://amr-review.org/sites/default/files/160525_Final%20paper_with%20cover.pdf (accessed on 29 May 2024).
13. Cosgrove, S.E.; Sakoulas, G.; Perencevich, E.N.; Schwaber, M.J.; Karchmer, A.W.; Carmeli, Y. Comparison of Mortality Associated with Methicillin-Resistant and Methicillin-Susceptible Staphylococcus Aureus Bacteremia: A Meta-Analysis. *Clin. Infect. Dis. Off. Publ. Infect. Dis. Soc. Am.* **2003**, *36*, 53–59. [CrossRef] [PubMed]
14. Seung, K.J.; Keshavjee, S.; Rich, M.L. Multidrug-Resistant Tuberculosis and Extensively Drug-Resistant Tuberculosis. *Cold Spring Harb. Perspect. Med.* **2015**, *5*, a017863. [CrossRef] [PubMed]
15. Bryce, A.; Hay, A.D.; Lane, I.F.; Thornton, H.V.; Wootton, M.; Costelloe, C. Global Prevalence of Antibiotic Resistance in Paediatric Urinary Tract Infections Caused by Escherichia Coli and Association with Routine Use of Antibiotics in Primary Care: Systematic Review and Meta-Analysis. *BMJ* **2016**, *352*, i939. [CrossRef] [PubMed]
16. Johnson, J.R.; Johnston, B.; Clabots, C.; Kuskowski, M.A.; Castanheira, M. Escherichia Coli Sequence Type ST131 as the Major Cause of Serious Multidrug-Resistant E. Coli Infections in the United States. *Clin. Infect. Dis. Off. Publ. Infect. Dis. Soc. Am.* **2010**, *51*, 286–294. [CrossRef] [PubMed]
17. Ioannou, P.; Plexousaki, M.; Dimogerontas, K.; Aftzi, V.; Drougkaki, M.; Konidaki, M.; Paschalidis, K.; Maraki, S.; Kofteridis, D.P. Characteristics of Urinary Tract Infections in Older Patients in a Tertiary Hospital in Greece. *Geriatr. Gerontol. Int.* **2020**, *20*, 1228–1233. [CrossRef] [PubMed]
18. Dejonckheere, Y.; Desmet, S.; Knops, N. A Study of the 20-Year Evolution of Antimicrobial Resistance Patterns of Pediatric Urinary Tract Infections in a Single Center. *Eur. J. Pediatr.* **2022**, *181*, 3271–3281. [CrossRef] [PubMed]
19. Sanchez, G.V.; Baird, A.M.G.; Karlowsky, J.A.; Master, R.N.; Bordon, J.M. Nitrofurantoin Retains Antimicrobial Activity against Multidrug-Resistant Urinary Escherichia Coli from US Outpatients. *J. Antimicrob. Chemother.* **2014**, *69*, 3259–3262. [CrossRef] [PubMed]
20. Poudel, A.N.; Zhu, S.; Cooper, N.; Little, P.; Tarrant, C.; Hickman, M.; Yao, G. The Economic Burden of Antibiotic Resistance: A Systematic Review and Meta-Analysis. *PLoS ONE* **2023**, *18*, e0285170. [CrossRef] [PubMed]

21. Meek, R.W.; Vyas, H.; Piddock, L.J.V. Nonmedical Uses of Antibiotics: Time to Restrict Their Use? *PLoS Biol.* **2015**, *13*, e1002266. [CrossRef]
22. Sihra, N.; Goodman, A.; Zakri, R.; Sahai, A.; Malde, S. Nonantibiotic Prevention and Management of Recurrent Urinary Tract Infection. *Nat. Rev. Urol.* **2018**, *15*, 750–776. [CrossRef]
23. Bishop, E.J.; Tiruvoipati, R. Management of Clostridioides Difficile Infection in Adults and Challenges in Clinical Practice: Review and Comparison of Current IDSA/SHEA, ESCMID and ASID Guidelines. *J. Antimicrob. Chemother.* **2022**, *78*, 21–30. [CrossRef] [PubMed]
24. Lou, J.; Cui, S.; Huang, N.; Jin, G.; Chen, C.; Fan, Y.; Zhang, C.; Li, J. Efficacy of Probiotics or Synbiotics in Critically Ill Patients: A Systematic Review and Meta-Analysis. *Clin. Nutr. ESPEN* **2024**, *59*, 48–62. [CrossRef] [PubMed]
25. Yenet, A.; Nibret, G.; Tegegne, B.A. Challenges to the Availability and Affordability of Essential Medicines in African Countries: A Scoping Review. *Clin. Outcomes Res.* **2023**, *15*, 443–458. [CrossRef]
26. Kováč, J.; Slobodníková, L.; Trajčíková, E.; Rendeková, K.; Mučaji, P.; Sychrová, A.; Bittner Fialová, S. Therapeutic Potential of Flavonoids and Tannins in Management of Oral Infectious Diseases—A Review. *Molecules* **2022**, *28*, 158. [CrossRef] [PubMed]
27. Górniak, I.; Bartoszewski, R.; Króliczewski, J. Comprehensive Review of Antimicrobial Activities of Plant Flavonoids. *Phytochem. Rev.* **2019**, *18*, 241–272. [CrossRef]
28. Fernández-Rojas, B.; Gutiérrez-Venegas, G. Flavonoids Exert Multiple Periodontic Benefits Including Anti-Inflammatory, Periodontal Ligament-Supporting, and Alveolar Bone-Preserving Effects. *Life Sci.* **2018**, *209*, 435–454. [CrossRef] [PubMed]
29. Justesen, U.; Knuthsen, P. Composition of Flavonoids in Fresh Herbs and Calculation of Flavonoid Intake by Use of Herbs in Traditional Danish Dishes. *Food Chem.* **2001**, *73*, 245–250. [CrossRef]
30. Roy, A.; Khan, A.; Ahmad, I.; Alghamdi, S.; Rajab, B.S.; Babalghith, A.O.; Alshahrani, M.Y.; Islam, S.; Islam, M.R. Flavonoids a Bioactive Compound from Medicinal Plants and Its Therapeutic Applications. *BioMed Res. Int.* **2022**, *2022*, 5445291. [CrossRef] [PubMed]
31. Chagas, M.D.S.S.; Behrens, M.D.; Moragas-Tellis, C.J.; Penedo, G.X.M.; Silva, A.R.; Gonçalves-de-Albuquerque, C.F. Flavonols and Flavones as Potential Anti-Inflammatory, Antioxidant, and Antibacterial Compounds. *Oxid. Med. Cell. Longev.* **2022**, *2022*, 9966750. [CrossRef]
32. Wittayathanarattana, T.; Wanichananan, P.; Supaibulwatana, K.; Goto, E. Enhancement of Bioactive Compounds in Baby Leaf Amaranthus Tricolor L. Using Short-Term Application of UV-B Irradiation. *Plant Physiol. Biochem.* **2022**, *182*, 202–215. [CrossRef]
33. Ferreyra, M.L.F.; Serra, P.; Casati, P. Recent Advances on the Roles of Flavonoids as Plant Protective Molecules after UV and High Light Exposure. *Physiol. Plant.* **2021**, *173*, 736–749. [CrossRef] [PubMed]
34. Dias, M.C.; Pinto, D.C.G.A.; Silva, A.M.S. Plant Flavonoids: Chemical Characteristics and Biological Activity. *Mol. Basel Switz.* **2021**, *26*, 5377. [CrossRef] [PubMed]
35. Miean, K.H.; Mohamed, S. Flavonoid (Myricetin, Quercetin, Kaempferol, Luteolin, and Apigenin) Content of Edible Tropical Plants. *J. Agric. Food Chem.* **2001**, *49*, 3106–3112. [CrossRef] [PubMed]
36. Hytti, M.; Piippo, N.; Korhonen, E.; Honkakoski, P.; Kaarniranta, K.; Kauppinen, A. Fisetin and Luteolin Protect Human Retinal Pigment Epithelial Cells from Oxidative Stress-Induced Cell Death and Regulate Inflammation. *Sci. Rep.* **2015**, *5*, 17645. [CrossRef] [PubMed]
37. Franza, L.; Carusi, V.; Nucera, E.; Pandolfi, F. Luteolin, Inflammation and Cancer: Special Emphasis on Gut Microbiota. *BioFactors* **2021**, *47*, 181–189. [CrossRef] [PubMed]
38. Jan, A.T.; Kamli, M.R.; Murtaza, I.; Singh, J.B.; Ali, A.; Haq, Q.M.R. Dietary Flavonoid Quercetin and Associated Health Benefits—An Overview. *Food Rev. Int.* **2010**, *26*, 302–317. [CrossRef]
39. Shu, Y.; Liu, Y.; Li, L.; Feng, J.; Lou, B.; Zhou, X.; Wu, H. Antibacterial Activity of Quercetin on Oral Infectious Pathogens. *Afr. J. Microbiol. Res.* **2011**, *5*, 5358–5361. [CrossRef]
40. Shalini, V.; Bhaskar, S.; Kumar, K.S.; Mohanlal, S.; Jayalekshmy, A.; Helen, A. Molecular Mechanisms of Anti-Inflammatory Action of the Flavonoid, Tricin from Njavara Rice (*Oryza sativa* L.) in Human Peripheral Blood Mononuclear Cells: Possible Role in the Inflammatory Signaling. *Int. Immunopharmacol.* **2012**, *14*, 32–38. [CrossRef] [PubMed]
41. Abdallah, H.M.; Almowallad, F.M.; Esmat, A.; Shehata, I.A.; Abdel-Sattar, E.A. Anti-Inflammatory Activity of Flavonoids from *Chrozophora tinctoria*. *Phytochem. Lett.* **2015**, *13*, 74–80. [CrossRef]
42. Fu, Y.; Chen, J.; Li, Y.-J.; Zheng, Y.-F.; Li, P. Antioxidant and Anti-Inflammatory Activities of Six Flavonoids Separated from Licorice. *Food Chem.* **2013**, *141*, 1063–1071. [CrossRef]
43. Das, T.; Mukherjee, S.; Chaudhuri, K. Effect of Quercetin on Vibrio Cholerae Induced Nuclear Factor-κB Activation and Interleukin-8 Expression in Intestinal Epithelial Cells. *Microbes Infect.* **2012**, *14*, 690–695. [CrossRef] [PubMed]
44. Li, B.Q.; Fu, T.; Dongyan, Y.; Mikovits, J.A.; Ruscetti, F.W.; Wang, J.M. Flavonoid Baicalin Inhibits HIV-1 Infection at the Level of Viral Entry. *Biochem. Biophys. Res. Commun.* **2000**, *276*, 534–538. [CrossRef] [PubMed]
45. Ono, K.; Nakane, H.; Fukushima, M.; Chermann, J.C.; Barré-Sinoussi, F. Inhibition of Reverse Transcriptase Activity by a Flavonoid Compound, 5,6,7-Trihydroxyflavone. *Biochem. Biophys. Res. Commun.* **1989**, *160*, 982–987. [CrossRef]
46. Khazeei Tabari, M.A.; Iranpanah, A.; Bahramsoltani, R.; Rahimi, R. Flavonoids as Promising Antiviral Agents against SARS-CoV-2 Infection: A Mechanistic Review. *Molecules* **2021**, *26*, 3900. [CrossRef] [PubMed]
47. Liu, A.-L.; Wang, H.-D.; Lee, S.M.; Wang, Y.-T.; Du, G.-H. Structure-Activity Relationship of Flavonoids as Influenza Virus Neuraminidase Inhibitors and Their in Vitro Anti-Viral Activities. *Bioorg. Med. Chem.* **2008**, *16*, 7141–7147. [CrossRef]

48. Clementi, N.; Scagnolari, C.; D'Amore, A.; Palombi, F.; Criscuolo, E.; Frasca, F.; Pierangeli, A.; Mancini, N.; Antonelli, G.; Clementi, M.; et al. Naringenin Is a Powerful Inhibitor of SARS-CoV-2 Infection in Vitro. *Pharmacol. Res.* **2021**, *163*, 105255. [CrossRef]
49. Martínez-Castillo, M.; Pacheco-Yepez, J.; Flores-Huerta, N.; Guzmán-Téllez, P.; Jarillo-Luna, R.A.; Cárdenas-Jaramillo, L.M.; Campos-Rodríguez, R.; Shibayama, M. Flavonoids as a Natural Treatment Against Entamoeba Histolytica. *Front. Cell. Infect. Microbiol.* **2018**, *8*, 209. [CrossRef]
50. Dsouza, D.; Nanjaiah, L. Antibacterial Activity of 3,3′,4′-Trihydroxyflavone from *Justicia wynaadensis* against Diabetic Wound and Urinary Tract Infection. *Braz. J. Microbiol. Publ. Braz. Soc. Microbiol.* **2018**, *49*, 152–161. [CrossRef]
51. Somerville, V.S.; Braakhuis, A.J.; Hopkins, W.G. Effect of Flavonoids on Upper Respiratory Tract Infections and Immune Function: A Systematic Review and Meta-Analysis. *Adv. Nutr.* **2016**, *7*, 488–497. [CrossRef]
52. Xie, Y.; Yang, W.; Tang, F.; Chen, X.; Ren, L. Antibacterial Activities of Flavonoids: Structure-Activity Relationship and Mechanism. *Curr. Med. Chem.* **2014**, *22*, 132–149. [CrossRef]
53. Cascioferro, S.; Totsika, M.; Schillaci, D. Sortase A: An Ideal Target for Anti-Virulence Drug Development. *Microb. Pathog.* **2014**, *77*, 105–112. [CrossRef] [PubMed]
54. Tsuchiya, H. Membrane Interactions of Phytochemicals as Their Molecular Mechanism Applicable to the Discovery of Drug Leads from Plants. *Molecules* **2015**, *20*, 18923–18966. [CrossRef] [PubMed]
55. Reygaert, W.C. The Antimicrobial Possibilities of Green Tea. *Front. Microbiol.* **2014**, *5*, 434. [CrossRef] [PubMed]
56. Sanver, D.; Murray, B.S.; Sadeghpour, A.; Rappolt, M.; Nelson, A.L. Experimental Modeling of Flavonoid-Biomembrane Interactions. *Langmuir ACS J. Surf. Colloids* **2016**, *32*, 13234–13243. [CrossRef] [PubMed]
57. Fathima, A.; Rao, J.R. Selective Toxicity of Catechin-a Natural Flavonoid towards Bacteria. *Appl. Microbiol. Biotechnol.* **2016**, *100*, 6395–6402. [CrossRef] [PubMed]
58. Nagle, D.G.; Ferreira, D.; Zhou, Y.-D. Epigallocatechin-3-Gallate (EGCG): Chemical and Biomedical Perspectives. *Phytochemistry* **2006**, *67*, 1849–1855. [CrossRef] [PubMed]
59. Kong, M.; Xie, K.; Lv, M.; Li, J.; Yao, J.; Yan, K.; Wu, X.; Xu, Y.; Ye, D. Anti-Inflammatory Phytochemicals for the Treatment of Diabetes and Its Complications: Lessons Learned and Future Promise. *Biomed. Pharmacother. Biomed. Pharmacother.* **2021**, *133*, 110975. [CrossRef] [PubMed]
60. Ravindranath, M.H.; Saravanan, T.S.; Monteclaro, C.C.; Presser, N.; Ye, X.; Selvan, S.R.; Brosman, S. Epicatechins Purified from Green Tea (Camellia Sinensis) Differentially Suppress Growth of Gender-Dependent Human Cancer Cell Lines. *Evid. Based Complement. Altern. Med. ECAM* **2006**, *3*, 237–247. [CrossRef] [PubMed]
61. Ikigai, H.; Nakae, T.; Hara, Y.; Shimamura, T. Bactericidal Catechins Damage the Lipid Bilayer. *Biochim. Biophys. Acta* **1993**, *1147*, 132–136. [CrossRef] [PubMed]
62. Budzyńska, A.; Rózalski, M.; Karolczak, W.; Wieckowska-Szakiel, M.; Sadowska, B.; Rózalska, B. Synthetic 3-Arylideneflavanones as Inhibitors of the Initial Stages of Biofilm Formation by Staphylococcus Aureus and Enterococcus Faecalis. *Z. Naturforschung C J. Biosci.* **2011**, *66*, 104–114. [CrossRef]
63. Cushnie, T.P.T.; Lamb, A.J. Antimicrobial Activity of Flavonoids. *Int. J. Antimicrob. Agents* **2005**, *26*, 343–356. [CrossRef] [PubMed]
64. Yan, Y.; Xia, X.; Fatima, A.; Zhang, L.; Yuan, G.; Lian, F.; Wang, Y. Antibacterial Activity and Mechanisms of Plant Flavonoids against Gram-Negative Bacteria Based on the Antibacterial Statistical Model. *Pharmaceuticals* **2024**, *17*, 292. [CrossRef] [PubMed]
65. Haraguchi, H.; Tanimoto, K.; Tamura, Y.; Mizutani, K.; Kinoshita, T. Mode of Antibacterial Action of Retrochalcones from Glycyrrhiza Inflata. *Phytochemistry* **1998**, *48*, 125–129. [CrossRef] [PubMed]
66. Gledhill, J.R.; Montgomery, M.G.; Leslie, A.G.W.; Walker, J.E. Mechanism of Inhibition of Bovine F1-ATPase by Resveratrol and Related Polyphenols. *Proc. Natl. Acad. Sci. USA* **2007**, *104*, 13632–13637. [CrossRef] [PubMed]
67. Chinnam, N.; Dadi, P.K.; Sabri, S.A.; Ahmad, M.; Kabir, M.A.; Ahmad, Z. Dietary Bioflavonoids Inhibit Escherichia Coli ATP Synthase in a Differential Manner. *Int. J. Biol. Macromol.* **2010**, *46*, 478–486. [CrossRef]
68. Plaper, A.; Golob, M.; Hafner, I.; Oblak, M.; Solmajer, T.; Jerala, R. Characterization of Quercetin Binding Site on DNA Gyrase. *Biochem. Biophys. Res. Commun.* **2003**, *306*, 530–536. [CrossRef]
69. Wu, D.; Kong, Y.; Han, C.; Chen, J.; Hu, L.; Jiang, H.; Shen, X. D-Alanine:D-Alanine Ligase as a New Target for the Flavonoids Quercetin and Apigenin. *Int. J. Antimicrob. Agents* **2008**, *32*, 421–426. [CrossRef] [PubMed]
70. Breidenstein, E.B.M.; de la Fuente-Núñez, C.; Hancock, R.E.W. Pseudomonas Aeruginosa: All Roads Lead to Resistance. *Trends Microbiol.* **2011**, *19*, 419–426. [CrossRef] [PubMed]
71. Sarbu, L.G.; Bahrin, L.G.; Babii, C.; Stefan, M.; Birsa, M.L. Synthetic Flavonoids with Antimicrobial Activity: A Review. *J. Appl. Microbiol.* **2019**, *127*, 1282–1290. [CrossRef] [PubMed]
72. Wrońska, N.; Szlaur, M.; Zawadzka, K.; Lisowska, K. The Synergistic Effect of Triterpenoids and Flavonoids-New Approaches for Treating Bacterial Infections? *Molecules* **2022**, *27*, 847. [CrossRef]
73. Gregoire, S.; Singh, A.P.; Vorsa, N.; Koo, H. Influence of Cranberry Phenolics on Glucan Synthesis by Glucosyltransferases and Streptococcus Mutans Acidogenicity. *J. Appl. Microbiol.* **2007**, *103*, 1960–1968. [CrossRef] [PubMed]
74. Gutiérrez-Venegas, G.; Gómez-Mora, J.A.; Meraz-Rodríguez, M.A.; Flores-Sánchez, M.A.; Ortiz-Miranda, L.F. Effect of Flavonoids on Antimicrobial Activity of Microorganisms Present in Dental Plaque. *Heliyon* **2019**, *5*, e03013. [CrossRef] [PubMed]
75. Srivastava, D.; Yadav, A.; Naqvi, S.; Awasthi, H.; Fatima, Z. Efficacy of Flavonoids in Combating Fluconazole Resistant Oral Candidiasis. *Curr. Pharm. Des.* **2022**, *28*, 1703–1713. [CrossRef] [PubMed]

76. Smiljković, M.; Kostić, M.; Stojković, D.; Glamočlija, J.; Soković, M. Could Flavonoids Compete with Synthetic Azoles in Diminishing Candida Albicans Infections? A Comparative Review Based on In Vitro Studies. *Curr. Med. Chem.* **2019**, *26*, 2536–2554. [CrossRef]
77. Srinivas, N.R. Combination of Flavonoids with Azole Drugs for Fungal Infections: Key Pharmacokinetic Challenges. *Future Microbiol.* **2019**, *14*, 733–738. [CrossRef] [PubMed]
78. da Silva, C.R.; de Andrade Neto, J.B.; de Sousa Campos, R.; Figueiredo, N.S.; Sampaio, L.S.; Magalhães, H.I.F.; Cavalcanti, B.C.; Gaspar, D.M.; de Andrade, G.M.; Lima, I.S.P.; et al. Synergistic Effect of the Flavonoid Catechin, Quercetin, or Epigallocatechin Gallate with Fluconazole Induces Apoptosis in Candida Tropicalis Resistant to Fluconazole. *Antimicrob. Agents Chemother.* **2014**, *58*, 1468–1478. [CrossRef] [PubMed]
79. Kohanski, M.A.; Dwyer, D.J.; Hayete, B.; Lawrence, C.A.; Collins, J.J. A Common Mechanism of Cellular Death Induced by Bactericidal Antibiotics. *Cell* **2007**, *130*, 797–810. [CrossRef] [PubMed]
80. Palacios, P.; Gutkind, G.; Rondina, R.V.; de Torres, R.; Coussio, J.D. Genus Baccharis. II. Antimicrobial Activity of B. Crispa and B. Notosergila. *Planta Med.* **1983**, *49*, 128. [CrossRef] [PubMed]
81. Oksüz, S.; Ayyildiz, H.; Johansson, C. 6-Methoxylated and C-Glycosyl Flavonoids from Centaurea Species. *J. Nat. Prod.* **1984**, *47*, 902–903. [CrossRef]
82. Basile, A.; Giordano, S.; López-Sáez, J.A.; Cobianchi, R.C. Antibacterial Activity of Pure Flavonoids Isolated from Mosses. *Phytochemistry* **1999**, *52*, 1479–1482. [CrossRef]
83. Basile, A.; Sorbo, S.; Giordano, S.; Ricciardi, L.; Ferrara, S.; Montesano, D.; Castaldo Cobianchi, R.; Vuotto, M.L.; Ferrara, L. Antibacterial and Allelopathic Activity of Extract from Castanea Sativa Leaves. *Fitoterapia* **2000**, *71* (Suppl. 1), S110–S116. [CrossRef] [PubMed]
84. Basile, A.; Conte, B.; Rigano, D.; Senatore, F.; Sorbo, S. Antibacterial and Antifungal Properties of Acetonic Extract of *Feijoa sellowiana* Fruits and Its Effect on *Helicobacter pylori* Growth. *J. Med. Food* **2010**, *13*, 189–195. [CrossRef] [PubMed]
85. Bakar, N.S.; Zin, N.M.; Basri, D.F. Synergy of Flavone with Vancomycin and Oxacillin against Vancomycin-Intermediate *Staphyloccus aureus*. *Pak. J. Pharm. Sci.* **2012**, *25*, 633–638.
86. Akilandeswari, K.; Ruckmani, K. Synergistic Antibacterial Effect of Apigenin with β-Lactam Antibiotics and Modulation of Bacterial Resistance by a Possible Membrane Effect against Methicillin Resistant *Staphylococcus aureus*. *Cell. Mol. Biol.* **2016**, *62*, 74–82. [CrossRef] [PubMed]
87. Usman Amin, M.; Khurram, M.; Khan, T.A.; Faidah, H.S.; Ullah Shah, Z.; Ur Rahman, S.; Haseeb, A.; Ilyas, M.; Ullah, N.; Umar Khayam, S.M.; et al. Effects of Luteolin and Quercetin in Combination with Some Conventional Antibiotics against Methicillin-Resistant *Staphylococcus aureus*. *Int. J. Mol. Sci.* **2016**, *17*, 1947. [CrossRef] [PubMed]
88. Lopes, L.A.A.; Dos Santos Rodrigues, J.B.; Magnani, M.; de Souza, E.L.; de Siqueira-Júnior, J.P. Inhibitory Effects of Flavonoids on Biofilm Formation by *Staphylococcus aureus* That Overexpresses Efflux Protein Genes. *Microb. Pathog.* **2017**, *107*, 193–197. [CrossRef] [PubMed]
89. Pepeljnjak, S.; Kosalec, I. Galangin Expresses Bactericidal Activity against Multiple-Resistant Bacteria: MRSA, Enterococcus Spp. and Pseudomonas Aeruginosa. *FEMS Microbiol. Lett.* **2004**, *240*, 111–116. [CrossRef] [PubMed]
90. Siriwong, S.; Thumanu, K.; Hengpratom, T.; Eumkeb, G. Synergy and Mode of Action of Ceftazidime plus Quercetin or Luteolin on *Streptococcus pyogenes*. *Evid. Based Complement. Altern. Med.* **2015**, *2015*, 759459. [CrossRef] [PubMed]
91. Eumkeb, G.; Sakdarat, S.; Siriwong, S. Reversing β-Lactam Antibiotic Resistance of Staphylococcus Aureus with Galangin from Alpinia Officinarum Hance and Synergism with Ceftazidime. *Phytomedicine Int. J. Phytother. Phytopharm.* **2010**, *18*, 40–45. [CrossRef]
92. Wang, S.-Y.; Sun, Z.-L.; Liu, T.; Gibbons, S.; Zhang, W.-J.; Qing, M. Flavonoids from *Sophora moorcroftiana* and Their Synergistic Antibacterial Effects on MRSA. *Phytother. Res.* **2014**, *28*, 1071–1076. [CrossRef]
93. Navrátilová, A.; Nešuta, O.; Vančatová, I.; Čížek, A.; Varela, M.R.E.; López-Abán, J.; Villa-Pulgarin, J.A.; Mollinedo, F.; Muro, A.; Žemličková, H.; et al. C-Geranylated Flavonoids from Paulownia Tomentosa Fruits with Antimicrobial Potential and Synergistic Activity with Antibiotics. *Pharm. Biol.* **2016**, *54*, 1398–1407. [CrossRef] [PubMed]
94. Yao, J.; Zhang, Y.; Wang, X.-Z.; Zhao, J.; Yang, Z.-J.; Lin, Y.-P.; Sun, L.; Lu, Q.-Y.; Fan, G.-J. Flavonoids for Treating Viral Acute Respiratory Tract Infections: A Systematic Review and Meta-Analysis of 30 Randomized Controlled Trials. *Front. Public Health* **2022**, *10*, 814669. [CrossRef] [PubMed]
95. El-Shiekh, R.A.; Elhemely, M.A.; Naguib, I.A.; Bukhari, S.I.; Elshimy, R. Luteolin 4′-Neohesperidoside Inhibits Clinically Isolated Resistant Bacteria In Vitro and In Vivo. *Molecules* **2023**, *28*, 2609. [CrossRef] [PubMed]
96. Manach, C.; Williamson, G.; Morand, C.; Scalbert, A.; Rémésy, C. Bioavailability and Bioefficacy of Polyphenols in Humans. I. Review of 97 Bioavailability Studies. *Am. J. Clin. Nutr.* **2005**, *81*, 230S–242S. [CrossRef] [PubMed]
97. Thilakarathna, S.H.; Rupasinghe, H.P.V. Flavonoid Bioavailability and Attempts for Bioavailability Enhancement. *Nutrients* **2013**, *5*, 3367–3387. [CrossRef] [PubMed]
98. Shen, Q.; Li, X.; Li, W.; Zhao, X. Enhanced Intestinal Absorption of Daidzein by Borneol/Menthol Eutectic Mixture and Microemulsion. *AAPS PharmSciTech* **2011**, *12*, 1044–1049. [CrossRef] [PubMed]
99. Zhang, Z.; Huang, Y.; Gao, F.; Gao, Z.; Bu, H.; Gu, W.; Li, Y. A Self-Assembled Nanodelivery System Enhances the Oral Bioavailability of Daidzein: In Vitro Characteristics and in Vivo Performance. *Nanomedicine* **2011**, *6*, 1365–1379. [CrossRef] [PubMed]

100. Nielsen, I.L.F.; Chee, W.S.S.; Poulsen, L.; Offord-Cavin, E.; Rasmussen, S.E.; Frederiksen, H.; Enslen, M.; Barron, D.; Horcajada, M.-N.; Williamson, G. Bioavailability Is Improved by Enzymatic Modification of the Citrus Flavonoid Hesperidin in Humans: A Randomized, Double-Blind, Crossover Trial. *J. Nutr.* **2006**, *136*, 404–408. [CrossRef] [PubMed]
101. Cao, H.; Jing, X.; Wu, D.; Shi, Y. Methylation of Genistein and Kaempferol Improves Their Affinities for Proteins. *Int. J. Food Sci. Nutr.* **2013**, *64*, 437–443. [CrossRef]
102. Foxman, B. Recurring Urinary Tract Infection: Incidence and Risk Factors. *Am. J. Public Health* **1990**, *80*, 331–333. [CrossRef]
103. Foxman, B.; Gillespie, B.; Koopman, J.; Zhang, L.; Palin, K.; Tallman, P.; Marsh, J.V.; Spear, S.; Sobel, J.D.; Marty, M.J.; et al. Risk Factors for Second Urinary Tract Infection among College Women. *Am. J. Epidemiol.* **2000**, *151*, 1194–1205. [CrossRef] [PubMed]
104. Hooton, T.M.; Scholes, D.; Hughes, J.P.; Winter, C.; Roberts, P.L.; Stapleton, A.E.; Stergachis, A.; Stamm, W.E. A Prospective Study of Risk Factors for Symptomatic Urinary Tract Infection in Young Women. *N. Engl. J. Med.* **1996**, *335*, 468–474. [CrossRef] [PubMed]
105. Fihn, S.D.; Boyko, E.J.; Normand, E.H.; Chen, C.L.; Grafton, J.R.; Hunt, M.; Yarbro, P.; Scholes, D.; Stergachis, A. Association between Use of Spermicide-Coated Condoms and Escherichia Coli Urinary Tract Infection in Young Women. *Am. J. Epidemiol.* **1996**, *144*, 512–520. [CrossRef] [PubMed]
106. Fihn, S.D.; Boyko, E.J.; Chen, C.L.; Normand, E.H.; Yarbro, P.; Scholes, D. Use of Spermicide-Coated Condoms and Other Risk Factors for Urinary Tract Infection Caused by Staphylococcus Saprophyticus. *Arch. Intern. Med.* **1998**, *158*, 281–287. [CrossRef] [PubMed]
107. Scholes, D.; Hooton, T.M.; Roberts, P.L.; Stapleton, A.E.; Gupta, K.; Stamm, W.E. Risk Factors for Recurrent Urinary Tract Infection in Young Women. *J. Infect. Dis.* **2000**, *182*, 1177–1182. [CrossRef] [PubMed]
108. Hooton, T.M.; Stapleton, A.E.; Roberts, P.L.; Winter, C.; Scholes, D.; Bavendam, T.; Stamm, W.E. Perineal Anatomy and Urine-Voiding Characteristics of Young Women with and without Recurrent Urinary Tract Infections. *Clin. Infect. Dis. Off. Publ. Infect. Dis. Soc. Am.* **1999**, *29*, 1600–1601. [CrossRef] [PubMed]
109. Schaeffer, A.J.; Jones, J.M.; Dunn, J.K. Association of in Vitro Escherichia Coli Adherence to Vaginal and Buccal Epithelial Cells with Susceptibility of Women to Recurrent Urinary-Tract Infections. *N. Engl. J. Med.* **1981**, *304*, 1062–1066. [CrossRef] [PubMed]
110. Källenius, G.; Winberg, J. Bacterial Adherence to Periurethral Epithelial Cells in Girls Prone to Urinary-Tract Infections. *Lancet* **1978**, *2*, 540–543. [CrossRef] [PubMed]
111. Fowler, J.E.; Stamey, T.A. Studies of Introital Colonization in Women with Recurrent Urinary Infections. VII. The Role of Bacterial Adherence. *J. Urol.* **1977**, *117*, 472–476. [CrossRef]
112. Stapleton, A.; Nudelman, E.; Clausen, H.; Hakomori, S.; Stamm, W.E. Binding of Uropathogenic Escherichia Coli R45 to Glycolipids Extracted from Vaginal Epithelial Cells Is Dependent on Histo-Blood Group Secretor Status. *J. Clin. Investig.* **1992**, *90*, 965–972. [CrossRef]
113. Lomberg, H.; Cedergren, B.; Leffler, H.; Nilsson, B.; Carlström, A.S.; Svanborg-Edén, C. Influence of Blood Group on the Availability of Receptors for Attachment of Uropathogenic Escherichia Coli. *Infect. Immun.* **1986**, *51*, 919–926. [CrossRef] [PubMed]
114. Bixler, B.R.; Anger, J.T. Updates to Recurrent Uncomplicated Urinary Tract Infections in Women: AUA/CUA/SUFU Guideline. *J. Urol.* **2022**, *208*, 754–756. [CrossRef] [PubMed]
115. Cymbal, M.; Chatterjee, A.; Baggott, B.; Auron, M. Management of Clostridioides Difficile Infection: Diagnosis, Treatment, and Future Perspectives. *Am. J. Med.* **2024**, *137*, 571–576. [CrossRef] [PubMed]
116. Tacconelli, E.; De Angelis, G.; Cataldo, M.A.; Mantengoli, E.; Spanu, T.; Pan, A.; Corti, G.; Radice, A.; Stolzuoli, L.; Antinori, S.; et al. Antibiotic Usage and Risk of Colonization and Infection with Antibiotic-Resistant Bacteria: A Hospital Population-Based Study. *Antimicrob. Agents Chemother.* **2009**, *53*, 4264–4269. [CrossRef]
117. Duarte, S.; Gregoire, S.; Singh, A.P.; Vorsa, N.; Schaich, K.; Bowen, W.H.; Koo, H. Inhibitory Effects of Cranberry Polyphenols on Formation and Acidogenicity of Streptococcus Mutans Biofilms. *FEMS Microbiol. Lett.* **2006**, *257*, 50–56. [CrossRef] [PubMed]
118. Guay, D.R.P. Cranberry and Urinary Tract Infections. *Drugs* **2009**, *69*, 775–807. [CrossRef] [PubMed]
119. Hisano, M.; Bruschini, H.; Nicodemo, A.C.; Srougi, M. Cranberries and Lower Urinary Tract Infection Prevention. *Clinics* **2012**, *67*, 661–668. [CrossRef] [PubMed]
120. Lynch, D.M. Cranberry for Prevention of Urinary Tract Infections. *Am. Fam. Physician* **2004**, *70*, 2175–2177.
121. Beachey, E.H. Bacterial Adherence: Adhesin-Receptor Interactions Mediating the Attachment of Bacteria to Mucosal Surface. *J. Infect. Dis.* **1981**, *143*, 325–345. [CrossRef]
122. Zafriri, D.; Ofek, I.; Adar, R.; Pocino, M.; Sharon, N. Inhibitory Activity of Cranberry Juice on Adherence of Type 1 and Type P Fimbriated Escherichia Coli to Eucaryotic Cells. *Antimicrob. Agents Chemother.* **1989**, *33*, 92–98. [CrossRef]
123. Howell, A.B.; Reed, J.D.; Krueger, C.G.; Winterbottom, R.; Cunningham, D.G.; Leahy, M. A-Type Cranberry Proanthocyanidins and Uropathogenic Bacterial Anti-Adhesion Activity. *Phytochemistry* **2005**, *66*, 2281–2291. [CrossRef]
124. Liu, Y.; Black, M.A.; Caron, L.; Camesano, T.A. Role of Cranberry Juice on Molecular-Scale Surface Characteristics and Adhesion Behavior of Escherichia Coli. *Biotechnol. Bioeng.* **2006**, *93*, 297–305. [CrossRef]
125. Ren, D.; Zuo, R.; González Barrios, A.F.; Bedzyk, L.A.; Eldridge, G.R.; Pasmore, M.E.; Wood, T.K. Differential Gene Expression for Investigation of Escherichia Coli Biofilm Inhibition by Plant Extract Ursolic Acid. *Appl. Environ. Microbiol.* **2005**, *71*, 4022–4034. [CrossRef] [PubMed]

126. Grace, M.H.; Massey, A.R.; Mbeunkui, F.; Yousef, G.G.; Lila, M.A. Comparison of Health-Relevant Flavonoids in Commonly Consumed Cranberry Products. *J. Food Sci.* **2012**, *77*, H176–H183. [CrossRef]
127. Howell, A.B.; Foxman, B. Cranberry Juice and Adhesion of Antibiotic-Resistant Uropathogens. *JAMA* **2002**, *287*, 3082–3083. [CrossRef] [PubMed]
128. Avorn, J.; Monane, M.; Gurwitz, J.H.; Glynn, R.J.; Choodnovskiy, I.; Lipsitz, L.A. Reduction of Bacteriuria and Pyuria after Ingestion of Cranberry Juice. *JAMA* **1994**, *271*, 751–754. [CrossRef]
129. Howell, A.B.; Botto, H.; Combescure, C.; Blanc-Potard, A.-B.; Gausa, L.; Matsumoto, T.; Tenke, P.; Sotto, A.; Lavigne, J.-P. Dosage Effect on Uropathogenic Escherichia Coli Anti-Adhesion Activity in Urine Following Consumption of Cranberry Powder Standardized for Proanthocyanidin Content: A Multicentric Randomized Double Blind Study. *BMC Infect. Dis.* **2010**, *10*, 94. [CrossRef]
130. Beerepoot, M.; Geerlings, S. Non-Antibiotic Prophylaxis for Urinary Tract Infections. *Pathogens* **2016**, *5*, 36. [CrossRef] [PubMed]
131. Luís, Â.; Domingues, F.; Pereira, L. Can Cranberries Contribute to Reduce the Incidence of Urinary Tract Infections? A Systematic Review with Meta-Analysis and Trial Sequential Analysis of Clinical Trials. *J. Urol.* **2017**, *198*, 614–621. [CrossRef] [PubMed]
132. Jepson, R.G.; Williams, G.; Craig, J.C. Cranberries for Preventing Urinary Tract Infections. *Cochrane Database Syst. Rev.* **2012**, *10*, CD001321. [CrossRef]
133. Gbinigie, O.A.; Spencer, E.A.; Heneghan, C.J.; Lee, J.J.; Butler, C.C. Cranberry Extract for Symptoms of Acute, Uncomplicated Urinary Tract Infection: A Systematic Review. *Antibiotics* **2020**, *10*, 12. [CrossRef] [PubMed]
134. Fu, Z.; Liska, D.; Talan, D.; Chung, M. Cranberry Reduces the Risk of Urinary Tract Infection Recurrence in Otherwise Healthy Women: A Systematic Review and Meta-Analysis. *J. Nutr.* **2017**, *147*, 2282–2288. [CrossRef] [PubMed]
135. Koradia, P.; Kapadia, S.; Trivedi, Y.; Chanchu, G.; Harper, A. Probiotic and Cranberry Supplementation for Preventing Recurrent Uncomplicated Urinary Tract Infections in Premenopausal Women: A Controlled Pilot Study. *Expert Rev. Anti Infect. Ther.* **2019**, *17*, 733–740. [CrossRef] [PubMed]
136. Stapleton, A.E.; Dziura, J.; Hooton, T.M.; Cox, M.E.; Yarova-Yarovaya, Y.; Chen, S.; Gupta, K. Recurrent Urinary Tract Infection and Urinary Escherichia Coli in Women Ingesting Cranberry Juice Daily: A Randomized Controlled Trial. *Mayo Clin. Proc.* **2012**, *87*, 143–150. [CrossRef] [PubMed]
137. Barbosa-Cesnik, C.; Brown, M.B.; Buxton, M.; Zhang, L.; DeBusscher, J.; Foxman, B. Cranberry Juice Fails to Prevent Recurrent Urinary Tract Infection: Results from a Randomized Placebo-Controlled Trial. *Clin. Infect. Dis. Off. Publ. Infect. Dis. Soc. Am.* **2011**, *52*, 23–30. [CrossRef] [PubMed]
138. Maki, K.C.; Kaspar, K.L.; Khoo, C.; Derrig, L.H.; Schild, A.L.; Gupta, K. Consumption of a Cranberry Juice Beverage Lowered the Number of Clinical Urinary Tract Infection Episodes in Women with a Recent History of Urinary Tract Infection. *Am. J. Clin. Nutr.* **2016**, *103*, 1434–1442. [CrossRef] [PubMed]
139. Bruyère, F.; Azzouzi, A.R.; Lavigne, J.-P.; Droupy, S.; Coloby, P.; Game, X.; Karsenty, G.; Issartel, B.; Ruffion, A.; Misrai, V.; et al. A Multicenter, Randomized, Placebo-Controlled Study Evaluating the Efficacy of a Combination of Propolis and Cranberry (Vaccinium Macrocarpon) (DUAB®) in Preventing Low Urinary Tract Infection Recurrence in Women Complaining of Recurrent Cystitis. *Urol. Int.* **2019**, *103*, 41–48. [CrossRef] [PubMed]
140. Vostalova, J.; Vidlar, A.; Simanek, V.; Galandakova, A.; Kosina, P.; Vacek, J.; Vrbkova, J.; Zimmermann, B.F.; Ulrichova, J.; Student, V. Are High Proanthocyanidins Key to Cranberry Efficacy in the Prevention of Recurrent Urinary Tract Infection? *Phytother. Res. PTR* **2015**, *29*, 1559–1567. [CrossRef] [PubMed]
141. Ledda, A.; Bottari, A.; Luzzi, R.; Belcaro, G.; Hu, S.; Dugall, M.; Hosoi, M.; Ippolito, E.; Corsi, M.; Gizzi, G.; et al. Cranberry Supplementation in the Prevention of Non-Severe Lower Urinary Tract Infections: A Pilot Study. *Eur. Rev. Med. Pharmacol. Sci.* **2015**, *19*, 77–80.
142. Kontiokari, T.; Sundqvist, K.; Nuutinen, M.; Pokka, T.; Koskela, M.; Uhari, M. Randomised Trial of Cranberry-Lingonberry Juice and Lactobacillus GG Drink for the Prevention of Urinary Tract Infections in Women. *BMJ* **2001**, *322*, 1571. [CrossRef]
143. Stothers, L. A Randomized Trial to Evaluate Effectiveness and Cost Effectiveness of Naturopathic Cranberry Products as Prophylaxis against Urinary Tract Infection in Women. *Can. J. Urol.* **2002**, *9*, 1558–1562. [PubMed]
144. Beerepoot, M.A.J.; ter Riet, G.; Nys, S.; van der Wal, W.M.; de Borgie, C.A.J.M.; de Reijke, T.M.; Prins, J.M.; Koeijers, J.; Verbon, A.; Stobberingh, E.; et al. Cranberries vs Antibiotics to Prevent Urinary Tract Infections: A Randomized Double-Blind Noninferiority Trial in Premenopausal Women. *Arch. Intern. Med.* **2011**, *171*, 1270–1278. [CrossRef] [PubMed]
145. Babar, A.; Moore, L.; Leblanc, V.; Dudonné, S.; Desjardins, Y.; Lemieux, S.; Bochard, V.; Guyonnet, D.; Dodin, S. High Dose versus Low Dose Standardized Cranberry Proanthocyanidin Extract for the Prevention of Recurrent Urinary Tract Infection in Healthy Women: A Double-Blind Randomized Controlled Trial. *BMC Urol.* **2021**, *21*, 44. [CrossRef] [PubMed]
146. Davis, J.A.; Freeze, H.H. Studies of Mannose Metabolism and Effects of Long-Term Mannose Ingestion in the Mouse. *Biochim. Biophys. Acta* **2001**, *1528*, 116–126. [CrossRef]
147. Fronzes, R.; Remaut, H.; Waksman, G. Architectures and Biogenesis of Non-Flagellar Protein Appendages in Gram-Negative Bacteria. *EMBO J.* **2008**, *27*, 2271–2280. [CrossRef] [PubMed]
148. Choudhury, D.; Thompson, A.; Stojanoff, V.; Langermann, S.; Pinkner, J.; Hultgren, S.J.; Knight, S.D. X-ray Structure of the FimC-FimH Chaperone-Adhesin Complex from Uropathogenic *Escherichia coli*. *Science* **1999**, *285*, 1061–1066. [CrossRef] [PubMed]

149. Zhou, G.; Mo, W.J.; Sebbel, P.; Min, G.; Neubert, T.A.; Glockshuber, R.; Wu, X.R.; Sun, T.T.; Kong, X.P. Uroplakin Ia Is the Urothelial Receptor for Uropathogenic *Escherichia coli*: Evidence from in Vitro FimH Binding. *J. Cell Sci.* **2001**, *114*, 4095–4103. [CrossRef] [PubMed]
150. Pak, J.; Pu, Y.; Zhang, Z.T.; Hasty, D.L.; Wu, X.R. Tamm-Horsfall Protein Binds to Type 1 Fimbriated *Escherichia coli* and Prevents *E. coli* from Binding to Uroplakin Ia and Ib Receptors. *J. Biol. Chem.* **2001**, *276*, 9924–9930. [CrossRef] [PubMed]
151. Eto, D.S.; Jones, T.A.; Sundsbak, J.L.; Mulvey, M.A. Integrin-Mediated Host Cell Invasion by Type 1-Piliated Uropathogenic Escherichia Coli. *PLoS Pathog.* **2007**, *3*, e100. [CrossRef]
152. Mydock-McGrane, L.K.; Cusumano, Z.T.; Janetka, J.W. Mannose-Derived FimH Antagonists: A Promising Anti-Virulence Therapeutic Strategy for Urinary Tract Infections and Crohn's Disease. *Expert Opin. Ther. Pat.* **2016**, *26*, 175–197. [CrossRef]
153. Michaels, E.K.; Chmiel, J.S.; Plotkin, B.J.; Schaeffer, A.J. Effect of D-Mannose and D-Glucose on Escherichia Coli Bacteriuria in Rats. *Urol. Res.* **1983**, *11*, 97–102. [CrossRef] [PubMed]
154. Schaeffer, A.J.; Chmiel, J.S.; Duncan, J.L.; Falkowski, W.S. Mannose-Sensitive Adherence of Escherichia Coli to Epithelial Cells from Women with Recurrent Urinary Tract Infections. *J. Urol.* **1984**, *131*, 906–910. [CrossRef] [PubMed]
155. Spaulding, C.N.; Klein, R.D.; Ruer, S.; Kau, A.L.; Schreiber, H.L.; Cusumano, Z.T.; Dodson, K.W.; Pinkner, J.S.; Fremont, D.H.; Janetka, J.W.; et al. Selective Depletion of Uropathogenic E. Coli from the Gut by a FimH Antagonist. *Nature* **2017**, *546*, 528–532. [CrossRef] [PubMed]
156. Mydock-McGrane, L.K.; Hannan, T.J.; Janetka, J.W. Rational Design Strategies for FimH Antagonists: New Drugs on the Horizon for Urinary Tract Infection and Crohn's Disease. *Expert Opin. Drug Discov.* **2017**, *12*, 711–731. [CrossRef]
157. Lenger, S.M.; Bradley, M.S.; Thomas, D.A.; Bertolet, M.H.; Lowder, J.L.; Sutcliffe, S. D-Mannose vs Other Agents for Recurrent Urinary Tract Infection Prevention in Adult Women: A Systematic Review and Meta-Analysis. *Am. J. Obstet. Gynecol.* **2020**, *223*, 265.e1–265.e13. [CrossRef]
158. Kyriakides, R.; Jones, P.; Somani, B.K. Role of D-Mannose in the Prevention of Recurrent Urinary Tract Infections: Evidence from a Systematic Review of the Literature. *Eur. Urol. Focus* **2021**, *7*, 1166–1169. [CrossRef] [PubMed]
159. Parazzini, F.; Ricci, E.; Fedele, F.; Chiaffarino, F.; Esposito, G.; Cipriani, S. Systematic Review of the Effect of D-Mannose with or without Other Drugs in the Treatment of Symptoms of Urinary Tract Infections/Cystitis (Review). *Biomed. Rep.* **2022**, *17*, 69. [CrossRef] [PubMed]
160. Kranjčec, B.; Papeš, D.; Altarac, S. D-Mannose Powder for Prophylaxis of Recurrent Urinary Tract Infections in Women: A Randomized Clinical Trial. *World J. Urol.* **2014**, *32*, 79–84. [CrossRef] [PubMed]
161. Porru, D.; Parmigiani, A.; Tinelli, C.; Barletta, D.; Choussos, D.; Di Franco, C.; Bobbi, V.; Bassi, S.; Miller, O.; Gardella, B.; et al. Oral D-Mannose in Recurrent Urinary Tract Infections in Women: A Pilot Study. *J. Clin. Urol.* **2014**, *7*, 208–213. [CrossRef]
162. Cooper, T.E.; Teng, C.; Howell, M.; Teixeira-Pinto, A.; Jaure, A.; Wong, G. D-Mannose for Preventing and Treating Urinary Tract Infections. *Cochrane Database Syst. Rev.* **2022**, *8*, CD013608. [CrossRef]
163. Fish, D.N.; Piscitelli, S.C.; Danziger, L.H. Development of Resistance during Antimicrobial Therapy: A Review of Antibiotic Classes and Patient Characteristics in 173 Studies. *Pharmacotherapy* **1995**, *15*, 279–291. [CrossRef] [PubMed]
164. Baraz, A.; Chowers, M.; Nevo, D.; Obolski, U. The Time-Varying Association between Previous Antibiotic Use and Antibiotic Resistance. *Clin. Microbiol. Infect. Off. Publ. Eur. Soc. Clin. Microbiol. Infect. Dis.* **2023**, *29*, 390.e1–390.e4. [CrossRef] [PubMed]
165. Kawalec, A.; Józefiak, J.; Kiliś-Pstrusińska, K. Urinary Tract Infection and Antimicrobial Resistance Patterns: 5-Year Experience in a Tertiary Pediatric Nephrology Center in the Southwestern Region of Poland. *Antibiotics* **2023**, *12*, 1454. [CrossRef] [PubMed]
166. Parry, C.M.; Taylor, A.; Williams, R.; Lally, H.; Corbett, H.J. Antimicrobial Resistance of Breakthrough Urinary Tract Infections in Young Children Receiving Continual Antibiotic Prophylaxis. *Eur. J. Pediatr.* **2023**, *182*, 4087–4093. [CrossRef] [PubMed]
167. Alsubaie, S.S.; Barry, M.A. Current Status of Long-Term Antibiotic Prophylaxis for Urinary Tract Infections in Children: An Antibiotic Stewardship Challenge. *Kidney Res. Clin. Pract.* **2019**, *38*, 441–454. [CrossRef]
168. Pothoven, R. Management of Urinary Tract Infections in the Era of Antimicrobial Resistance. *Drug Target Insights* **2023**, *17*, 126–137. [CrossRef]

antibiotics

Review

Novel Siderophore Cephalosporin and Combinations of Cephalosporins with β-Lactamase Inhibitors as an Advancement in Treatment of Ventilator-Associated Pneumonia

Szymon Viscardi [1,*], Ewa Topola [1], Jakub Sobieraj [1] and Anna Duda-Madej [2,*]

1 Faculty of Medicine, Wroclaw Medical University, Ludwika Pasteura 1, 50-367 Wrocław, Poland; ewa.topola@student.umw.edu.pl (E.T.); jakub.sobieraj@student.umw.edu.pl (J.S.)
2 Department of Microbiology, Faculty of Medicine, Wroclaw Medical University, Chałubińskiego 4, 50-368 Wrocław, Poland
* Correspondence: szymon.viscardi@student.umw.edu.pl (S.V.); anna.duda-madej@umw.edu.pl (A.D.-M.)

Abstract: In an era of increasing antibiotic resistance among pathogens, the treatment options for infectious diseases are diminishing. One of the clinical groups especially vulnerable to this threat are patients who are hospitalized in intensive care units due to ventilator-associated pneumonia caused by multidrug-resistant/extensively drug-resistant Gram-negative bacteria. In order to prevent the exhaustion of therapeutic options for this life-threatening condition, there is an urgent need for new pharmaceuticals. Novel β-lactam antibiotics, including combinations of cephalosporins with β-lactamase inhibitors, are proposed as a solution to this escalating problem. The unique mechanism of action, distinctive to this new group of siderophore cephalosporins, can overcome multidrug resistance, which is raising high expectations. In this review, we present the summarized results of clinical trials, in vitro studies, and case studies on the therapeutic efficacy of cefoperazone-sulbactam, ceftolozane-tazobactam, ceftazidime-avibactam, and cefiderocol in the treatment of ventilator-associated pneumonia. We demonstrate that treatment strategies based on siderophore cephalosporins and combinations of β-lactams with β-lactamases inhibitors show comparable or higher clinical efficacy than those used with classic pharmaceuticals, like carbapenems, colistin, or tigecycline, and are often associated with a lower risk of adverse events.

Keywords: cefoperazone-sulbactam; ceftolozane-tazobactam; ceftazidime-avibactam; cefiderocol; ventilator-associated pneumonia; hospital-associated pneumonia; nosocomial infections; multidrug-resistant; Gram-negative bacteria

Citation: Viscardi, S.; Topola, E.; Sobieraj, J.; Duda-Madej, A. Novel Siderophore Cephalosporin and Combinations of Cephalosporins with β-Lactamase Inhibitors as an Advancement in Treatment of Ventilator-Associated Pneumonia. *Antibiotics* **2024**, *13*, 445. https://doi.org/10.3390/antibiotics13050445

Academic Editor: Marc Maresca

Received: 25 April 2024
Revised: 10 May 2024
Accepted: 12 May 2024
Published: 14 May 2024

1. Introduction

Ventilator-associated pneumonia (VAP) is a major health problem occurring in intensive care units (ICUs), the importance of which may increase due to arising antibiotic resistance. VAP is defined as a pneumonia that arises more than 48–72 h after endotracheal intubation [1]. Half of all cases of hospital-acquired pneumonia (HAP) are due to ventilation, with 9% to 27% of intubated patients at risk of developing this condition and with an event risk of around 1.2 to 8.5 per 1000 ventilation days. The excess cost of VAP treatment is estimated to be over USD 40,000 per patient [1,2]. However, the prevalence may vary depending on the country; a large retrospective study in the USA estimated that 1.81% of mechanically ventilated patients developed VAP [3], while in a study conducted in India, it reached 34% of patients [4]. Also, the mortality of VAP may be different and has been reported to be up to 50%, but this may be distorted by confounding factors, as patients admitted to the ICU have initially serious conditions. Therefore, the attributable mortality is estimated to be 9% to 13% [5], which still makes VAP an important clinical issue.

VAP is usually a bacterial infection. The main pathogens responsible for 80% of ventilator-associated bacterial pneumonia (VABP) are ESKAPE group pathogens: *Enterococcus faecium*, *Staphylococcus aureus*, *Klebsiella pneumoniae*, *Acinetobacter baumannii*, *Pseudomonas aeruginosa*, and *Enterobacter* spp. [6]. Viruses as a sole cause of VAP in the ICU are rare [7]. The most frequently detected are HSV-1 (*Herpes simplex* virus-1) infections (31% of cases) [7] and rhinovirus, influenza A virus and cytomegalovirus (CMV) (5.1% of patients with VAP) [7,8]. However, acute respiratory distress syndrome (ARDS) due to SARS-CoV2 (severe acute respiratory syndrome coronavirus-2) infection can lead to bacterial superinfection and VABP in 21–64% of patients [9]. Even though detection of *Candida* spp. respiratory tract colonization in patients with VAP is associated with higher mortality (45.45% vs. 28.67% in VAP without confirmed colonization, $p < 0.05$), the role of fungi in the pathogenesis of VAP is controversial [10]. Therefore, although other pathogens may contribute to the progress and outcome of VAP, bacteria remain the most important etiological factor.

Because of the fact that bacteria are the main etiological agent associated with VAP, the problem of drug resistance is a real-life risk for these patients. Around 50% of antibiotics used in ICUs are administered because of VABP, but they are often ineffective due to the high proportion of resistant bacteria [11]. They can exhibit a panel of resistance to antimicrobial categories (AMCs): (i) multidrug-resistant (MDR)—nonsusceptible to at least one agent in three or more AMCs; (ii) extensively drug-resistant (XDR)—nonsusceptible to at least one agent in all but two or fewer AMCs; and (iii) pan-drug-resistant (PDR)—nonsusceptible to all antimicrobial agents [12]. Meta-analyses performed showed that *P. aeruginosa* MDR was the etiologic agent of VABP in 33% of cases [13]. However, *A. baumannii* isolates showed resistance rates of 13.3%, 68.3%, and 18.3% for MDR, XDR, and PDR, respectively [14]. In such cases, empiric antibiotic therapy may not be sufficient to completely eradicate the bacterial agent. According to 2016 guidelines from the American Society of Infectious Diseases/American Thoracic Society, a local epidemiological panel should be considered before initiating treatment. However, due to the easy distribution of resistance genes in the hospital environment, appropriate antibiotics may still be limited [15]. In the calculation of this problem, new therapeutic options from the group of β-lactams are promising. There are reports of the high effectiveness of these preparations against carbapenem-resistant *Enterobacterales*, which are an increasing danger for patients hospitalized in the ICU [16,17]. In turn, therapies with cephalosporins, β-lactam antibiotics, whose mechanism of action is to inhibit bacterial cell wall synthesis, are showing promising results. However, as a result of evolution and drug abuse, bacteria have developed resistance to β-lactams, mainly through the production of β-lactamases that inactivate these antibiotics. Recent drugs can combine these antibiotics with β-lactamase inhibitors, reducing bacterial resistance to therapy. These combinations include ceftazidime-avibactam (CAZ-AVI), cefoperazone-sulbactam (CFP-SBT), and ceftolozane-tazobactam (CEF-TAZ). One group of cephalosporins that may contribute to more effective treatment of VABP are the newly developed siderophore cephalosporins, such as cefiderocol (CFD). Table 1 summarizes the abilities of formulations to overcome multidrug resistance among bacteria according to the Ambler classification.

The aim of this study was to evaluate the therapeutic potential of novel drugs in VABP treatment. In this review, we carefully compiled the data on the clinical efficiency of the mentioned cephalosporins combinations with β-lactamase inhibitors and cefiderocol, as well as their in vitro activity against clinical isolates derived from patients with confirmed VABP. The collection of this information is not only important for researchers focused on β-lactams, but, more importantly, we believe that our review will help clinicians offer accurate therapeutic options for treating patients in the ICU.

Table 1. Ability of siderophore cephalosporins and selected cephalosporin-inhibitor combinations to overcome multidrug resistance among bacteria.

Cephalosporin-Inhibitor Combination/Antibiotic	Spectrum of Action According to Ambler Classification	Examples of β-Lactamases	References
Cefoperazone-sulbactam	class A	narrow spectrum: TEM-1 and TEM-2	[18]
	class C	AmpC	
Ceftolozane-tazobactam	class A	extended spectrum (ESβL): SHV-2 and CTX-M-15	[19,20]
	class C	AmpC	
Ceftazidime-avibactam	class A	narrow spectrum: TEM-1	[21,22]
		extended spectrum: SHV and CTX-M	
		KPC-2 and KPC-3	
	class C	AmpC	
	class D	OXA-48	
Cefiderocol	class A	extended spectrum, e.g., SHV type	[23]
		KPC	
	class B	MBL: VIM, IMP, and NDM	
	class C	AmpC	
	class D	OXA-48 and OXA-23	

AmpC—cephalosporinases encoded on the chromosomes of many of the *Enterobacteriaceae*; CTX-M-15—CTX-M-type ESβL; ESβL—extended-spectrum β-lactamase; KPC—*K. pneumoniae* carbapenemase; IMP—metallo-β-lactamase type imipenemase; MBL—metallo-β-lactamase; NDM—New Delhi MBL; OXA—oxacillinase-type β-lactamase; SHV—sulfhydryl variable penicillinase; TEM—plasmid-encoded β-lactamase in Gram-negative bacteria; VIM—Verona integron-encoded MBL.

2. Characteristics of Cefoperazone-Sulbactam

CFP-SBT is a combination of the antibiotic cefoperazone, a IIIrd-generation cephalosporin, and a class A β-lactamase inhibitor, sulbactam [18]. It has significant activity against Gram-negative bacteria, e.g., *Enterobacterales*, *Acinetobacter* spp., and *P. aeruginosa* [24]. This combination increases the effectiveness of the antibiotic against MDR bacteria, except *P. aeruginosa* resistant to carbapenems. CFP-SBT is used at antibiotic-to-inhibitor ratios of 2:1, 1:1, and 1:2, which lead to increasing reductions in the MIC (minimal inhibitory concentration) [25]. Moreover, bacterial resistance to the combination in the 1:1 ratio is lower compared to the tested combination in the 2:1 ratio [26]. Sulbactam addition in the 1:1 ratio reduced the MIC_{90} > 128 mg/L for *Enterobacter cloacae* (MIC_{90} = 32 mg/L; MIC_{90}—the lowest concentration of the antibiotic at which 90% of the isolates were inhibited); *Serratia marcescens* (MIC_{90} = 32 mg/L), *K. pneumoniae* (ESβL: 32 mg/L; non-ESβL: 4 mg/L); and *A. baumannii* (imipenem resistant: 32 mg/L; imipenem susceptible: 16 mg/L). The 2:1 ratio also reduced the bacterial MIC [27]. In vitro tests showed that the tested combination is active against *Acinetobacter* spp., with an MIC value of 1.0 μg/mL [28]. Potential uses of this drug include infections of the lower and upper respiratory systems, infections of the lower and upper urinary systems, intra-abdominal infections, sepsis, skin and soft tissue infections, joint and bone infections, and bacterial infections of the genitals, e.g., gonorrhea. The indications listed are part of Pfizer's guidelines. The combination should be administered in a 1:1 or 1:2 ratio. For a 1:1 ratio, 1–2 g of sulbactam and 1–2 g of cefoperazone are used, and for a 1:2 ratio, 0.5–1 g of sulbactam and 1–2 g of cefoperazone. Doses should be administered to the patient every 12 h. The described mode of administration is intended for adults and is part of Pfizer's guidelines. A very important issue is the degree of drug penetration into the lungs. A group of patients (121 people) with infections caused by *A. baumannii* obtained CFP-SBT at a dose of 3 g every 8 h, but the effectiveness of the drug decreased over time. It is possible to administer a higher dose (4 g), which may improve the quality of treatment [29]. Adverse events connected with CFP-SBT treatment include nephrotoxicity, thrombocytopenia, leukopenia, increased liver enzyme activity, prolonged prothrombin time (PT), and increased international normalized ratio (INR) [30]. However, the most common side effect is a rash (which occurs in 10.1% of patients), as well as vomiting (4.4%) [31].

Cefoperazone

Cefoperazone (CFP)—(6R,7R)-7-[[(2R)-2-[(4-ethyl-2,3-dioxopiperazine-1-carbonyl) amino]-2-(4-hydroxyphenyl)acetyl]amino]-3-[(1-methyltetrazol-5-yl)sulfanylmethyl]-8-oxo-5-thia-1-azabicyclo[4.2.0]oct-2-ene-2-carboxylic acid (IUPAC name according to PubChem) is a semisynthetic, IIIrd-generation cephalosporin. It is an antibiotic intended for intramuscular or intravenous (i.v.) infusions. In contrast to Ist- and IInd-generation cephalosporins, it is active against *P. aeruginosa* and shows greater stability to hydrolysis by bacterial enzymes. Its action also covers Gram-positive and Gram-negative bacteria, including *Enterobacteriaceae* and anaerobes [32]. The combination of sulbactam with this cephalosporine expands its spectrum to, among others, *Bacteroides* spp. and *Acinetobacter* spp. [33]. The structural model of cefoperazone is shown in Figure 1A.

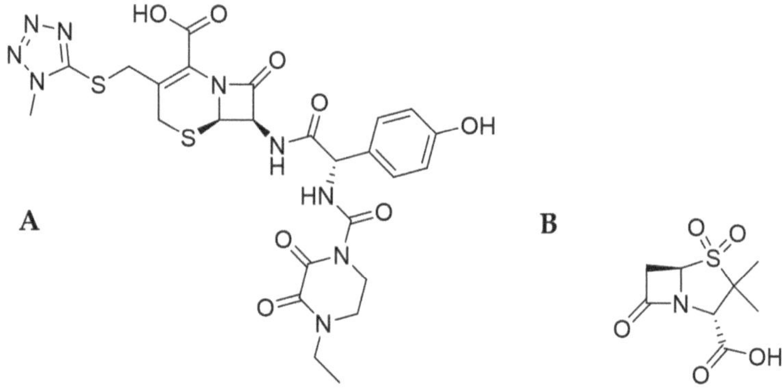

Figure 1. Structural models of cefoperazone (**A**) and sulbactam (**B**).

Sulbactam

Sulbactam (SBT)—(2S,5R)-3,3-dimethyl-4,4,7-trioxo-4lambda6-thia-1-azabicyclo [3.2.0] heptane-2-carboxylic acid (IUPAC name according to PubChem), acts on bacteria with an ESβL-resistance mechanism (such as *E. coli* or *K. pneumoniae*) and AmpC (cephalosporinases encoded on the chromosomes of many of the *Enterobacteriaceae*), as well as against *Acinetobacter* resistant to carbapenems [18,26]. On the molecular level, SBT contains a β-lactam ring in its structure. Sulbactam presented significant activity on bacterial enzymes transferred by plasmids [33]. Studies on *A. baumannii* showed that sulbactam also inhibits penicillin-binding proteins (PBPs) (mainly PBP1 and PBP3), which are key proteins in bacterial cell wall synthesis. This action leads to the death of bacterial cells [34,35]. Resistance to sulbactam is influenced by TEM β-lactamases, OXA-23 enzymes, and ADC (*Acinetobacter*-derived cephalosporinase). Mutations in PBPs can significantly increase sulbactam's MIC [29]. Importantly, SBT has an impressive effect on plasmid-mediated β-lactamases in contrast to CFP alone [28]. The structural model of sulbactam is shown in Figure 1B.

2.1. Clinical Efficiency of Cefoperazone-Sulbactam

2.1.1. In Vitro Studies

The Namiduru et al. study evaluated the in vitro activity of available antibiotics against 230 clinical strains of bacteria isolated from patients with confirmed VABP between 2001 and 2003. The predominant pathogens were *P. aeruginosa* (33.9%), *S. aureus* (30%), *A. baumannii* (26.1%), and *Enterobacter* spp. (4.3%). The study showed that the highest efficacy against *P. aeruginosa* and *A. baumannii* isolates was obtained by CFP-SBT (80.7% vs. 80%, respectively) and imipenem (*A. baumannii* 83.3%) [36]. Similar results come from the Xia et al. study conducted in a population of pediatric patients (N = 94) hospitalized in an ICU with a confirmed diagnosis of VABP. Dominant pathogens isolated from the respiratory tract were identified as MDR isolates of *K. pneumoniae* and *A. baumannii* (78%). The highest activity against *A. baumannii* was detected (similarly to the Namiduru et al. study above) in the case of CFP-SBT and imipenem [37]. Similar reports to those from Xia et al. and Namiduru et al. come from the Capoor et al. study, in which sensitivity to antibiotics was assessed among 128 clinical isolates (derived from patients diagnosed with VABP). The predominant isolated pathogens were *Acinetobacter calcoaceticus-Acinetobacter baumannii* complex. CFP-SBT was found to be the most effective against the pathogen, with 95.6% of strains being effectively treated with it, along with MER and PIP-TAZ [38].

Another study evaluated the in vitro susceptibility to CFP separately and in combination with SBT against bacterial isolates from patients hospitalized (N = 1796 Gram (−) isolates and 476 Gram (+)) in several hospitals in Taiwan in 2012. The most frequently isolated pathogens identified as etiological factors of VABP/HAP in ICU wards were MSSA (methicillin-susceptible *S. aureus*) (10.65%), *E. coli* (11%), *K. pneumoniae* (10.17%), *A. baumannii* (16.51%), *P. aeruginosa* (7.83%), and *Stenotrophomonas maltophila* (3.83%). The study showed a higher activity of the combination of CFP-SBT over CFP alone for MSSA (MIC_{50} = 2 μg/mL vs. 4 μg/mL; MIC_{50}—the lowest concentration of the antibiotic at which 50% of the isolates were inhibited) for Gram (+) pathogens. Pathogens of the ESKAPE group were also characterized by a higher sensitivity to the combination with the inhibitor *K. pneumoniae* and *P. aeruginosa* (MIC_{90} = 32 vs. 128/mL), *A. baumannii* (MIC_{90} = 64 vs. >128 μg/mL), and *S. maltophila* (MIC_{50} = 64 vs. 128 μg/mL). The CFP-SBT showed elevated activities against *K. pneumoniae* and *E. coli* ESβL strains, as well as against CRAB and CRPA (carbapenem-resistant *P. aeruginosa*) in comparison to CFP alone [39].

In another study, Huang et al. evaluated the in vitro efficacy of combinations of azithromycin (AZT) with CFP-SBT for clinical isolates of MDR *P. aeruginosa* collected from 151 ICU patients diagnosed with VABP. The strains were characterized by a high degree of resistance to, among others, aztreonam (ATM) or imipenem. The study included a total of four schemes of empirical therapy containing CFP-SBT, as follows: CFP-SBT + AZT, CFP-SBT + amikacin (AKC), CFP-SBT + levofloxacin (LEV), and CFP-SBT + AZT + LEV/AKC combination. The highest effectiveness was achieved by combining CFP-STB with AKC (70%) and AZT (60%). The combination with AZT allowed for a decrease in the MIC_{50} for CFP-SBT from 64 to 16 μg/mL and MIC_{90} from 128 to 64 μg/mL. It has been shown that both the combinations of the preparation with AZT and AZT + AKC provide an additive effect. The results indicate that the empirically tested combination of BL + BLI (β-lactam + β-lactamase inhibitors) with AZT may have potential benefits in treating VABP caused by *P. aeruginosa* [40].

Xia et al. evaluated the in vitro efficacy of CFP-SBT therapy in combination with other antibiotics against CRAB isolates derived from patients hospitalized in the ICU (N = 71, of which 45 were subjected to mechanical ventilation). Patients qualified for the study, due to the lack of a possibility of receiving polymyxin therapy (renal failure), were randomized into the following two groups: those receiving and not receiving a CPF-SBT therapy-based scheme. A higher 30-day survival ratio was demonstrated in the CFP-SBT-receiving population (96.4% vs. 73.3%). In vitro studies showed a high percentage of expression of the following *A. baumannii* carbapenemases: OXA-23 and OXA-51. The evaluation of antibiotic sensitivity involved choosing multiple combinations with additive/synergistic properties, and the combination of CFP-SBT + MER was revealed to be the most active connection (MIC_{50} = 16 μg/mL; MIC_{90} = 64 μg/mL) [41]. The Sader et al. study evaluated the in vitro susceptibility of *Enterobacteriaceae* isolates (including *P. aeruginosa* (3818) and *A. baumannii* (1310)) to CFP-SBT. A total of 28.3% of all isolates were collected from patients diagnosed with HAP/VABP. The susceptibility profile of *P. aeruginosa* to the preparation was defined in the range of 59.5–83%, of which the highest activity was recorded in relation to strains from Western Europe and Latin America (83%). In turn, the sensitivity of *A. baumannii* to the preparation was relatively lower at 43–73.8% [24]. A summary of the in vitro antimicrobial activity of cefoperazone-sulbactam is presented in Table 2.

Table 2. In vitro antimicrobial activity of CFP-STB and its combinations against VABP clinical isolates.

Pathogen	Antibiotic Scheme Used In Vitro	MIC Value	Reference
MSSA	CFP-SBT	MIC_{50} = 2 μg/mL	[39]
	CFP	MIC_{50} = 4 μg/mL	
K. pneumoniae	CFP-SBT	MIC_{90} = 32 μg/mL	
	CFP	MIC_{90} = 128 μg/mL	
P. aeruginosa	CFP-SBT	MIC_{90} = 32 μg/mL	
	CFP	MIC_{90} = 128 μg/mL	
A. baumannii	CFP-SBT	MIC_{90} = 64 μg/mL	
	CFP	MIC_{90} > 128 μg/mL	
S. maltophila	CFP-SBT	MIC_{50} = 64 μg/mL	
	CFP	MIC_{50} = 128 μg/mL	
MDR *P. aeruginosa*	CFP-SBT	MIC_{50} = 64 μg/mL	[42]
		MIC_{90} = 128 μg/mL	
	CFP-SBT + AZT	MIC_{50} = 16 μg/mL	
		MIC_{90} = 64 μg/mL	
Carbapenem-resistant *A. baumannii*	CFP-SBT + MER	MIC_{50} = 16 μg/mL	[37]
		MIC_{90} = 64 μg/mL	

AZT—azithromycin; CFP-SBT—cefoperazone-sulbactam; MDR—multidrug-resistant; MER—meropenem; MSSA—methicillin-susceptible *S. aureus*.

2.1.2. Clinical Trials

The randomized noninferiority clinical trial was conducted to evaluate the therapeutic efficacy of CFP-SBT vs. cefepime (CPM) for the treatment of HAP and healthcare-associated pneumonia (HCAP). Ultimately, 166 patients were randomized into the following two groups: those treated with CFP-SBT (N = 79) and with CPM (N = 87). Patients were assessed twice during the test-of-cure (TOC) in the following phases: intention-to-treat (ITT) and per protocol (PP). Clinical success has been shown for patients receiving CFP-SBT and CPM treatment after approximately 10 days in both the ITT and PP analyses. In turn, the percentages of clinical cures with CFP-SBT therapy obtained better parameters in contrast to CPM, as follows: clinical cure at ITT analysis = 73.1% vs. 56.8% and clinical cure at PP analysis = 74.2% vs. 56.8%. Among the sputum-isolated pathogens obtained from patients treated with CFP-SBT, the presences of *P. aeruginosa* (100% effective therapy) and *A. baumannii* (75% effectiveness) were found. Both formulations achieved a similar adverse event profile (CFP-SBT 58/79, of which 20 were directly drug related; CPM 58/87, of which 15 were directly drug related). It was also shown that in the case of patients suffering from HAP, CFP-SBT showed a lower effectiveness than CPM in terms of the percentage of clinical cure (~43% vs. ~88%). The study proves that the preparation is noninferior to CPM activity and is an interesting option for HAP/VABP therapy [43].

Another study compared the efficacy and safety profile of CFP-SBT and piperacillin-tazobactam (PIP-TAZ) in elderly patients hospitalized due to, among others, HAP/VABP. Patients were randomly assigned into the following two groups: those receiving PIP-TAZ (N = 150) and those receiving CFP-SBT (N = 167); the most commonly isolated pathogens were *K. pneumoniae*, *P. aeruginosa*, and *A. baumannii*. Clinical cure was achieved in both groups at 85 vs. 81%, respectively and clinical failure at 14 and 17%, respectively. The study also revealed the in-hospital mortality ratio (including pneumonia-related mortality), it was 40/167 (including 24 due to pneumonia) and 32/150 (of which 14 related to pneumonia) for CFP-SBT and PIP-TAZ, respectively. In the scope of TEAE (treatment-emergent adverse event), both preparations obtained comparable outcomes. Diarrhea was the predominant type in the PIP-TAZ group (N = 2). In turn, patients treated with CFP-SBT developed hepatitis, rash, and hemorrhage (in total, N = 3) [42]. The Chen et al. study compared the effectiveness of therapy, among others, in VABP in patients treated with CFP-SBT (N = 37) or PIP-TAZ (N = 28). The most commonly isolated pathogens in both groups were *A. baumannii*, *P. aeruginosa*, and *P. aeruginosa*. In the study, similar results were obtained for clinical cure in patients hospitalized due to VABP for both CFP-SBT and PIP-TAZ at 78.4 vs. 71.4% and treatment failures at 16.2 vs. 28.6%. However, the mortality ratio due to pneumonia was higher for CFP-SBT therapy vs. PIP-TAZ (13.5 vs. 3.6%). Nevertheless, the results of the study clearly show that cefoperazone sulbactam is equally effective in VABP therapy as its comparator, PIP-TAZ [44].

The Guclu et al. study evaluated the efficacy of CFP-SBT and PIP-TAZ therapy in the treatment of nosocomial infections in patients hospitalized at the ICU of Sakarya University Training and Research Hospital in the years 2017–2018. Patients were randomly assigned to the following two groups: those receiving empirical therapy based on CFP-SBT or PIP-TAZ (number in each group, N = 154). The percentages of patients hospitalized for HAP/VABP in both groups were 17.5% and 22.5%, respectively. Pathogens classified as MDR accounted for more than 70% of all etiological factors of diseases, the predominant species included *K. pneumoniae*, *E. coli*, *P. aeruginosa*, and *A. baumannii*. Both preparations had similar therapeutic parameters, as follows: treatment success—50% vs. 51.2%; 14-day mortality ratio—29.2% vs. 35%; 28-day mortality ratio—46.1% vs. 42.8%; and TEAEs frequency—50.6% vs. 46.1%. The most common adverse events were increased INR value, thrombocytopenia, hepatotoxicity, and nephrotoxicity. As it was demonstrated by other studies in this field [42,44], CFP-SBT had a noninferior activity profile compared to PIP-TAZ in the empirical therapy of HAP/VABP [30].

Kara et al. evaluated the efficacy of colistin (COL) monotherapy and its combinations with other antibiotics in VABP therapy. Between 2009 and 2014, 134 ICU patients were enrolled in the study, and *A. baumannii* was the most common etiological factor for VABP. One of the tested combinations was the combination of high doses of CFP-SBT with COL, which resulted in a relatively low effectiveness (median duration of therapy: 6 days; treatment failure: >70%; and mortality ratio: 15/17 of the patients); however, the results were comparable to the effectiveness of the combination of COL + tigecycline (TGC) or COL + ampicillin-sulbactam [45]. The Qin et al. study evaluated the effect of a high-dose combination of CFP-SBT with TGC on the efficacy of VABP therapy with the etiology of XDR *A. baumannii*. Patients were randomized into the following two groups: receiving a combination of both drugs and receiving TGC as monotherapy. The study showed a higher efficacy of CFP-SBT + TGC scheme vs. TGS therapy separately (85.7% vs. 47.6% clinical success ratio) and also demonstrated the ability of CFP-SBT to lower the MIC of TGC. In both study groups, there were no significant differences in the incidence of TEAE [46]. Another study presented interesting reports on the effectiveness of VABP therapy of etiology of carbapenem-resistant *A. baumannii* (CRAB) using CFP-SBT. A total of 80 patients were enrolled in the study and randomized into the following two groups: those who received CFP-SBT as an adjunct to COL, meropenem (MER), or TGC (N = 52); those who received BAT (best available therapy; drug scheme without CFP-SBT), N = 38. Significantly, differences in mortality were found in both groups, as follows: 14-day mortality ratio—17 vs. 39%; 30-day mortality ratio—35 vs. 61%; in-hospital mortality ratio—39 vs. 68%. A similar frequency of adverse events has been demonstrated, and AKI (acute kidney injury) and bone marrow aplasia were the most common AEs (adverse events). According to the study, adjuvant CFP-SBT therapy has a significant impact on survival among patients with VABP with a CRAB etiology [47].

The efficacy of TGC monotherapy or in combination with CFP-SBT in patients (N = 114) with a diagnosis of lower respiratory tract infection (LRTI) with the etiology of MDR *A. baumannii* was assessed in the Qin et al. study. Patients were randomly divided into the following two groups: those who received CFP-SBT monotherapy and those treated with a combination therapy (both groups equally, N = 57). The end-point of the study was a decrease in the concentration of inflammatory factors in the serum on the 14th day after the start of therapy (PCT—procalcitonin; CRP—C-reactive protein; IL-6—interleukine-6; and TNF-α—tumor necrosis factor α) and evaluation on the APACHE II scale (Acute Physiology and Chronic Health Evaluation II). The combination of TGC with CFP-SBT has been shown to reduce systemic inflammation (PCT, CRP, IL-6, and TNF-α had lower values in this group of subjects). Similarly, improvements in patients were reported on the APACHE II scale. The results clearly indicate the advantage of combination therapy over CFP-SBT monotherapy in the treatment of MDR *A. baumannii* etiological VABP [48].

A summary of clinical efficiency of cefoperazone-sulbactam treatment among patients with VABP is presented in Table 3.

Table 3. Efficacy of CFP-SBT in the treatment of VABP.

Number of Patients	Treatment Scheme in VABP Population	Outcomes of Trial		Etiology of Infection	Reference
N = 166	CFP-SBT (N = 79)	CC:	73.1%	*P. aeruginosa* *A. baumannii*	[43]
	CPM (N = 87)		56.8%		
N = 317	CFP-SBT (N = 167)	CC	85%	*K. pneumoniae* *P. aeruginosa* *A. baumannii*	[42]
			81%		
	PIP-TAZ (N = 150)	IHMR	24%		
			23%		
N = 65	CFP-SBT (N = 37)	CC in VABP group:	78.4%	*A. baumannii*	[44]
			71.4%	*P. aeruginosa*	
	PIP-TAZ (N = 28)	Total mortality ratio:	13.5%	*K. pneumoniae*	
			3.6%	*P. aeruginosa*	
N = 308	CFP-SBT (N = 154)	CC:	50%	MDR (70%): *K. pneumoniae* *E. coli* *P. aeruginosa* *A. baumannii*	[30]
			51.2%		
		14-MR:	29.2%		
			35%		
	PIP-TAZ (N = 154)	28-MR:	46.1%		
			42.8%		
N = 42	CFP-SBT + TGC	CC:	85.7%	XDR *A. baumannii*	[46]
	TGC		47.6%		

Table 3. *Cont.*

Number of Patients	Treatment Scheme in VABP Population	Outcomes of Trial		Etiology of Infection	Reference
N = 80	Combination therapy with CFP-SBT (N = 52)	14-MR	17%	Carbapenem-resistant *A. baumannii*	[47]
			39%		
		30-MR	35%		
			61%		
	Combination therapy without CFP-SBT (N = 38)	IHMR	39%		
			68%		

14/28/30-MR—14/28/30-day mortality ratio; CC—clinical cure; CFP-SBT—cefoperazone-sulbactam; CPM—cefepime; IHMR—in-hospital mortality ratio; MDR—multidrug-resistant; PIP-TAZ—piperacillin-tazobactam; TGC—tigecycline; VABP—ventilator-associated bacterial pneumonia; XDR—extensively drug-resistant.

3. Characteristics of Ceftolozane-Tazobactam

CEF-TAZ is a novel combination of the β-lactam antibiotic-ceftolozane and the β-lactamase inhibitor-tazobactam. It presents a promising effect in the treatment of MDR bacteria of the *Enterobacteriaceae* genus, including ESβL strains [49]. Furthermore, CEF-TAZ is also highly efficient against MDR *P. aeruginosa* and become an important option of treatment for infection of this etiology and may be used empirically [50]. According to the EMA (European Medicines Agency) guidelines, the indications for treatment using the combination are complicated urinary tract infections (cUTIs), including acute pyelonephritis (AP) and cystitis; complicated intra-abdominal infections (cIAIs); lung infections associated with mechanical ventilation (VABP); and HAP. Similar indications are presented by the FDA (Food and Drug Administration), taking into account mainly cUTIs and cIAIs. The drug should be administered every 8 h for 1 h lasting i.v. infusion at a dose of 1.5 g (in the ratio of 1 g of CEF and 0.5 g of tazobactam). The duration of treatment for cUTIs should be 7 days and for cIAIs 4–14 days. CEF-TAZ is capable of overcoming the efflux pump's resistance mechanism and also porins mutations, which are increasingly frequent among MDR bacteria. However, it is inactivated by some β-lactamase enzymes such as carbapenemases [51]. Lung penetration by CEF-TAZ was assessed compared to PIP-TAZ. The concentration in epithelial lining fluid (ELF) was more than 8 mg/L, indicating that the growth in the microorganism in this case, *P. aeruginosa,* should be inhibited [52]. In the Sheffield et al. study, a lack of serious adverse events were reported among patients treated with CEF-TAZ, except for one patient, who experienced a worsening of a gout attack. The other most common side effects were headaches and digestive system problems, e.g., nausea and vomiting [51,53].

Ceftolozane

Ceftolozane (CEF) is a new β-lactam antibiotic from the group of Vth-generation cephalosporins, and its name, according to IUPAC, is (6R,7R)-3-[[3-amino-4-(2-aminoethylcarbamoylamino)-2-methylpyrazol-1-ium-1-yl]methyl]-7-[[(2Z)-2-(5-amino-1,2,4-thiadiazol-3-yl)-2-(2-carboxypropane-2-yloxyimino)acetyl]amino]-8-oxo-5-thia-1-azabicyclo[4.2.0]oct-2-ene-2-carboxylate (PubChem). This preparation showed a promising activity profile against MDR strains and the relatively rare presence of adverse events [54]. It is highly active against Gram-negative bacteria (GNB) and presented an 8–16 times stronger activity against MDR strains than, for example, CPM or CAZ. CEF is also characterized by greater affinity to PBPs: 1b; 1c 2 and 3. The preparation was also shown to be superior to CAZ's eradication activity against numerous bacterial strains [55]. The structural model of ceftolozane is shown in Figure 2A.

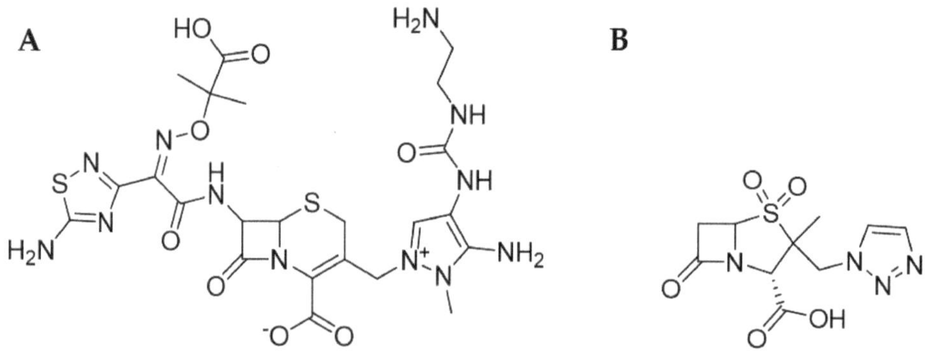

Figure 2. The structural models of ceftolozane (**A**) and tazobactam (**B**).

Tazobactam

Tazobactam (TAZ) is a β-lactamase inhibitor with the following IUPAC name: (2S,3S,5R)-3-methyl-4,4,7-trioxo-3-(triazol-1-ylmethyl)-4lambda6-thia-1-azabicyclo[3.2.0]heptane-2-carboxylic acid (PubChem). Molecule of TAZ is characterized by presence of a β-lactam ring in its structure. Tazobactam is a β-lactamase inhibitor that acts on the active site of the enzyme by irreversibly binding to it [52]. The combination of CEF with TAZ is highly effective against MDR microorganisms producing Ambler A class β-lactamases like ESβL, e.g., CTX-M-14 and CTX-M-15 [52]. However, it is inactive against bacteria with the expression of KPC and Ambler B class enzymes such as metallo-β-lactamases [19]. Research by Haidar et al. reported that among a group of 21 patients with MDR *P. aeruginosa* infections treated with CEF-TAZ, 71% achieved clinical cure. However, a worrying fact is that in three cases, the development of a resistance mechanism to the preparation was observed, and an increased expression of AmpC was also detected [56]. The structural model of tazobactam is shown in Figure 2B.

3.1. Clinical Efficiency of Ceftolozane-Tazobactam

3.1.1. In Vitro Studies

Perez and coworkers presented findings of a study conducted as part of the MagicBullet trial, which was performed to assess the antibiotic resistance and efficacy of treatment for infections caused by XDR/MDR *P. aeruginosa* strains that occurred in several European countries, such as Greece, Italy, and Spain. In 2012–2015, a total of 121 *P. aeruginosa* isolates were collected, the highest numbers of PDR, MDR, and XDR isolates were identified in Greece, and the isolates were characterized by 22.6% resistance to CEF-TAZ and 24.5% resistance to CAZ-AVI. Resistant isolates expressed the VIM-2 MBL gene. In general, it was shown that the most active anti-*Pseudomonas* drugs in the study were COL (94.3% activity), CEF-TAZ (77.4%), and CAZ-AVI (75.5%) [57]. Sader et al. assessed the in vitro effectiveness of murepavadine against clinical isolates of XDR *P. aeruginosa* from 21 European countries as part of the SENTRY Antimicrobial Surveillance Program 2016–2017. Bacteria that were isolated from patients, including HAP/VABP, accounted for 63% of all isolates. In the general population of bacterial isolates, CEF-TAZ tested as a comparator showed an average MIC_{50} and MIC_{90} at the levels of 2 and >32 mg/mL, respectively, presenting the activity against 70.6% of isolates according to EUCAST (European Committee on Antimicrobial Susceptibility Testing) (COL MIC_{50} = 1 mg/mL; MIC_{90} = 2 mg/mL; activity = 93.6%). In the group of isolates from North America (N = 432), CEF-TAZ showed a higher activity of 86.8%, with MIC_{50} and MIC_{90} values of 1 and 8 mg/mL, respectively [58].

Carvalhaes et al. examined the in vitro susceptibility of, among others, VABP-derived clinical isolates of *P. aeruginosa* (N = 1531) and other *Enterobacteriaceae* (N = 2373) on CEF-TAZ using the microdilution method. The preparation exhibited an MIC_{50} and MIC_{90} in relation to the general population of *P. aeruginosa* at the levels of 0.5 and 2 mg/L and in the cases of the MDR and XDR strains with MIC_{50}/MIC_{90} values of 1/8 and 2/16 mg/L. In turn, the activity against non-CRE (noncarbapenem-resistant *Enterobacterales*) *E. coli* and *K. pneumoniae* was as follows: 0.5/2 mg/L and 1/8 mg/L. According to EUCAST/CLSI (Clinical and Laboratory Standards Institute) guidelines, CEF-TAZ showed activity against 97.5% of *P. aeruginosa* isolates, making it the second most active after COL at 99.9%. The activities of CEF-TAZ against MDR and XDR *Pseudomonas* strains were similar at 87.9% and 82.9%. The activities against non-CRE *E. coli* and *K. pneumoniae* isolates oscillated around 80 and 90%, respectively [19]. This is reflected in a study conducted by Pfaller et al., which showed that CEF-TAZ is active against ESβL non-CRE *Enterobacterales* isolates between 81.4% and 89.4% (MIC_{50}/MIC_{90} in the range of 0.5–8 mg/L). However, there was no significant effect on CRE isolates [59]. It is important is that among all tested BL and BL + BLI, CEF-TAZ showed the highest activity against MDR/XDR *P. aeruginosa* isolated from patients with VABP. This is a significant premise in the era of increasing antibiotic resistance among pathogens in the ESKAPE group [19].

The Idowu et al. study evaluated the in vitro efficacy of CEF-TAZ alone and in combination with a TOB homodimer against MDR and XDR *P. aeruginosa* strains. The study proved that the combination results in an increase in CEF-TAZ activity against *Pseudomonas* isolates and also prevents the development of resistance to the preparation, surpassing CAZ-AVI in this aspect [60].

The Karlowsky et al. study evaluated the susceptibility of bacterial isolates, including MDR *P. aeruginosa*, *K. pneumoniae*, and *E. coli*, from patients with LRTI (SMART—Study for Monitoring Antimicrobial Resistance Trends, conducted in the USA in 2018–2019). A total of 1237 *P. aeruginosa* isolates were tested. A high activity of CEF-TAZ against the pathogens was demonstrated (96%), which exceeded all other tested antibiotics, except AKC (96%). There was also a significant activity against *E. coli* and *K. pneumoniae* isolates (97.2% and 92.6%, respectively). Importantly, the preparation in these cases showed a superior activity profile than typical anti-*Pseudomonas* preparations, e.g., PIP-TAZ and CAZ. The medicament was active at the level of 81.6% and 86.8% against *P. aeruginosa* strains resistant to CAZ and PIP-TAZ, respectively. For ESβL *K. pneumoniae* and *E. coli* strains, the preparation was less active (67.1% and 86.5%). In turn, CEF-TAZ showed the highest activity against the following MDR strains: *E. coli* at 82.2% and *P. aeruginosa* at 76.2%, among other tested comparators. It was also observed that CEF-TAZ by its activity (96%) is equal to AKC and significantly exceeds PIP-TAZ (72.3%), CPM (78%), CAZ (76.6%), and even MER (72.9%) against *P. aeruginosa* isolates derived from ICU patients. In general, the treatment was equally effective when compared to AKC, which, when considering the risk of aminoglycoside nephrotoxicity, supports the CEF-TAZ treatment option [61]. Similar conclusions can be drawn from the Shortridge et al. study, which demonstrated that the in vitro activity against 1345 *P. aeruginosa* isolates derived from HAP/VABP patients was the highest in the monotherapies based on COL (>99%), AKC (98.1%), and CEF-TAZ (96.5%) [62]. These results correlate with another evaluation of *P. aeruginosa*'s sensitivity to CEF-TAZ and comparators [63]. Furthermore, CEF-TAZ monotherapy has been shown to be superior to CPM, CAZ, MER, and PIP-TAZ alone. Moreover, these BLs presented as comparable to the CEF-TAZ activity only in combination with AKC/COL [62].

The SMART study also assessed the susceptibility of *Enterobacteriaceae* (720 *P. aeruginosa* bacterial isolates, 338 *K. pneumoniae* isolates, and 291 *E. coli* isolates derived from ICU patients with confirmed LRTI (including VABP)) to, among others, CEF-TAZ. The study showed that *P. aeruginosa* isolates collected from ICU ward patients were characterized by the highest sensitivity to CEF-TAZ among the tested comparators, with activity inferior only to AKC (94% vs. 96%). In addition, relatively rare resistance to the preparation was noted in contrast to other BL and BL + BLI comparators (CEF-TAZ = 6%; CAZ = 27.4%; and PIP-TAZ = 31.7% resistance rates). It is worth emphasizing that CEF-TAZ showed significant activity against strains resistant to, among others, MER (83.1%), PIP-TAZ (82.5%), or CAZ (78.2%). It is also worth noting, that the preparation had relatively high activity against MDR *P. aeruginosa* (71.4% of strains from ICU) and pan-β-lactam-non-susceptible

P. aeruginosa (resistant to IIIrd- and IVth-generation cephalosporins, ATM, carbapenems, and PIP-TAZ), where it was effective against 65.6% of isolates (collected from ICU patients) [64].

The Candel et al. study evaluated the in vitro susceptibility of bacterial pathogens, including isolates from patients suffering from VABP to cefiderocol and comparators, including CEF-TAZ. A total of 20,911 isolates were collected, of which 34.4% were from VABP infections. *Enterobacteriaceae* accounted for 51.2% of all isolates, the most numerous species were *K. pneumoniae* and *E. coli*. Nonfermenting pathogens constituted 48.8% of isolates, the most numerous were *P. aeruginosa*, *A. baumannii*, and *S. maltophila*. There was no activity against *A. baumannii* and *S. maltophila* from the CEF-TAZ, while the preparation showed significant activity in relation to isolates sensitive to the carbapenems *E. coli*, *K. pneumoniae*, and *P. aeruginosa* (97%, 86.9%, and 97.5%, respectively). Among the CRE isolates, a lower activity was observed. A moderate susceptibility rate was observed only in the case of *P. aeruginosa* (CRPA), at the level of 48.1%. The study proved the increasing resistance to CEF-TAZ among ESKAPE pathogens, while cefiderocol is a promising response to this trend (with a high activity against carbapenem-resistant *S. maltophila* and *P. aeruginosa* of the order of >90%). This well reflects the MIC_{50}/MIC_{90} values for the pathogens mentioned, respectively, for cefiderocol vs. CEF-TAZ: *P. aeruginosa*—0.25/0.5 mg/L vs. 0.5/8 mg/L and *K. pneumoniae*—0.25/2 mg/L vs. 0.5/> 64 mg/L [65]. A summary of the in vitro antimicrobial activity of ceftolozane-tazobactam is presented in Table 4.

Table 4. In vitro antimicrobial activity of CEF-TAZ and its combinations against VABP clinical isolates.

Pathogen	Antibiotic Scheme Used In Vitro	MIC Value	Reference
XDR *P. aeruginosa*	CEF-TAZ	MIC_{50} = 2 mg/mL MIC_{90} > 32 mg/mL	[58]
		MIC_{50} = 1 mg/mL MIC_{90} = 8 mg/mL	
MDR *P. aeruginosa*	CEF-TAZ	MIC_{50} = 1 mg/L	[19]
		MIC_{90} = 8 mg/L	
XDR *P. aeruginosa*	CEF-TAZ	MIC_{50} = 2 mg/L	
		MIC_{90} = 16 mg/L	
Non-CRE *E. coli*	CEF-TAZ	MIC_{50} = 0.5 mg/L	
		MIC_{90} = 2 mg/L	
Non-CRE *K. pneumoniae*	CEF-TAZ	MIC_{50} = 1 mg/L	
		MIC_{90} = 8 mg/L	
ESβL non-CRE *Enterobacterales*	CEF-TAZ	0.5–8 mg/L	[59]
P. aeruginosa	CEF-TAZ	MIC_{50} = 0.5 mg/L	[65]
		MIC_{90} = 8 mg/L	
K. pneumoniae	CEF-TAZ	MIC_{50} = 0.5 mg/L	
		MIC_{90} > 64 mg/L	

CEF-TAZ—ceftolozane-tazobactam; CRE—carbapenem-resistant *Enterobacterales;* MDR—multidrug-resistant; XDR—extensively drug-resistant.

3.1.2. Clinical Trials

The effectiveness and safety of the treatment for HAP/VABP with CEF-TAZ vs. MER were assessed in a phase 3 randomized, double-blind, and controlled trial (ASPECT-NP). The primary end-point of the study was, among others, 28-day mortality for any cause. The secondary end-point was the clinical response at TOC (test-of-cure: 7–14 days after EOT (end of treatment)) in the microbiological ITT population (mITT—microbiological intention to treat population). The patients were randomized into two groups, as follows: treated with CEF-TAZ (N = 362) and treated with MER (N = 364). Those in the first group received 3 g of CEF-TAZ (2 g of CEF and 1 g of TAZ) and the second 1 g of MER as 1-h i.v. infusions every 8 h for 8–14 days. In the general population, the predominant form of the disease was VABP (71%). The dominant isolated pathogens were represented by *K. pneumoniae*, *E. coli*, and *P. aeruginosa* (including strains of ESβL). The 28-day mortality rates were as follows: 24% for CEF-TAZ vs. 25.3% for MER treatment. In turn, the obtained clinical responses at TOC, respectively, were 54% vs. 53%. Another relevant end-point was microbial eradication assessed in the mITT population at TOC (73.1% vs. 68%, respectively). In both cases, a similar eradication activity of the preparations was observed against MDR *P. aeruginosa* (~54%), XDR *P. aeruginosa* (40%), and against ESβL *Enterobacteriaceae* (57.1% vs. 61.6%). The adverse events profile (TEAE) was also similar in both groups at 11% vs. 8% (of which severe TEAE accounted for 8/38 and 2/27 of the total amount of TEAE, respectively). According to the study, CEF-TAZ has a comparable antimicrobial profile to MER in VABP therapy, indicating a potential for its application in an empirical therapy [66–68].

Pogue et al. revealed the effectiveness of CEF-TAZ therapy in relation to the comparators, as follows: aminoglycosides (gentamicin [GEN], tobramycin [TOB], and AKC) and COL in the treatment of MDR/XDR *P. aeruginosa*. In this multicenter, retrospective, and observational cohort study, patients were randomly divided (N = 200) into two (I and II) groups (both equally, N = 100). The dominant type of infection was VABP (63% of patients were mechanically ventilated, and VABP developed in 52%). The need to switch to combination therapy (adding antibiotic—most often MER, β-lactams, and PIP-TAZ in I or COL in II) occurred significantly more frequently in the cohort receiving aminoglycoside or polymyxin than CEF-TAZ (75% vs. 15%). Importantly, better clinical cure values were achieved in the CEF-TAZ cohort (81% vs. 61%), as was the issue of hospital mortality (20% vs. 25%). The difference was also related to the safety profile of the therapy, especially nephrotoxicity understood as AKI (6% vs. 34%). Nearly 7% of patients treated with aminoglycosides/polymyxins required renal replacement therapy. The study highlighted the benefits of CEF-TAZ in the treatment of VABP. The CEF-TAZ therapy was found to have lower kidney toxicity than the comparator, as well as superior effectiveness against MDR/XDR *P. aeruginosa* [69].

Mogyorodi et al. compared the efficacy of CEF-TAZ or COL therapy in the treatment of VABP with the etiology of XDR *P. aeruginosa* among patients hospitalized in an ICU ward. A total of 51 patients were enrolled in the trial and were randomly assigned to the following two groups: those receiving CEF-TAZ (N = 18) and those receiving COL (N = 33). The end-points of the study were clinical cure (understood as the resolution of symptoms and signs of infection), microbiological cure (eradication or persistent colonization), and 28-day mortality ratio and frequency of AEs. The median duration of treatment was 7 days in the CEF-TAZ group and 9 days in the COL group. The need for combination therapy was observed in 32/33 (97%) cases in the COL group (mainly the addition of inhaled COL) and in 8/18 (44%) cases in the CEF-TAZ group. Clinical success was achieved to a greater extent in the CEF-TAZ group than in the COL group (72.2% vs. 30.3%). Microbiological eradication occurred in 44.4% of patients in the first group and in only 15.2% of patients treated with COL. There were no significant differences in the 28-day mortality rate (27.8% vs. 33.3%). The AEs were significantly more frequent in the COL population (72.7%) vs. CEF-TAZ (55.5%). The development of AKI was more frequent in patients treated with COL than CEF-TAZ (48.5% vs. 11.1%) [70]. In a retrospective clinical study conducted in 2016–2018 in 22 Italian hospitals, the therapeutic effectiveness of CEF-TAZ in the treatment of, among others, nosocomial infections with the etiology of *P. aeruginosa* was assessed. A total of 101 patients were enrolled in the study (VABP cases accounted for 12, in turn HAP = 20). Overall, 75% of patients presenting symptoms of LRTI achieved clinical success, 8 patients from the VABP group developed clinical failure, which accounted for 8/17 of all cases of failure. Side effects were reported in only 3/101 patients. Particularly important is that it was shown that out of the tested bacterial isolates, as much as 50.5% were classified as XDR

P. aeruginosa (among others, high resistance to fosfomycin (FOS = 89.1%), MER (77.2%), PIZ-TAZ (77.2%), and GEN and TOB (50.5%)) [71].

A study in the Canadian Leadership on Antimicrobial Real-Life use (CLEAR) registry conducted in Canada, in 2019–2020, evaluated the clinical effectiveness of CEF-TAZ in the treatment of, among others, patients suffering from HAP and VABP and hospitalized on and off in the ICU. The study included 51 patients, of which 8 were diagnosed with VABP and 19 were treated for HAP. Based on the in vitro isolates tested, the average activity of the preparation was determined to be 88.2%. Patients treated for VABP received CEF-TAZ due to the resistance of bacteria to previous therapy (8/8), all of them had microbiologically confirmed *P. aeruginosa* infection. In three cases, combination therapy with fluoroquinolone and aminoglycoside was implemented. There were three deaths; in addition, one of the patients developed clinical failure in the course of therapy. Clinical success (improvement and clinical cure) was observed in the remaining patients. The duration of treatment was usually in the range of 7–10 days. Adverse events were reported in 1/8 cases, and the development of neutropenia occurred [72].

In another retrospective multicenter clinical trial, Gallagher et al. investigated the efficacy of the treatment of MDR *P. aeruginosa* infections with CEF-TAZ. The primary end-points of the study were 30-day mortality and the in-hospital mortality rate. For secondary points, the following were chosen: clinical cure (resolution of the signs and symptoms of infection) and microbiological eradication (negative culture in EOT). A total of 205 patients were enrolled, of which 121 were suffering from pneumonia (with VABP cases accounting for 63). *P. aeruginosa* isolates were shown to have a high rate of antibiotic resistance (%) to carbapenems (96.8%), PIP-TAZ (94.2%), ATM (92.8%), and CPM/CAZ (85.6%). In the general population, the total mortality was 19%, the VABP population achieved clinical success in 50%, microbiological eradication was confirmed in 53.4% of patients, 22 out of 39 deaths related to patients with VABP [73].

In the multicenter retrospective study by Holger et al., the efficacy of CEF-TAZ therapy vs. BAT (including the use of PIP-TAZ, CPM, MER, CAZ/AVI, CAZ, or COL) was evaluated in the treatment of LRTI with the etiology of MDR/XDR *P. aeruginosa*. The study enrolled 206 patients, who were then randomly assigned into two groups: those who received CEF-TAZ (N = 118) and those treated with BAT (N = 88). The percentage of VABP in both groups was 52.5% and 38.6%, respectively. In the general populations of CEF-TAZ and BAT, the primary end-points were as follows: clinical failure—23.7% vs. 48.9%; 30-day mortality rate—15.3% vs. 20.5%; and TEAE frequency—10.2% vs. 33%. Taking into account these results and a much higher percentage of VABP in the CEF-TAZ treatment group, it can be concluded that the preparation showed a significant advantage over BAT in the treatment of MDR/XDR *P. aeruginosa* etiology LRTI. The therapy was also characterized by greater safety and a lower percentage of TEAEs [74].

A summary of the clinical efficiency of ceftolozane-tazobactam therapy among patients with VABP is presented in Table 5.

Table 5. Efficacy of CEF-TAZ in the treatment of VABP.

Number of Patients	Treatment Scheme in VABP Population	Outcomes of Trial		Etiology of Infections	Reference
N = 726 (71% VABP)	CEF-TAZ (N = 362)	28-MR	24%	*K. pneumoniae*, *E. coli* *P. aeruginosa* (including ESβL)	[66]
			25.3%		
		Clinical response at TOC	54%		
	MER (N = 364)		53%		
		Microbiological eradication ratio	73.1%		
			68%		

Table 5. *Cont.*

Number of Patients	Treatment Scheme in VABP Population	Outcomes of Trial		Etiology of Infections	Reference
N = 200 (52% VABP)	CEF-TAZ (N = 100)	CC	81%	MDR/XDR *P. aeruginosa*	[69]
	Aminoglycosides (TOB, GEN, AKC)/COL (N = 100)		61%		
		IHMR	20%		
			25%		
N = 51	CEF-TAZ (N = 18)	CC	72.2%	XDR *P. aeruginosa*	[70]
			30.3%		
		Microbiological eradication ratio	44.4%		
			15.2%		
	COL (N = 33)	28-MR	27.8%		
			33.3%		
		Frequency of AEs	55.5%		
			72.7%		
N = 205 (63/205 VABP)	CEF-TAZ (N = 63)	30-MR	35%	MDR *P. aeruginosa*	[73]
		CC	50%		
		Microbiological eradication in the EOT	53.4%		
N = 206 (46.6% VABP)	CEF-TAZ (N = 118)	Clinical failure	23.7%	MDR/XDR *P. aeruginosa*	[74]
			48.9%		
	BAT (N = 88)	30-MR	15.3%		
			20.5%		
		Frequency of TEAEs	10.2%		
			33%		

28/30-MR—28/30-day mortality ratio; AE—adverse event; AKC—amikacin; BAT—best available treatment (therapy included combinations of, among others, piperacillin-tazobactam, cefepime, meropenem, ceftazidime-avibactam, ceftazidime, and colistin); CC—clinical cure; CEF-TAZ—ceftolozane-tazobactam; COL—colistin; EOT—end of treatment; GEN—gentamicin; IHMR—in-hospital mortality ratio; MDR—multidrug-resistant; MER—meropenem; TEAE—treatment emergent adverse event; TOB—tobramycin; TOC—test of cure; VABP—ventilator-associated bacterial pneumonia; XDR—extensively drug-resistant.

4. Characteristics of Ceftazidime-Avibactam

Ceftazidime (CAZ) is a IIIrd-generation cephalosporin which, in combination with the non-β-lactam β-lactamase inhibitor avibactam (AVI), is active against GNB, especially *Enterobacteriaceae* [21]. Worse activity is observed against nonfermenting *P. aeruginosa* due to its resistance of 2.9–18%, as well as against *A. baumannii*, which has a resistance to CAZ-AVI that reaches 50% [75]. However, AVI, in combination with antibiotics, decreases the MIC value for *P. aeruginosa* by nearly four times. The effect of CAZ-AVI has been demonstrated also against anaerobic bacteria, e.g., *Prevotella* spp., *Bacterioides fragilis*, *Clostridium perfringens*, and *Porphyromonas* spp. [76]. The combination of CAZ-AVI with other antibiotics achieved promising results in the terms of the eradication of *K. pneumoniae* strains colonizing intestines. Success was achieved in 11/12 cases despite the presence of a KPC resistance mechanism among pathogens [77]. CAZ-AVI, according to EMA indications, can be used in the following cases: cIAIs, cUTIs, including AP; nosocomial pneumonia (HAP); and VABP, as well as in the case of bacteremia associated with the above disease states. FDA guidelines also indicate the use of this combination in the conditions mentioned. Moreover, CAZ-AVI can also be used in pediatric patients over 3 months of age in treatment of cUTIs and cIAIs. The drug is administered at a ratio of 2 g of CAZ to 0.5 g of AVI in i.v. infusions lasting 2 h every 8 h in adult patients with normal renal function (CrCl > 50 mL/min and eGFR > 50 mL/min/1.73 m^2). For pediatric patients, the dosage is altered, as follows: children between 3 and 6 months of age should receive 40 mg/kg of CAZ and 10 mg/kg of AVI. However, above 6 months of age, 50 mg/kg of CAZ and 12.5 mg/kg of AVI should be administered. The most common adverse events revealed in clinical trials included gastrointestinal problems, such as abdominal pain, diarrhea, and constipation, as well as headaches, injection site reactions, fever, and increased levels of transaminases. Renal failure and diarrhea have been reported rarely [78,79]. In Drwiega et al.'s study, the effects of different ways of administering CAZ-AVI were assessed. Patients received a bolus of the drug at a dose of 20 mg/kg body weight, and then after 60 min a continuous infusion at a dose of 60 mg/kg body weight per day. Pulmonary penetration of CAZ in studies among patients hospitalized due to VABP achieved better values during continuous infusion in contrast to bolus infusion. The median ELF concentration was 12 μg/mL, and the ELF-to-plasma penetration ratio was 0.42 [80].

Ceftazidime

Ceftazidime (CAZ) is a β-lactam antibiotic belonging to the group of IIIrd-generation cephalosporins with the IUPAC name (6R,7R)-7-[[(2Z)-2-(2-amino-1,3-thiazol-4-yl)-2-(2-carboxypropane-2-yloxyimino)acetyl]amino]-8-oxo-3-(pyridin-1-ium-1-ylmethyl)-5-thia - 1-azabicyclo[4.2.0]oct-2-ene-2-carboxylate (PubChem). Similarly to other cephalosporines, the CAZ molecule binds to PBP proteins preventing the cross-linking of peptidoglycan in the microorganism wall, resulting in cell death [21]. The structural model of ceftazidime is shown in Figure 3A.

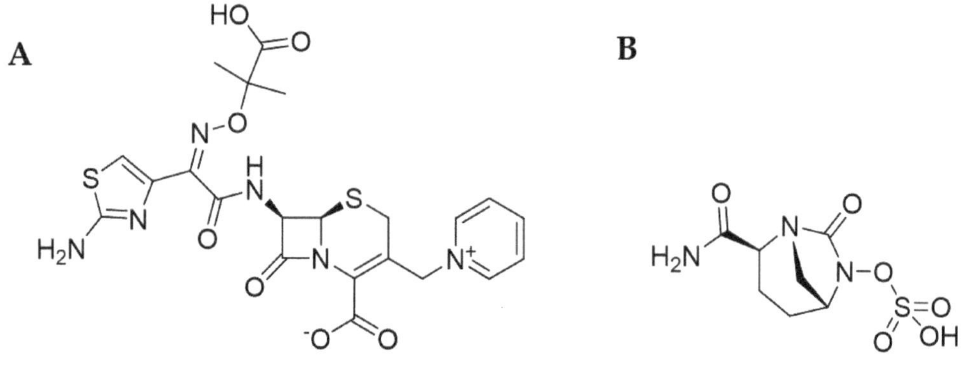

Figure 3. The structural models of ceftazidime (**A**) and avibactam (**B**).

Avibactam

Avibactam (AVI) is a non-β-lactam β-lactamase inhibitor, and its IUPAC name is [(2S,5R)-2-carbamoyl-7-oxo-1,6-diazabicyclo[3.2.1]acetan-6-yl] hydrogen sulfate (PubChem). The inhibitor molecule has the ability to reversibly covalently attach to the active site of serine β-lactamases [52,81]. It is active against *Enterobacteriaceae* producing: ESβL, KPC, AmpC, and OXA-48. However, it is ineffective against metallo-β-lactamases (group B, according to Ambler, e.g., VIM, IMP, and NDM). Therefore, it is active against Ambler classes A, C, and some enzymes from group D producers [21,78]. The resistance to CAZ-AVI is connected with structural point mutations in β-lactamase molecules, as well as changes in the membrane proteins and efflux pump expression [75]. The structural model of avibactam is shown in Figure 3B.

4.1. Clinical Efficiency of Ceftazidime-Avibactam

4.1.1. In Vitro Studies

The Sader et al. study evaluated the activity of CAZ-AVI against Gram (−) bacterial isolates collected from patients (ICU and other hospital wards) presenting, among others, symptoms of VABP. In total, 18,864 isolates were tested, of which 435 derived from VABP infections. The results obtained by CAZ-AVI were very promising. The growth of 99.9% of the *Enterobacteriaceae* was inhibited at MIC levels lower than 8 mg/L. In turn, the MIC_{50}/MIC_{90} determined for isolates derived from VABP were 0.12/0.5 mg/L. The KPC isolates and carbapenem-resistant *Enterobacteriaceae* also had high susceptibility to CAZ-AVI at MIC_{50}/MIC_{90} = 0.5/2.0 mg/L. The CAZ-AVI activity against *P. aeruginosa* isolates collected from VABP was also significant at MIC_{50}/MIC_{90} = 2.0/4.0 mg/L, respectively. However, individuals resistant to CAZ or MER and MDR isolates were less susceptible (4.0/16.0 mg/L). XDR *P. aeruginosa* achieved higher MIC values similar to *A. baumannii* (8.0/32.0 and 16.0/>32 mg/L, respectively). Importantly, AVI significantly increased the sensitivity of *P. aeruginosa* to CAZ (especially with ICU-ward-derived isolates), where this pathogen was initially sensitive in ~78% and in combination with AVI it increased to 95.6% [82]. Perez et al. evaluated the susceptibility of clinical isolates of MDR and XDR *P. aeruginosa* (collected from patients suffered from VABP) to a number of antibiotics including CAZ-AVI. The preparation was characterized by an MIC_{50}/MIC_{90} at the level of 2.0/16.0 mg/L (sensitivity of 75.5%), which makes it comparable to the CEF-TAZ described above (77.4%) [57].

Another study evaluated the CAZ-AVI activity separately and in combination with ATM against *K. pneumoniae* CRE, NDM, and KPC isolates in vitro, as well as in vivo in a murine model of infection. A total of 47 *Klebsiella* isolates were collected, of which 16 were KPC-2, 1 was OXA-232, 28 were identified as NDM, and 2 were identified as KPC-2 + NDM. CAZ-AVI was highly active in vitro against KPC-2 isolates (MIC = 0.4–0.8 mg/dL) and OXA (0.2 mg/dL). The preparation showed lower activity against NDM isolates (MIC in the range of 0.5 to 256 mg/L) and against KPC and NDM isolates (MIC value of 8 and 128 mg/L, respectively). Significantly, in most cases (90%) synergistic interactions between CAZ-AVI and ATM were demonstrated. The combination led to a decrease in the MICs of both compounds and restored ATM activity against KPC and NDM isolates [83].

The in vitro susceptibilities of *Enterobacteriaceae* clinical isolates (including *P. aeruginosa*) collected from a series of studies (e.g., REPROVE) were assessed for many antibiotics, including CAZ-AVI. The combination obtained MIC_{50}/MIC_{90} values for individual pathogens causing VABP as follows: MDR *P. aeruginosa* (8/64 mg/L, strain sensitivity 34.8%) and MDR *K. pneumoniae* (0.5/1.0 mg/L, sensitivity 75%). The REPROVE study confirmed previous reports on the superiority to comparators of the antimicrobial activity profile of CAZ-AVI against MDR GNB causing VABP [84,85].

In the next study, the in vitro activity of CAZ-AVI was evaluated in relation to GNB bacterial isolates collected from patients hospitalized due to pneumonia (including VABP) in 2011–2015. A total of 11185 isolates were tested, of which 1097 were collected from patients suffering from VABP. Subsequently, the sensitivity profiles to the drugs CAZ-AVI and comparators were determined. CAZ-AVI showed high activity against *P. aeruginosa* (MIC_{50}/MIC_{90} = 2/4 mg/L), including MER-resistant strains (4/16 mg/L) and PIP-TAZ-resistant MDR strains (4/16 mg/L in both strains). The XDR pathogens were characterized (similarly to Sader et al.'s 2015 study) by higher MICs (8/32 mg/L). This trend was observed also among *A. baumannii* isolates (16/>32 mg/L). Pathogenic CREs were highly susceptible to CAZ-AVI, with MIC_{50}/MIC_{90} at 0.5/2 mg/L. Among ESβL strains, there was also high sensitivity to combinations, as follows: *K. pneumoniae* (0.25/1 mg/L) and *E. coli* (0.12/0.5 mg/L). Overall, a high percentage of CAZ-AVI susceptibility was demonstrated for most GNB isolates derived from VABP, e.g., *P. aeruginosa* (97.8% including MDR/XDR = 87.5%), ESβL *K. pneumoniae* (93.1%), and ESβL *E. coli* (100%). The study showed noninferior activity of the preparation to comparators against *Enterobacteriaceae* and significantly weaker in relation to *A. baumannii* (e.g., COL = 1/2 mg/L) [86].

The Candel et al. study evaluated the in vitro activity of cefiderocol and comparators, including CAZ-AVI against GNB isolates obtained in the years 2013–2018 as part of the SIDERO-WT and SIDERO-Proteeae studies. A total of 20,911 isolates were collected, of which 34% were collected from patients with VABP (mainly *K. pneumoniae*, *P. aeruginosa*, and *A. baumannii*). The activity of CAZ-AVI against isolates sensitive to carbapenems (derived mainly from VABP infections) was at the level of 98–99.6%. CRE pathogens were less sensitive to CAZ-AVI (*K. pneumoniae* at 70.5% and *P. aeruginosa* at 46.1%). Cefiderocol or COL presented activity profiles against *A. baumannii* and *S. maltophila* superior to CAZ-AVI [65]. The study showed some limitations of CAZ-AVI activity in relation to *A. baumannii* or *S. maltophila*, which were also reported by other researchers [82,86]. A summary of the in vitro antimicrobial activity of ceftazidime-avibactam is presented in Table 6.

Table 6. In vitro antimicrobial activity of CAZ-AVI and its combinations against VABP clinical isolates.

Pathogens	Antibiotic Scheme Used In Vitro	MIC Value	Reference
KPC *Enterobacteriaceae*	CAZ-AVI	MIC_{50} = 0.5 mg/L	[82]
		MIC_{90} = 2 mg/L	
CRE *Enterobacteriaceae*		MIC_{50} = 0.5 mg/L	
		MIC_{90} = 2 mg/L	
P. aeruginosa		MIC_{50} = 2 mg/L	
		MIC_{90} = 4 mg/L	
P. aeruginosa (MER-NS, CAZ-NS, or MDR strains)		MIC_{50} = 4 mg/L	
		MIC_{90} = 16 mg/L	
XDR *P. aeruginosa*		MIC_{50} = 8 mg/L	
		MIC_{90} = 32 mg/L	
A. baumannii		MIC_{50} = 16 mg/L	
		MIC_{90} > 32 mg/L	
MDR *P. aeruginosa*	CAZ-AVI	MIC_{50} = 2 mg/L	[57]
		MIC_{90} = 16 mg/L	
KPC-2 *K. pneumoniae*	CAZ-AVI	MIC = 0.4–0.8 mg/L	[83]
OXA-232 *K. pneumoniae*		MIC = 0.2 mg/L	
NDM *K. pneumoniae*		MIC = 0.5–256 mg/L	
KPC-2 + NDM *K. pneumoniae*		MIC = 8–128 mg/L	
MDR *P. aeruginosa*	CAZ-AVI	MIC_{50} = 8 mg/L	[84]
		MIC_{90} = 64 mg/L	
MDR *K. pneumoniae*		MIC_{50} = 0.5 mg/L	
		MIC_{90} = 1 mg/L	
P. aeruginosa	CAZ-AVI	MIC_{50} = 2 mg/L	[86]
		MIC_{90} = 4 mg/L	
MER-NS *P. aeruginosa*		MIC_{50} = 4 mg/L MIC_{90} = 16 mg/L	
PIP-TAZ-NS *P. aeruginosa*			
MDR *P. aeruginosa*			

Table 6. *Cont.*

Pathogens	Antibiotic Scheme Used In Vitro	MIC Value	Reference
XDR *P. aeruginosa*	CAZ-AVI	MIC_{50} = 8 mg/L	[86]
		MIC_{90} = 32 mg/L	
A. baumannii		MIC_{50} = 16 mg/L	
		MIC_{90} > 32 mg/L	
ESβL *K. pneumoniae*		MIC_{50} = 0.25 mg/L	
		MIC_{90} = 1 mg/L	
ESβL *E. coli*		MIC_{50} = 0.12 mg/L	
		MIC_{90} = 0.5 mg/L	

CAZ-AVI—ceftazidime-avibactam; CRE—carbapenem-resistant *Enterobacterales*; MDR—multidrug-resistant; MER-NS—meropenem-nonsusceptible; PIP-TAZ-NS—piperacillin-tazobactam-nonsusceptible; XDR—extensively drug-resistant.

4.1.2. Clinical Trials

In a multicenter, randomized, double-blind, and phase 3 study—REPROVE (Randomized Trial of ceftazidime-avibactam Versus Meropenem for Treatment of Hospital-Acquired and Ventilator-Associated Bacterial Pneumonia)—Torres et al. evaluated the CAZ-AVI therapeutic efficiency vs. MER for the treatment of nosocomial infections (including VABP). Patients were randomly divided into the following two groups: those who received CAZ-AVI (N = 356, CAZ-AVI 2000 + 500 mg every 8 h) and those treated with MER (N = 370, MER 1000 mg every 8 h). The percentage of VABP infections per group were 33 vs. 35%. The predominant pathogens causing infections were identified as: *K. pneumoniae* and *P. aeruginosa* (total ~70% of isolates). Clinical cure (the primary end-point of the trial) was achieved in 68.8% (CAZ-AVI) vs. 73% (MER) of the mITT population (clinical modified intention-to-treat—patients meeting entry trial criteria with confirmed GNB etiology infection), in the clinically evaluable population (CEP: mITT patients after receiving treatment), the results for clinical cure were as follows: 77.4% vs. 78.15%. Patients with confirmed VABP achieved clinical success in the mITT and CEP populations for the CAZ-AVI and MER treatments, respectively, at 70.3% vs. 74.2% and 77.5% vs. 75.9%. Both preparations were characterized by similar activities against pathogens isolated from patients, both Gram (−) and (+). Similarly, the 28-day mortality ratio was 9 vs. 7%, and the frequency of AEs was 75% vs. 74% (the most common were diarrhea, hypokalemia, and anemia). According to the results of the study, CAZ-AVI showed noninferior activity to MER, which is a prerequisite for its use in carbapenem-saving therapy for VABP and HAP [87,88].

Interesting insights also come from the Shi et al. study, which assessed the effectiveness of CAZ-AVI vs. TGC therapy in patients hospitalized in an ICU ward due to HAP/VABP with the etiology of CRE *K. pneumoniae*. The study included 105 patients (% of confirmed VABP = 71.4) who were randomized into two groups: those obtaining TGC (N = 62, the scheme included 200 mg in loading dose and then a maintenance dose) and treated CAZ-AVI (N = 43, regimen of 2000 + 500 mg every 8 h). Clinical success was defined as the resolution of clinical symptoms in combination with the normalization of nonmicrobiological components, for example, laboratory parameters. Microbiological success was defined as the eradication/negative cultures at the end of the patient's hospitalization. CAZ-AVI therapy has been shown to be associated with a higher percentage of clinical cure rates, as follow: 51.2% vs. 29%, respectively. Similarly, the issue of microbiological success was presented, as follows: 74.4% vs. 33.9%. There was no significant difference in 28-day survival (69.8% vs. 66.1%). Importantly, CAZ-AVI therapy was also associated with a lower percentage of AEs (mainly diarrhea) than TGC treatment, as follows: 7% vs. 27.4%, respectively. The study proved that CAZ-AVI could be an alternative treatment option for critically ill ICU ward patients with VABP of CRE pathogens etiology [89].

A retrospective observational cohort study conducted in Greece in a population of critically ill ICU patients (diagnosed with VABP) demonstrated the superiority of CAZ-AVI therapy over other treatment regimens for infections with a CRE etiology. Patients were randomly assigned into the following two groups: those receiving CAZ-AVI (N = 41) and those receiving BAT (N = 36). A total of 26 cases of VABP were reported (19 vs. 7 among populations). CAZ-AVI presented better results in the field of clinical cure among patients (80.5% vs. 52.8% for BAT). Moreover, it also achieved a microbiological eradication ratio superior to its comparator (94.3% vs. 67.7%).

The 28-day survival rate also favored CAZ-AVI therapy (85.4% vs. 61.1%), in turn, there was no difference in the frequency of AEs. CAZ-AVI was shown to have a higher efficacy rate than COL-based regimens, which is crucial because of colistin's nephrotoxicity and limited therapeutic use in patients with impaired renal function [90]. These conclusions are supported by van Duin et al.'s study, in which the superiority of CAZ-AVI therapy over colistin was demonstrated in the field of in-hospital mortality (CAZ-AVI treated patients—9%; COL treated patients—32%) in the therapy of CRE etiology infections [91].

Zheng et al., in turn, investigated the efficacy of CAZ-AVI monotherapy and CAZ-AVI-based combination therapy in the treatment of CRE etiology infections. Patients (N = 62) were randomly divided into the following two groups: those treated with CAZ-AVI alone (N = 21) and those receiving combination therapy (N = 41). The primary end-point of the study was the 30-day mortality of patients from any cause. Mortality in the first population was 47.6% vs. 24.4% in the combination group. A similar advantage of polytherapy was demonstrated in the evaluation of microbiological eradication ratio: 42.9% vs. 61%. Particularly promising clinical parameters (related to the reduction in 30-day mortality) were obtained by schemes based on the combination of CAZ-AVI with carbapenems (MER and IMI), TGC, FOS, and ATM [92]. There are also reports on the potential effectiveness of combined therapies that consisted of CAZ-AVI + ATM or CAZ-AVI + MER + ertapenem (double-carbapenem therapy) in the treatment of VABP with the etiology of PDR *K. pneumoniae*. The therapeutic regimens used led to the alleviation of parameters on both the SOFA (Sequential Organ Failure Assessment) and CPIS (Clinical Pulmonary Infection Scale) scales. This is an important premise in the topic of using CAZ-AVI + ATM in VABP empirical therapy as a beneficial option to increase the chances of survival for patients [93].

The Burastero et al. reported the potential use of CAZ-AVI plus FOS/AKC/trimethoprim-sulfamethoxazole (TRM-STX)/MER therapy regimen in the treatment of VABP with an etiology of *P. aeruginosa* DTT (DTT—difficult-to-treat, strains nonsusceptible to β-lactams and fluoroquinolones), *Burkholderia cepacia*, and *S. maltophila* complicated by SARS-CoV2 infection. It was reported that in critically ill patients with mentioned viral coinfection, such a treatment scheme led to significant microbiological eradication (achieved in 14/23 of the cases studied). The problem remaining is a high mortality ratio, recorded at the level above 60% despite ICU hospitalization. Because of the challenges of treating infections of this etiology, the completion of the microbiological eradication in 3/6 cases of *S. maltophila* infection is a significant premise for the continuation of research in this field [94].

The effectiveness of CAZ-AVI therapy in the treatment of infections with the etiology of MDR/XDR *P. aeruginosa* was also assessed in a study by Corbella et al. The study population (N = 61) included patients presenting, among others, LRTI (34.4%). It was confirmed that predominant etiological factor of VABP infections was *P. aeruginosa* (91.8% of isolates were identified as MDR and 8.2% as XDR). The median duration of therapy was 7 days, and in 47% of cases an additional drug was administered, including MER, COL, and ATM. Clinical cure was achieved on day 14 of therapy in 54.1% of cases, while the 30-day mortality rate was 13.1%, recurrence by the day 90 occurred in 12.5% of cases. Promising results were also obtained when CAZ-AVI was combined with ATM (total absence of recurrent infections) and in combination with AKC (100% 30-day survival ratio vs. 93.8%

in case of CAZ-AVI monotherapy). Interestingly, the combination of CAZ-AVI with COL did not achieve better parameters than CAZ-AVI monotherapy [95].

Another multicenter, retrospective, and cohort clinical trial evaluated the therapeutic efficacy of CAZ-AVI in the treatment of carbapenem-resistant *Enterobacteriaceae* and *P. aeruginosa* etiology infections. The study achieved similar results to Corbella et al., as follows: 30-day mortality ratio of 17.2%, recurrence by the day 30 ratio of 5.9%, and clinical success reached in 70.9% of patients (N = 203). Particularly important is that these results were similar among patients in the general population, those infected with *P. aeruginosa*, and among patients with CRE infection. The LRTI accounted for 37.4% of all infections, and *K. pneumoniae* accounted for as much as 43.8% of all isolates (including 63.2% of all CRE isolates). The MIC_{50}/MIC_{90} values of CAZ-AVI were *K. pneumoniae* CRE 2/4 mg/L and 2/6 mg/L for *P. aeruginosa*. Both studies demonstrate the high efficacy of CAZ-AVI monotherapy in the treatment of GNB MDR infections, including mechanically ventilated patients. This is an important premise in the era of increasing resistance to carbapenems and anti-*Pseudomonas* preparations [96].

A summary of the clinical efficiency of ceftazidime-avibactam therapy among patients with VABP is presented in Table 7.

Table 7. Efficacy of CAZ-AVI in the treatment of VABP.

Number of Patients	Treatment Scheme in VABP Population	Outcomes of Trial		Etiology of Infection	Reference
N = 726	CAZ-AVI (N = 356) (33% VABP)	Clinical success in mITT population	70.3%	*K. pneumoniae* *P. aeruginosa*	[87]
			74.2%		
		Clinical success in CEP population	77.5%		
			75.9%		
	MER (N = 370) (35% VABP)	28-MR	9%		
			7%		
		Frequency of AEs	75%		
			74%		
N = 105 (71.4% VABP)	CAZ-AVI (N = 43)	Clinical success	51.2%	CRE *K. pneumoniae*	[89]
			29%		
		Microbiological success	74.4%		
	TGC (N = 62)		33.9%		
		28-MR	69.8%		
			66.1%		
N = 77 (33.8% VABP)	CAZ-AVI (N = 41, 19/41 VABP)	CC	80.5%	CRE: *P. aeruginosa* *K. pneumoniae* *E. coli*	[90]
			52.8%		
		Microbiological eradication ratio	94.3%		
	BAT (N = 36, 7/36 VABP)		67.7%		
		28-day survival ratio	85.4%		
			61.1%		
N = 62	CAZ-AVI monotherapy (N = 21)	30-MR	47.6%	CRE: *P. aeruginosa* *K. pneumoniae* *E. coli*	[92]
			24.4%		
	CAZ-AVI combined therapy (N = 41)	Microbiological eradication ratio	42.9%		
			61%		

Table 7. *Cont.*

Number of Patients	Treatment Scheme in VABP Population	Outcomes of Trial		Etiology of Infection	Reference
N = 61	CAZ-AVI in monotherapy 53% CAZ-AVI in combined therapy 47% (scheme obtained, e.g., COL/MER/ATM)	CC by the day 14	54.1%	*P. aeruginosa*: MDR—91.8% XDR—8.2%	[95]
		30-MR	13.1%		
		Recurrence by the day 90	12.5%		
		30-day survival ratio	93.8% (for CAZ-AVI monotherapy)		
N = 203 (37.4% LRTI include VABP)	CAZ-AVI monotherapy (N = 203)	Clinical success	70.9%	CRE: *P. aeruginosa* *K. pneumoniae*	[96]
		Recurrence by the day 30 ratio	5.9%		
		30-MR	17.2%		

28/30-MR—28/30-day mortality ratio; AE—adverse event; ATM—aztreonam; BAT—best available therapy; CAZ-AVI—ceftazidime-avibactam; CC—clinical cure; CEP—clinical evaluable population (mITT patients after receiving treatment); COL—colistin; CRE—carbapenem-resistant *Enterobacterales;* GNB—Gram-negative bacteria; LRTI—lower respiratory tract infection; MDR—multidrug-resistant; MER—meropenem; mITT—modified intention to treat population (patients meeting entry trial criteria with confirmed Gram (−) bacteria etiology infection); TGC—tigecycline; VABP—ventilator-associated bacterial pneumonia; XDR—extensively drug-resistant.

5. Characteristics of Cefiderocol

Cefiderocol (CFD) is a newly synthetized siderophore cephalosporin antibiotic (IUPAC name: (6R,7R)-7-[[(2Z)-2-(2-amino-1,3-thiazol-4-yl)-2-(2-carboxypropane- 2-yloxyimino) acetyl]amino]-3-[[1-[2-[(2-chloro-3,4-dihydroxybenzoyl)amino]ethyl] pyrrolidin-1-ium-1-yl]methyl]-8-oxo-5-thia-1-azabicyclo[4.2.0]oct-2-ene-2-carboxylate (PubChem)) (Figure 4). CFD presents the unique mechanism of action which uses bacteria's ability to collect iron ions from the external environment and transport them into the cells. The siderophore antibiotic molecule after active transport through iron ions carriers into periplasmatic space attach to PBP proteins and then disrupt the structure of the bacterial wall, which ends in cell lysis. It was revealed that CFD can also be taken into bacterial cell via porines, which is the mechanism of penetration into the cell similar to that exhibited by classical β-lactam preparations [97]. Figure 5 shows the mechanism of action of cefiderocol in comparison to the β-lactam β-lactamase inhibitor combination against a bacterial cell on molecular level. The spectrum of action of CFD includes nonfermenting *P. aeruginosa*, *A. baumannii*, *S. maltophila*, and other *Enterobacteriaceae*, as well as MDR bacteria [98]. The MIC_{90} value for *Enterobacterales*, nonfermenting bacilli and *Proteaceae* obtained were in the range of 0.12 to 2 mg/L. Moreover, 95% to even 100% of the strains remained susceptible [22]. According to EMA, CFD is indicated for the treatment of infections caused by Gram (−) aerobic bacteria and should be used when previous therapeutic options have failed. The FDA recommendations describe that the drug can be used to treat HAP, VABP, and cUTIs including AP in patients over 18 years of age. Doses of 2 g should be administered every 8 h as an i.v. infusion lasting 3 h in patients with normal renal function (defined as CrCl in the range of 60–119 mL/min). Resistance to CFD on molecular level may be associated with the Ton-B-dependent iron transporter mutations and mutations in *ampC* genes. It was also observed that the MIC value for this preparation may increase when the microorganism retains NDM resistance (it acts only on 58% of *Enterobacterales* NDM isolates; considering that the MIC value $\leq 4\ \mu g/mL$ indicates the susceptibility of the bacteria) [99]. CFD molecule is stable against hydrolysis by serine β-lactamases like KPC and also against MBL [100]. The penetration of CFD into the lungs has been also investigated. Patients suffering from VABP received CFD at a dose of 2 g every 8 h in a 3 h infusion, achieving ELF values of 7.63 mg/L immediately after the infusion and 10.40 mg/L for 2 h after the end of the drug's administration. The ratio of ELF to the unbound drug concentration in plasma was 0.212 at the end of the 3 h of continuous infusion and 0.547 in 2 h after the end of the infusion. The above results demonstrate that the permeability of the drug is adequate for the treatment of pneumonia [101]. The most common AEs connected with the treatment include skin changes, such as rashes, headaches, abdominal pain, throat and mouth pain, fever, hypertension, nausea, vomiting, diarrhea, constipation, cough, and an increase in liver enzymes and creatine phosphokinase plasma activity and leukocytosis [102,103].

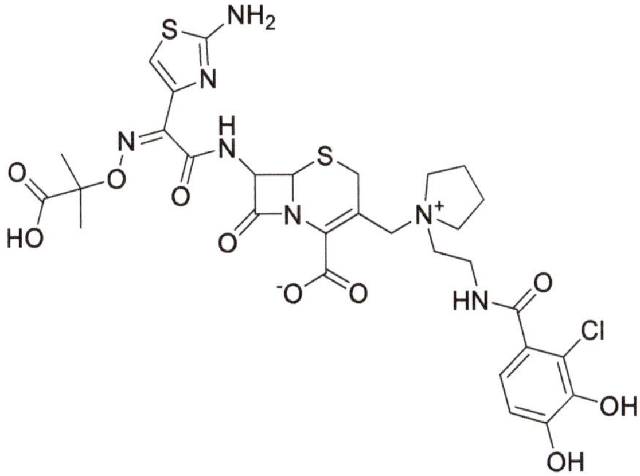

Figure 4. The structural model of cefiderocol.

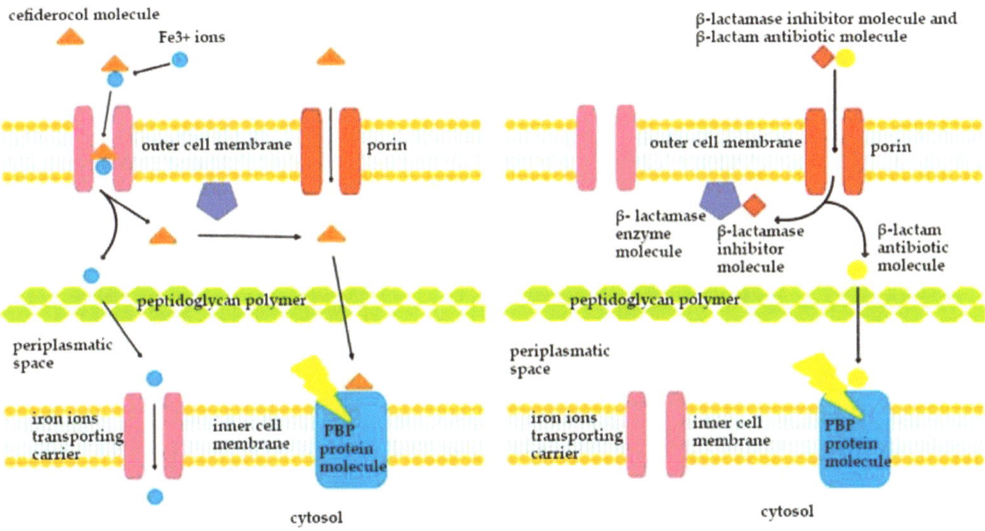

Figure 5. Comparison of the molecular mechanism of action of cefiderocol and the β-lactam β-lactamase inhibitor combination on the bacterial cell wall. Inhibition of peptidoglycan synthesis due to β-lactam antibiotic molecule's interaction with PBP is marked with a lightning bolt symbol.

5.1. Clinical Efficiency of Cefiderocol

5.1.1. In Vitro Studies

The in vitro activity of CFD on GNB isolates was assessed in the course of the SIDERO-WT study conducted in 2014–2017. A total of 19,119 isolates were assessed, among which *E. coli*, *K. pneumoniae*, *P. aeruginosa*, and *A. baumannii* were dominant. Overall, a promising CFD activity profile was demonstrated for the isolates tested; the MIC_{90} for *K. pneumoniae* over the years was assessed as 1 µg/mL, while those of *P. aeruginosa* at 0.5 µg/mL and *A. baumannii* in the range 1–4 µg/mL. It is worth emphasizing that the preparation was also highly active against *S. maltophila* (MIC_{90} = 0.25–0.5 µg/mL). The CFD was also shown to be highly active against MER-resistant isolates, as follows: *P. aeruginosa* (MIC_{90} = 0.5–1 μg/mL) and *A. baumannii* (MIC_{90} = 1–4 μg/mL). The preparation within 3 years of the study showed excellent activity against CRE isolates, as follows: *P. aeruginosa* (99.7–100%) and *A. baumannii* (91%-96.9%). The activity of CFD was superior to the evaluated comparators, such as CEF-TAZ and CAZ-AVI, or even COL (MIC_{90} > 8 μg/mL) against the tested MER-resistant *Enterobacteriaceae* isolates [20,104]. In the SIDERO-CR study, the CFD activity against carbapenem-nonsusceptible clinical isolates of *Enterobacteriaceae* and *P. aeruginosa*, *A. baumannii* and *S. maltophila* (the study included MDR isolates) was evaluated. The preparation showed the highest activity against *Enterobacteriaceae* (97% for CFD vs. 77% for CAZ-AVI vs. 77.8% for COL), similarly for nonfermenting bacilli: *P. aeruginosa* (99.2% CFD vs. 99.6% COL), *A. baumannii* (90.9% CFD vs. 94.6% COL). There was a 100% activity against *S. maltophila* isolates insensitive to carbapenems, which made CFD the most active of the tested compounds. The MIC_{90} for CFD was determined in study at the value of 4 μg/mL, however the compound inhibited the growth of 97% of isolates already at subinhibitory concentration [105].

Another in vitro study evaluated the activity of CFD and comparators against GNB isolates obtained from patients in the SIDERO-WT and SIDERO–Proteeae studies. Of the 20911 isolates evaluated, nearly 34.4% were collected from patients diagnosed with VABP. Among the isolates predominant were *Enterobacteriaceae*, especially *Klebsiella* spp., nonfermenting bacilli represented 48.8% of VABP isolates (mainly *A. baumannii* and emphP. aeruginosa). The CFD activity against VABP isolates was assessed. The percentages of susceptible bacterial strains to CFD are presented as follows: *E. coli*—99.6%; *Klebsiella* spp. [CR]—63.8%; carbapenem-sensitive [CS]—97.9%,; *P. aeruginosa* (CR: 97.1%, CS: 99.6%), *A. baumannii* (CR: 91%, CS: 94.4%), *S. maltophila* (CR: 99.6%). CFD showed higher activity than comparators: CAZ-AVI, CEF-TAZ. Only COL showed activity against *A. baumannii* and *S. maltophila* outside CFD [65]. Another in vitro study evaluated the activity of CFD and comparators (BL + BLI) against GNB fermenting and nonfermenting isolates (including those derived from HAP/VABP positive patients) under the SENTRY Antimicrobial Surveillance Program for 2020. CFD was the most active of the tested preparations (including COL) against *P. aeruginosa* (MIC_{50}/MIC_{90} = 0.12/0.5 mg/L) and including XDR and MER-VAB resistant isolates (0.12/1 mg/L for both isolates). In addition, it demonstrated, similar to COL, activity against isolates resistant to all tested BL + BLI connections (CEF-TAZ, CAZ-AVI, PIP-TAZ, and imipenem-relebactam). Moreover, it was also the most active formulation against *A. baumannii* (0.25/1 mg/L), including MER-resistant isolates (0.5/2 mg/L). Interestingly, CFD showed the same activity against *S. maltophila* isolates as TRM-STX (0.12/0.5 mg/L), which makes these two preparations the most active among the tested substances [106]. A summary of the in vitro antimicrobial activity of cefiderocol is presented in Table 8.

Table 8. In vitro antimicrobial activity of CFD and its combinations against VABP clinical isolates.

Pathogens	Antibiotic Scheme Used In Vitro	MIC Value	References
K. pneumoniae	CFD	MIC_{90} = 1 μg/mL	[20,104]
P. aeruginosa		MIC_{90} = 0.5 μg/mL	
A. baumannii		MIC_{90} = 1–4 μg/mL	
S. maltophila		MIC_{90} = 0.25–0.5 μg/mL	
MER-NS *P. aeruginosa*		MIC_{90} = 0.5–1 μg/mL	
MER-NS *A. baumannii*		MIC_{90} = 1–4 μg/mL	
carbapenem-NS *Enterobacteriaceae*	CFD	MIC_{90} = 4 μg/mL	[105]

Table 8. *Cont.*

Pathogens	Antibiotic Scheme Used In Vitro	MIC Value	References
P. aeruginosa	CFD	MIC_{50} = 0.12 mg/L	[106]
		MIC_{90} = 0.5 mg/L	
XDR *P. aeruginosa*		MIC_{50} = 0.12 mg/L	
		MIC_{90} = 1 mg/L	
MER-NS *P. aeruginosa*		MIC_{50} = 0.12 mg/L	
		MIC_{90} = 1 mg/L	
A. baumannii		MIC_{50} = 0.25 mg/L	
		MIC_{90} = 1 mg/L	
MER-NS *A. baumannii*		MIC_{50} = 0.5 mg/L	
		MIC_{90} = 2 mg/L	
S. maltophila		MIC_{50} = 0.12 mg/L	
		MIC_{90} = 0.5 mg/L	

CFD—cefiderocol; MER-NS—meropenem-nonsusceptible; XDR—extensively drug-resistant.

5.1.2. Clinical Trials

A randomized, double-blind, and phase 3 study—APEKS-NP—evaluated the efficacy of i.v. CFD therapy (2000 mg every 8 h) vs. MER (2000 mg every 8 h) in the treatment of nosocomial infections (including HAP and VABP). Both preparations were administered in combination with linezolid. Patients (N = 300) were randomized into the following two groups: those treated with CFD and those treated with MER (N = 148 and N = 152, respectively). The percentages of VABP infections in both groups was 41% vs. 44%. Among these patients, the most common etiological factors of infection were *K. pneumoniae*, *P. aeruginosa*, and *A. baumannii*. Moreover, *A. baumannii* was shown to be predominant carbapenemase (70%) and EsβL producer (43–67%). The MIC range for *P. aeruginosa* and *E. coli* ($\leq$0.03–1 g/mL) indicates a promising CFD activity profile against *Enterobacterales* and nonfermenting bacteria; higher values were recorded for *K. pneumoniae* and *A. baumannii* ($\leq$0.03–4 g/mL and $\leq$0.03–> 6 4 g/mL, respectively). Both treatments lasted comparatively around 10 days, and the primary end-point defined as 14-day mortality rate was also similar among patients with VABP (15% vs. 13%). The 28-day mortality rate was similar in both groups among patients with VABP (23% vs. 22%). Clinical cure and microbiological eradication was achieved in 66% and 42% of cases for the CFD treated group and 56% and 34% for the MER group, respectively. The safeties of the CFD and MER therapies were also evaluated; the overall frequency of AEs was similar (88% vs. 86%), and the most common were UTIs, hypokalemia, diarrhea, and anemia. The preparation showed noninferior effectiveness in relation to MER; it also did not generate numerous severe TEAEs. In summary, it is a potential therapeutic option in the case of VABP with GNB etiology, including *A. baumannii* and CRE [107].

In a multicenter, randomized, and phase 3 study—CREDIBLE-CR—Bassetti et al. evaluated the efficacy of CFD therapy among patients with carbapenem-resistant GNB etiology infections. The study enrolled 152 patients who were randomly divided into the following two groups: those treated with CFD (N = 101, 2000 mg every 8 h) and those receiving BAT (N = 51). The percentages of patients presenting VABP were as follows: 24% vs. 27% in both groups. Importantly, the dominant pathogen isolated from patients suffering from VABP was CRAB (65% vs. 53%). The study showed a noninferior effect of CFD vs. BAT (primary end-point: clinical cure in TOC were 50% vs. 53% for the nosocomial infections group). Microbiological eradication in TOC accounted for 23% vs. 21%, respectively. Treatment-emergent adverse events occurred in 15% vs. 22%, respectively. Significantly, both 14- and 28-day mortalities were higher in the CFD vs. BAT groups (24%

vs. 14% and 31% vs. 18%, respectively). This was probably due to the more frequent CRAB etiology of VABP/HAP among patients in the CFD-treated group than among those receiving BAT [108].

Falcone et al. evaluated the efficacy of CFD as a salvage therapy for the infection of non-fermenting NDM or CRE pathogens. The study included 10 patients, 4 of whom presented symptoms of VABP complicated by SARS-CoV2 infection. Each of the described cases had a different microbiological etiology, as follows: *A. baumannii*, *A. baumannii* + *K. pneumoniae* NDM, *K. pneumoniae* NDM, and *K. pneumoniae* NDM + *S. maltophila*. Significantly, three-quarters of patients showed clinical success with 30 days of therapy, and only one case of death was reported (30-day mortality ratio). In all cases, CFD monotherapy was performed at the site of previous combination regimens that did not cause clinical improvement in patients (including COL, ATM, TGC, CAZ-AVI, or FOS). The MIC values for *K. pneumoniae* NDM were in the range of 1–2 μg/mL in patients with VABP. The study demonstrated the benefits of CFD monotherapy as a salvage therapy for severe infections, e.g., VABP in patients who had exhausted other treatment options [109].

Another study by Falcone et al. compared the therapeutic efficacy of CFD vs. COL therapy in the treatment of infections with an etiology of CRAB. The study included 124 patients who were randomized into two groups: those treated with CFD (N = 47) and those treated with COL (N = 77). A total of 35 patients were clinically and laboratory confirmed to have a VABP infection with CRAB etiology (12 vs. 23 cases, respectively). Among the 12 patients treated with CFD (MIC = 0.12–2 mg/L), only 2 received the preparation as a monotherapy, and the others received CFD combined with TGC, FOS, or meropenem-vaborbactam (MER-VAB). The median duration of treatment with CFD vs. COL was comparable (between 15 and 13 days). The 30-day mortality ratio achieved a lower score among those treated with CFD vs. COL (34% vs. 54%, risk ratio of 0.44, 95% confidence interval of 0.22–0.66, $p < 0.001$) in the general population. Among patients treated for VABP, there was no significant difference in the 30-day mortality (58.3% vs. 56.5%). In terms of therapy safety, the advantage of CFD therapy in a COL-based regimen has been demonstrated. The incidence of TEAEs was higher in the COL-treated group at 21.1% (100% was AKI) vs. CFD at 2.1% (rash) [110].

Similar reports come from Rando et al. These researchers evaluated the efficacy of CFD therapy (mainly in combination therapy) in 13 patients with severe infections of multi-bacterial etiology including CRAB (10/13 developed VABP complicated by SARS-CoV2 infection). The median duration of treatment was 10 days, 7/13 patients achieved clinical cure, the remaining patients died despite treatment (30-day mortality ratio = 46%). No TEAEs associated with CFD treatment were reported. Although there was a considerable amount of mortality, it is noteworthy that the patients were critically ill, had numerous comorbidities, and the infection was multibacterial, complicated by SARS-CoV2 [111]. The effectiveness of various treatment options for VABP with the etiology of CRAB complicated by SARS-CoV2 infection in ICU was also evaluated. Russo et al. compared the efficacy of therapeutic regimens based on COL and CFD. The study enrolled 73 patients who were randomly assigned to appropriate subgroups treated with the appropriate drug regimen. All tested strains were classified as XDR or PDR. The population treated with CFD in combination with other drugs had lower 14- and 30-day mortality ratios than the population treated with COL-based regimens (5.2% and 31.5% vs. 75.9% and 98.1%, respectively). Special benefits in the 30-day survival ratio were attributed, especially, to the therapy with a combination of CFD + FOS [112].

Conclusions similar to those from Russo et al.'s study were derived from Rando et al.'s study, which compared the effectiveness of therapeutic regimens based on the combination of CFD + COL ± TGC or, among others, COL + TGC + FOS in VABP therapy in the etiology of CRAB complicated by SARS-CoV2 infection. The 28-day mortality ratio showed that the CFD-containing therapy was associated with a significantly lower risk of death than other regimens without CFD at 44% vs. 67%, respectively. As in the previous study, the significant benefits of CFD use in VABP with CRAB etiology in the field of increased patient survival were demonstrated [113]. The efficacy of CFD-based monotherapy and combination therapy in XDR and DTR (difficult-to-treat resistant) *P. aeruginosa* etiology infections was also evaluated. The study included 17 patients, and symptoms of VABP were presented by 7 patients (5/7 had SARS-CoV2 coinfection). Only one patient received CFD as monotherapy, the others were treated with combination therapy (COL inhaled, FOS, moxifloxacin, or CAZ-AVI). In 4/7 cases, the infection was caused by *P. aeruginosa*; in other cases, *K. pneumoniae* KPC, *S. maltophila*, and *A. baumannii* PDR was additionally found. In the VABP group, no significant TEAEs were found during treatment, and 3/7 patients died, while the remaining patients showed clinical success and microbiological eradication, with only 1/7 patients developing a recurrent infection. The study proves the effectiveness of CFD therapy in critically ill patients whose previous therapy, including TGC, CAZ-AVI, or MER, was unsuccessful [114].

In the single-center prospective observational study by Dalfino et al., different therapeutic regimens (including CFD or COL) were evaluated for CRAB etiology VABP suffered patients. The patient population (N = 90) was randomized into two groups: treated with a CFD-based regimen with inhaled COL (N = 40) and treated with a COL-based regimen with additional inhaled COL (N = 50). Clinical failure occurred less frequently in patients treated with CFD (25% vs. 48%). Microbiological failure was also shown more frequently among treated with COL combinations (30% vs. 60%). The 14-day mortality ratio was significantly lower for patients treated with CFD (10% vs. 38%). Of the CFD-based regimens, the highest percentage of clinical cures was recorded with the combination of CFD + FOS + COL inhaled [115].The findings are highly consistent with previously described studies [111–113].

The study by de la Fuente et al. evaluated the efficacy of CFD-based therapy in the treatment of carbapenem-resistant (CR) GNB infections in a population (N = 13) of patients hospitalized in an ICU. Eleven patients were confirmed clinically and in the laboratory to have VABP, showing the following etiological factors: 5/10 CR *P. aeruginosa*, 3/10 CR *Burkholderia cepacia*, 1/10 *S. maltophila*, and 1/10 CR *K. pneumoniae* KPC. Moreover, 4/10 patients received CFD alone, and the others were treated with CFD combination regimens with LEV/MER + TRM-STX/ciprofloxacin/COL/LEV + TRM-STX. In the group of patients treated for VABP, the 28-day mortality ratio reached 30%, and the remaining patients presented clinical success. All patients, except for one, developed microbiological eradication. According to the study, the treated group did not present severe TEAEs, and the high effectiveness of VABP therapy with CFD for a CR-GNB etiology was demonstrated [116].

A summary of the clinical efficiency of the cefiderocol therapy among patients with VABP is presented in Table 9.

Table 9. Efficacy of CFD in the treatment of VABP.

Number of Patients	Treatment Scheme in VABP Population	Outcomes of Trial		Etiology of Infection	Reference
N = 300	CFD (N = 148) (41% VABP)	14-MR	15%	*K. pneumoniae* *P. aeruginosa* *A. baumannii* (CRE strains: 70% ESβL producers: 43–67%)	[107]
			13%		
		28-MR	23%		
			22%		
	MER (N = 152) (44% VABP)	CC	66%		
			56%		
		Microbiological eradication	42%		
			34%		

Table 9. *Cont.*

Number of Patients	Treatment Scheme in VABP Population	Outcomes of Trial		Etiology of Infection	Reference
N = 152	CFD (N = 101) (24% VABP)	CC at TOC	50%	CR-GNB With a predominance of CRAB (65% of isolates in the CFD and 53% in the BAT groups)	[108]
	BAT (N = 51) (27% VABP)		53%		
	CFD (N = 101) (24% VABP)	Microbiological eradication at TOC	23%		
	BAT (N = 51) (27% VABP)		21%		
	CFD (N = 101) (24% VABP)	14-MR	24%		
	BAT (N = 51) (27% VABP)		31%		
	CFD (N = 101) (24% VABP)	28-MR	14%		
	BAT (N = 51) (27% VABP)		18%		
N = 35 (VABP group)	12/35 CFD-based therapy	30-MR	58.3%	CRAB	[110]
	23/35 COL-based therapy		56.5%		
N = 73	CFD (N = 19)-based therapy	14-MR	5.2%	CRAB complicated with SARS-CoV2	[112]
	COL (N = 54)-based therapy		75.9%		
	CFD (N = 19)-based therapy	28-MR	31.5%		
	COL (N = 54)-based therapy		98.1%		
N = 90	CFD-based therapy with additional inh. COL (N = 40)	Clinical failure	25%	CRAB	[115]
	COL-based therapy with additional inh. COL (N = 50)		48%		
	CFD-based therapy with additional inh. COL (N = 40)	Microbiological failure	30%		
	COL-based therapy with additional inh. COL (N = 50)		60%		
	CFD-based therapy with additional inh. COL (N = 40)	14-MR	10%		
	COL-based therapy with additional inh. COL (N = 50)		38%		

14/28/30-MR—14/28/30-day mortality ratio; CC—clinical cure; CFD—cefiderocol; COL—colistin; CRAB—carbapenem-resistant *A. baumannii*; CRE—carbapenem-resistant *Enterobacterales*; CR-GNB—carbapenem-resistant Gram negative bacteria; inh.—inhaled; MER—meropenem; TOC—test of cure; VABP—ventilator-associated bacterial pneumonia.

5.1.3. Case Studies

Mercadante et al. described the use of CFD as a salvage therapy in an 8-month-old infant treated for VABP with an etiology of *P. aeruginosa* VIM. The patient's severe condition led to the need for therapy with ECMO (extra-corporeal membrane oxygenation) and the use of CRRT (continuous-renal-replacement Therapy). The previous therapy, which included CAZ-AVI + ATM, did not lead to an improvement. After a week of treatment, the patient's condition deteriorated. Because of the sensitivity of *P. aeruginosa* to CFD (MIC = 1 µg/mL), the decision was made to use the off-label preparation. The therapy lasted for 14 days and led to the patient's recovery; after 4 days from the beginning of treatment, the condition improved so much that ECMO and CRRT were discontinued, and after a week of treatment, negative blood and BAL (bronchoalveolar lavage) cultures were obtained. The study demonstrates the effectiveness of CFD-based salvage therapy in infants undergoing invasive intensive care procedures, as follows: ECMO and CRRT [117].

Another case study presented the successful use of CFD in the treatment of necrotizing pneumonia caused by *P. aeruginosa* associated with mechanical ventilation. The pathogen developed resistances to CEF-TAZ and MER during previous antibiotic therapy. The combination of CFD with TOB and ciprofloxacin led to the patient's recovery and confirmed microbiological eradication. The evidence suggests that CFD is effective in salvage therapy for people with VABP caused by CEF-TAZ-resistant MDR P. aeruginosa [118].

5.1.4. Mechanism of Action of β-Lactam Drugs

The mechanism of action of the cephalosporins described in this manuscript is to block the final stage of cell wall biosynthesis. These antibiotics inhibit the activity of transpeptidases (PBPs)—protoplasmic proteins responsible for cross-linking the bacterial cell wall. As a result of their action, bonds among its components—mucopolysaccharides—are not formed, causing lysis of the bacterial cell.

The combinations we looked at closely in this review increase the sensitivity of strains to the associated cephalosporin. This is due to the fact that β-lactamase inhibitors bind to some PBPs, thus protecting the antibiotic from enzymatic degradation in the presence of β-lactamase-producing bacteria. In addition, the higher the class of cephalosporins used in combination with the inhibitor, the broader the activity against β-lactamases. Therefore, wider use in the treatment of infections with a G-negative bacterial etiology in order from the least to most active is as follows: CFP-SBT, CEF-TAZ, and CAZ-AVI.

The highest activity is shown by cefiderocol, a siderophore cephalosporin, which has been confirmed against WHO priority pathogens, as follows: *A. baumannii*, *P. aeruginosa*, and *Enterobacteriaceae* resistant to carbapenems. Moreover, its efficacy against ESβL-, AmpC-, KPC-, and MBL-positive strains was confirmed. The mechanism of cefiderocol is based on the inhibition of the synthesis of the bacterial cell wall by binding to PBP proteins, mainly PBP3. This cephalosporin, in the first step, binds iron, which is necessary for bacteria to function, and then, by active transport, enters the periplasmic space. During this process, it reduces the number of porins in the outer membrane and the expression of MDR pumps responsible for the efflux [100].

6. Material and Methods

This review refers to articles from the Scopus, Web of Science, PubMed and Google Scholar databases. In total, 115 papers were included. The following keywords were used in the search of the titles and abstracts of the articles: 'ventilator-associated pneumonia', 'ceftolozane tazobactam', 'ceftazidime avibactam', 'cefoperazone sulbactam', and 'cefiderocol'. Table 1 presents ability of cefiderocol and selected cephalosporin-inhibitor combinations to overcome multidrug resistance among bacteria. Tables 3, 5, 7, and 9 refer to the clinical efficiency of the described preparations in VABP therapy. Tables 2, 4, 6, and 8 show the in vitro antimicrobial activity of the mentioned drugs against clinical isolates of bacteria derived from patients suffering from VABP. Figures 1–4 present two-dimensional structures of the revealed preparations. Figure 5 depicts the scheme of the molecular action of the described β-lactams inside the bacterial cell.

7. Conclusions

New combinations of cephalosporins with β-lactamase inhibitors are a useful therapeutic option in the treatment of serious nosocomial infections, e.g., HAP and VABP. The addition of β-lactamase inhibitors extend the spectrum of action of classical β-lactam antibiotics by bacteria with a high potential of resistance, especially KPC and ESβL *Enterobacteriaceae*. The analyzed clinical trials report that the BL + BLI combination scan be used to optimize treatment times, minimize the risk of adverse events, and improve patient survival. The use of the described pharmaceuticals allowed, in many cases, for the limitation of the use of multidrug combination therapies based on colistin, a last resort antibiotic, which is connected with a high risk of nephrotoxicity. In the studies mentioned above, it was shown that the drugs used obtained a generally noninferior profile of action compared to polymyxins or carbapenems, which is an important premise in the era of increasing antibiotic resistance, especially in ICU wards. Special hopes are, therefore, raised by the new generation of β-lactam antibiotics: siderophore cephalosporins and their representative—cefiderocol. In both in vitro studies and clinical trials, the preparation displayed an impressive level of action, demonstrating that it has therapeutic efficacy comparable or superior to colistin and meropenem. Particularly important is the high activity of cefiderocol against resistance to the carbapenems *A. baumannii* and *S. maltophila* (pathogens especially dangerous to critically ill ICU patients), which has been repeatedly demonstrated in clinical trials. In conclusion, while VABP is still a challenge for clinicians, the cefiderocol and BL + BLI combinations described above can be used as salvage therapy in patients treated in the ICU ward when the other treatment options have been exhausted.

Limited treatment options for patients developing VABP should also encourage the search for innovative treatments for this life-threatening infection. Because of the increase in antibiotic resistance, it is necessary to develop products that can combine the properties of classic drugs (e.g., inhibition of PBP protein) with a novel mechanism to break down antibiotic resistance (e.g., siderophore mechanism). Future directions of research should oscillate around the search for new ways of influencing bacterial cells. High hopes in this field may be raised by substances of natural origin (e.g., plant extracts and alkaloids), which due to their cytostatic properties have already been used in, among others, the therapy of oncological diseases. It also seems particularly important to search for new combinations of classic antibiotics with antibacterial substances of natural origin. The detection of additive or synergistic interactions among these compounds could reduce the use of last resort drugs (e.g., colistin) in favor of preparations with, among others, lower toxicity. Although more research is needed, the combination of innovative therapeutic strategies and rapid molecular diagnostics systems (MALDI-TOF MS) is a worthwhile option in the face of the growing global problem of antibiotic resistance.

However, this study has some limitations. Various clinical studies had significant differences, including the number of VABP patients in each cohort, which could lead to different statistical power values and, sometimes, bias in the outcomes. In addition, individuals who qualified for specific groups and received a particular treatment regimen could differ disproportionately based on their severity of concomitant diseases, age, or the number of previous hospitalizations (a factor that may increase the risk of colonization with MDR pathogens). The treatment effects may have been different, although not due to drug action directly but to the worsening prognosis of specific patients. It was also difficult to determine if the clinical success was caused by a specific drug or combined preparations in multiple situations in which the evaluated drugs were used in a polytherapy. The antibiotic resistance profile of pathogens, which are the underlying causes of VABP, can be different depending on the patient's location or the type of ward they are in during hospitalization. Conducting a meta-analysis study, which would enable us to develop the topic in depth, seems worthwhile for the described reasons.

Author Contributions: Conceptualization, S.V. and A.D.-M.; software, E.T. and J.S.; investigation, S.V., E.T. and J.S.; resources, S.V., E.T. and J.S.; data curation, A.D.-M., S.V., E.T. and J.S.; writing—original draft preparation, S.V., E.T. and J.S.; writing—review and editing, A.D.-M., S.V., E.T. and J.S.; visualization, A.D.-M., S.V., E.T. and J.S.; supervision, A.D.-M. and S.V. All authors have read and agreed to the published version of the manuscript.

Funding: This research received no external funding.

Institutional Review Board Statement: Not applicable.

Informed Consent Statement: Not applicable.

Data Availability Statement: Not applicable.

Conflicts of Interest: The authors declare no conflicts of interest.

Abbreviations

14/28/30-MR—14/28/30-day mortality ratio; ADC—*Acinetobacter*-derived cephalosporinase; AE—adverse event; AKC—amikacin; AKI—acute kidney injury; AmpC—cephalosporinases encoded on the chromosomes of many of the *Enterobacteriaceae*; AP—acute pyelonephritis; APACHE II—Acute Physiology and Chronic Health Evaluation II; ATM—aztreonam; AZT—azithromycin; BAL—bronchoalveolar lavage; BAT—best available treatment; BL + BLI—β-lactam + β-lactamase inhibitor; CAZ-AVI—ceftazidime-avibactam; CC—clinical cure; CEF-TAZ—ceftolozane-tazobactam; CEP—clinical evaluable population; CFD—cefiderocol; CFP-SBT—cefoperazone-sulbactam; cIAI—complicated intra-abdominal infection; CLSI—Clinical and Laboratory Standards Institute; COL—colistin; CPIS—clinical pulmonary infection score; CPM—cefepime; CR—carbapenem-resistant; CRAB—carbapenem-resistant *A. baumannii*; CrCl—creatinine clearance; CRE—carbapenem-resistant *Enterobacterales*; CRP—C-reactive protein; CRPA—carbapenem-resistant *P. aeruginosa*; CRRT—continuous renal replacement treatment; CS—carbapenem-sensitive; CTX-M—CTX-M-type β-lactamase; cUTI—complicated urinary tract infection; DTT—difficult-to-treat; ECMO—extracorporeal membrane oxygenation; eGFR—estimated glomerular filtration rate; ELF—epithelial lung fluid; EMA—European Medicines Agency; EOT—end-of-treatment; ESKAPE—acronym related to the most commonly isolated bacteria in ICU wards; ESβL—extended-spectrum β-lactamase; EUCAST—European Committee on Antimicrobial Susceptibility Testing; FDA—Food Drug Administration; FOS—fosfomycin; GEN—gentamycin; GNB—Gram-negative bacteria; h—hours; HAP—hospital-acquired pneumonia; HCAP—healthcare-acquired pneumonia; i.v.—intravenous; ICU—intensive care unit; IHMR—in-hospital mortality ratio; IL—interleukin; IMI—imipenem; IMP—imipenemase-type MBL; Inh.—inhaled; INR—international normalized ratio; ITT—intention-to-treat; IUPAC—International Union of Pure and Applied Chemistry; KPC—*K. pneumoniae* carbapenemase; LEV—levofloxacin; LRTI—lower respiratory tract infection; MALDI-TOF MS—matrix-assisted laser desorption/ionization-time of flight mass spectrometry; MBL—metallo-β-lactamase; MDR—multidrug-resistant; MER—meropenem; MIC—minimum inhibitory concentration; MIC_{50}—the lowest concentration of the antibiotic at which 50% of the isolates were inhibited; MIC_{90}—the lowest concentration of the antibiotic at which 90% of the isolates were inhibited; mITT—modified intention to treat; MSSA—methicillin-susceptible *S. aureus*; N—number of; NDM—New-Delhi-type MBL; NS—nonsusceptible; OXA—oxacillinase-type β-lactamase; PBP—penicillin-binding protein; PCT—procalcitonin; PDR—pan-drug-resistant; PIP-TAZ—piperacillin-tazobactam; PP—per protocol; PT—prothrombin time; SARS-CoV2—severe acute respiratory syndrome coronavirus-2; SHV—sulfhydryl-variable penicillinase; TEAE—treatment emergent adverse event; TEM—plasmid-encoded β-lactamase in Gram-negative bacteria; TGC—tigecycline; TNF-α—tumor necrosis factor α; TOB—tobramycin; TOC—test of cure; TRM-STX—trimethoprim-sulfamethoxazole; VABP—ventilator-associated bacterial pneumonia; VAP—ventilator-associated pneumonia; VIM—Verona integron-encoded MBL; vs.—versus; XDR—extensively drug-resistant

References

1. American Thoracic Society; Infectious Diseases Society of America. Guidelines for the Management of Adults with Hospital-Acquired, Ventilator-Associated, and Healthcare-Associated *Pneumonia. Am. J. Respir. Crit. Care Med.* **2005**, *171*, 388–416. [CrossRef]
2. Kalanuria, A.A.; Zai, W.; Mirski, M. Ventilator-Associated Pneumonia in the ICU. *Crit. Care* **2014**, *18*, 208. [CrossRef]
3. Shah, H.; Ali, A.; Patel, A.A.; Abbagoni, V.; Goswami, R.; Kumar, A.; Velasquez Botero, F.; Otite, E.; Tomar, H.; Desai, M.; et al. Trends and Factors Associated with Ventilator-Associated Pneumonia: A National Perspective. *Cureus* **2022**, *14*, e23634. [CrossRef]
4. Gunalan, A.; Sastry, A.S.; Ramanathan, V.; Sistla, S. Early-vs Late-Onset Ventilator-Associated Pneumonia in Critically Ill Adults: Comparison of Risk Factors, Outcome, and Microbial Profile. *Indian J. Crit. Care Med.* **2023**, *27*, 411–415. [CrossRef]
5. Papazian, L.; Klompas, M.; Luyt, C.E. Ventilator-Associated Pneumonia in Adults: A Narrative Review. *Intensive Care Med.* **2020**, *46*, 888–906. [CrossRef]
6. Sandiumenge, A.; Rello, J. Ventilator-Associated Pneumonia Caused by ESKAPE Organisms: Cause, Clinical Features, and Management. *Curr. Opin. Pulm. Med.* **2012**, *18*, 187–193. [CrossRef]
7. Daubin, C.; Vincent, S.; Vabret, A.; Du Cheyron, D.; Parienti, J.J.; Ramakers, M.; Freymuth, F.; Charbonneau, P. Nosocomial Viral Ventilator-Associated Pneumonia in the Intensive Care Unit: A Prospective Cohort Study. *Intensive Care Med.* **2005**, *31*, 1116–1122. [CrossRef]
8. Shorr, A.F.; Ilges, D.T.; Micek, S.T.; Kollef, M.H. The Importance of Viruses in Ventilator-Associated Pneumonia. *Infect. Control Hosp. Epidemiol.* **2023**, *44*, 1137–1142. [CrossRef]
9. Fumagalli, J.; Panigada, M.; Klompas, M.; Berra, L. Ventilator-Associated Pneumonia among SARSCoV-2 Acute Respiratory Distress Syndrome Patients. *Curr. Opin. Crit. Care* **2022**, *28*, 74–82. [CrossRef]
10. Huang, Y.; Jiao, Y.; Zhang, J.; Xu, J.; Cheng, Q.; Li, Y.; Liang, S.; Li, H.; Gong, J.; Zhu, Y.; et al. Microbial Etiology and Prognostic Factors of Ventilator-Associated Pneumonia: A Multicenter Retrospective Study in Shanghai. *Clin. Infect. Dis.* **2018**, *67*, S146–S152. [CrossRef]

11. Vincent, J.L.; Bihari, D.J.; Suter, P.M.; Bruining, H.A.; White, J.; Nicolas Chanoin, M.H.; Wolff, M.; Spencer, R.C.; Hemmer, M. The Prevalence of Nosocomial Infection in Intensive Care Units in Europe: Results of the European Prevalence of Infection in Intensive Care (EPIC) Study. *JAMA* **1995**, *274*, 639–644. [CrossRef]
12. Magiorakos, A.P.; Srinivasan, A.; Carey, R.B.; Carmeli, Y.; Falagas, M.E.; Giske, C.G.; Harbarth, S.; Hindler, J.F.; Kahlmeter, G.; Olsson-Liljequist, B.; et al. Multidrug-Resistant, Extensively Drug-Resistant and Pandrug-Resistant Bacteria: An International Expert Proposal for Interim Standard Definitions for Acquired Resistance. *Clin. Microbiol. Infect.* **2012**, *18*, 268–281. [CrossRef]
13. Li, Y.; Roberts, J.; Walker, M.; Aslan, A.; Harris, P.; Sime, F. The Global Epidemiology of Ventilator-Associated Pneumonia Caused by Multi-Drug Resistant *Pseudomonas aeruginosa*: A Systematic Review and Meta-Analysis. *Int. J. Infect. Dis.* **2023**, *139*, 78–85. [CrossRef]
14. Čiginskienė, A.; Dambrauskienė, A.; Rello, J.; Adukauskienė, D. Ventilator-Associated Pneumonia Due to Drug-Resistant *Acinetobacter baumannii*: Risk Factors and Mortality Relation with Resistance Profiles, and Independent Predictors of in-Hospital Mortality. *Medicina* **2019**, *55*, 49. [CrossRef]
15. Awad, L.S.; Abdallah, D.I.; Mugharbil, A.M.; Jisr, T.H.; Droubi, N.S.; El-Rajab, N.A.; Moghnieh, R.A. An Antibiotic Stewardship Exercise in the ICU: Building a Treatment Algorithm for the Management of Ventilator-Associated Pneumonia Based on Local Epidemiology and the 2016 Infectious Diseases Society of America/American Thoracic Society Guidelines. *Infect. Drug Resist.* **2018**, *11*, 17–28. [CrossRef]
16. Sader, H.S.; Carvalhaes, C.G.; Kimbrough, J.H.; Mendes, R.E.; Castanheira, M. Activity of Aztreonam-Avibactam against Enterobacterales Resistant to Recently Approved Beta-Lactamase Inhibitor Combinations Collected in Europe, Latin America, and the Asia-Pacific Region (2020–2022). *Int. J. Antimicrob. Agents* **2024**, *63*, 107113. [CrossRef] [PubMed]
17. Duda-Madej, A.; Viscardi, S.; Topola, E. Meropenem/Vaborbactam: β-Lactam/β-Lactamase Inhibitor Combination, the Future in Eradicating Multidrug Resistance. *Antibiotics* **2023**, *12*, 1612. [CrossRef]
18. Ku, Y.-H.; Yu, W.-L. Cefoperazone/Sulbactam: New Composites against Multiresistant Gram Negative Bacteria? *Infect. Genet. Evol.* **2021**, *88*, 104707. [CrossRef] [PubMed]
19. Carvalhaes, C.G.; Castanheira, M.; Sader, H.S.; Flamm, R.K.; Shortridge, D. Antimicrobial Activity of Ceftolozane-Tazobactam Tested against Gram-Negative Contemporary (2015–2017) Isolates from Hospitalized Patients with Pneumonia in US Medical Centers. *Diagn. Microbiol. Infect. Dis.* **2019**, *94*, 93–102. [CrossRef] [PubMed]
20. Karlowsky, J.A.; Hackel, M.A.; Tsuji, M.; Yamano, Y.; Echols, R.; Sahm, D.F. In Vitro Activity of Cefiderocol, a Siderophore Cephalosporin, Against Gram-Negative Bacilli Isolated by Clinical Laboratories in North America and Europe in 2015-2016: SIDERO-WT-2015. *Int. J. Antimicrob. Agents* **2019**, *53*, 456–466. [CrossRef] [PubMed]
21. Shirley, M. Ceftazidime-Avibactam: A Review in the Treatment of Serious Gram-Negative Bacterial Infections. *Drugs* **2018**, *78*, 675–692. [CrossRef] [PubMed]
22. Syed, Y.Y. Cefiderocol: A Review in Serious Gram-Negative Bacterial Infections. *Drugs* **2021**, *81*, 1559–1571. [CrossRef]
23. Ong'uti, S.; Czech, M.; Robilotti, E.; Holubar, M. Cefiderocol: A New Cephalosporin Stratagem Against Multidrug-Resistant Gram-Negative Bacteria. *Clin. Infect. Dis.* **2022**, *74*, 1303–1312. [CrossRef] [PubMed]
24. Sader, H.S.; Carvalhaes, C.G.; Streit, J.M.; Castanheira, M.; Flamm, R.K. Antimicrobial Activity of Cefoperazone-Sulbactam Tested against Gram-Negative Organisms from Europe, Asia-Pacific, and Latin America. *Int. J. Infect. Dis.* **2020**, *91*, 32–37. [CrossRef] [PubMed]
25. Lai, C.-C.; Chen, C.-C.; Lu, Y.-C.; Lin, T.-P.; Chuang, Y.-C.; Tang, H.-J. Appropriate Composites of Cefoperazone–Sulbactam against Multidrug-Resistant Organisms. *Infect. Drug Resist.* **2018**, *11*, 1441–1445. [CrossRef] [PubMed]
26. Chang, P.-C.; Chen, C.-C.; Lu, Y.C.; Lai, C.-C.; Huang, H.-L.; Chuang, Y.-C.; Tang, H.-J. The Impact of Inoculum Size on the Activity of Cefoperazone-Sulbactam against Multidrug Resistant Organisms. *J. Microbiol. Immunol. Infect.* **2018**, *51*, 207–213. [CrossRef] [PubMed]
27. Kuo, H.-Y.; Wang, F.-D.; Yen, Y.-F.; Lin, M.-L.; Liu, C.-Y. In Vitro Activities of Piperacillin or Cefoperazone Alone and in Combination with Beta-Lactamase Inhibitors against Gram-Negative Bacilli. *New Microbiol.* **2009**, *32*, 49–55. [PubMed]
28. Jones, R.N.; Wilson, H.W.; Thornsberry, C.; Barry, A.L. In Vitro Antimicrobial Activity of Cefoperazone-Sulbactam Combinations against 554 Clinical Isolates Including a Review and β-Lactamase Studies. *Diagn. Microbiol. Infect. Dis.* **1985**, *3*, 489–499. [CrossRef] [PubMed]
29. Wang, Q.; Huang, M.; Zhou, S. Observation of Clinical Efficacy of the Cefoperazone/Sulbactam Anti-infective Regimen in the Treatment of Multidrug-resistant *Acinetobacter baumannii* Lung Infection. *J. Clin. Pharm. Ther.* **2022**, *47*, 1020–1027. [CrossRef]
30. Guclu, E.; Kaya, G.; Ogutlu, A.; Karabay, O. The Effect of Cefoperazone Sulbactam and Piperacillin Tazobactam on Mortality in Gram-Negative Nosocomial Infections. *J. Chemother.* **2020**, *32*, 118–123. [CrossRef]
31. Lan, S.-H.; Chang, S.-P.; Lai, C.-C.; Lu, L.-C.; Tang, H.-J. Efficacy and Safety of Cefoperazone-Sulbactam in Empiric Therapy for Febrile Neutropenia. *Medicine* **2020**, *99*, e19321. [CrossRef] [PubMed]
32. Brogden, R.N.; Carmine, A.; Heel, R.C.; Morley, P.A.; Speight, T.M.; Avery, G.S. Cefoperazone: A Review of Its In Vitro Antimicrobial Activity, Pharmacological Properties and Therapeutic Efficacy. *Drugs* **1981**, *22*, 423–460. [CrossRef]
33. Williams, J.D. β-Lactamase Inhibition and In Vitro Activity of Sulbactam and Sulbactam/Cefoperazone. *Clin. Infect. Dis.* **1997**, *24*, 494–497. [CrossRef]

34. Penwell, W.A.; Shapiro, A.B.; Giacobbe, R.A.; Gu, R.-F.; Gao, N.; Thresher, J.; McLaughlin, R.E.; Huband, M.; DeJonge, B.L.M.; Ehmann, D.E.; et al. Molecular Mechanisms of Sulbactam Antibacterial Activity and Resistance Determinants in *Acinetobacter baumannii*. *Antimicrob. Agents Chemother.* **2015**, *59*, 1680–1689. [CrossRef]
35. Moussa, S.H.; Shapiro, A.B.; Mcleod, S.M.; Iyer, R.; Carter, N.M.; Tsai, Y.; Siu, L.K.; Alita, A. Contemporary Clinical Isolates of *Acinetobacter baumannii*. *Antimicrob. Agents Chemother.* **2021**, *10*, 28.
36. Namiduru, M.; Güngör, G.; Karaoğlan, I.; Dikensoy, Ö. Antibiotic Resistance of Bacterial Ventilator-Associated Pneumonia in Surgical Intensive Care Units. *J. Int. Med. Res.* **2004**, *32*, 78–83. [CrossRef]
37. Xia, J.; Zhang, D.; Xu, Y.; Gong, M.; Zhou, Y.; Fang, X. A Retrospective Analysis of Carbapenem-Resistant *Acinetobacter baumannii*-Mediated Nosocomial Pneumonia and the in Vitro Therapeutic Benefit of Cefoperazone/Sulbactam. *Int. J. Infect. Dis.* **2014**, *23*, 90–93. [CrossRef] [PubMed]
38. Capoor, M.R.; Nair, D.; Srivastava, L.; Gupta, B.; Aggarwal, P. Characterization and Changing Minimum Inhibitory Concentration (MIC) of Acinetobacter Species from a Tertiary Care Setup. *J. Commun. Dis.* **2005**, *37*, 99–107. [PubMed]
39. Chiang, T.-T.; Tang, H.-J.; Chiu, C.-H.; Chen, T.-L.; Ho, M.-W.; Lee, C.-H.; Sheng, W.-H.; Yang, Y.-S. Antimicrobial Activities of Cefoperazone-Sulbactam in Comparison to Cefoperazone against Clinical Organisms from Medical Centers in Taiwan. *J. Med. Sci.* **2016**, *36*, 229. [CrossRef]
40. Huang, Y.; Wang, W.; Huang, Q.; Wang, Z.; Xu, Z.; Tu, C.; Wan, D.; He, M.; Yang, X.; Xu, H.; et al. Clinical Efficacy and In Vitro Drug Sensitivity Test Results of Azithromycin Combined with Other Antimicrobial Therapies in the Treatment of MDR P. Aeruginosa Ventilator-Associated Pneumonia. *Front. Pharmacol.* **2022**, *13*, 944965. [CrossRef]
41. Xia, Y.-L.; Ge, M.; Wang, Z. Pathogenic Analysis of Ventilator-Associated Pneumonia in the Pediatric Intensive Care Unit in High-Altitude Areas. *Zhongguo Dang Dai Er Ke Za Zhi* **2014**, *16*, 787–790. [PubMed]
42. Huang, C.-T.; Chen, C.-H.; Chen, W.-C.; Wang, Y.-T.; Lai, C.-C.; Fu, P.-K.; Kuo, L.-K.; Chen, C.-M.; Fang, W.-F.; Tu, C.-Y.; et al. Clinical Effectiveness of Cefoperazone-Sulbactam vs. Piperacillin-Tazobactam for the Treatment of Pneumonia in Elderly Patients. *Int. J. Antimicrob. Agents* **2022**, *59*, 106491. [CrossRef] [PubMed]
43. Liu, J.-W.; Chen, Y.-H.; Lee, W.-S.; Lin, J.-C.; Huang, C.-T.; Lin, H.-H.; Liu, Y.-C.; Chuang, Y.-C.; Tang, H.-J.; Chen, Y.-S.; et al. Randomized Noninferiority Trial of Cefoperazone-Sulbactam versus Cefepime in the Treatment of Hospital-Acquired and Healthcare-Associated Pneumonia. *Antimicrob. Agents Chemother.* **2019**, *63*, e00023-19. [CrossRef] [PubMed]
44. Chen, C.-H.; Tu, C.-Y.; Chen, W.-C.; Kuo, L.-K.; Wang, Y.-T.; Fu, P.-K.; Ku, S.-C.; Fang, W.-F.; Chen, C.-M.; Lai, C.-C. Clinical Efficacy of Cefoperazone-Sulbactam versus Piperacillin-Tazobactam in the Treatment of Hospital-Acquired Pneumonia and Ventilator-Associated Pneumonia. *Infect. Drug Resist.* **2021**, *14*, 2251–2258. [CrossRef]
45. Kara, İ.; Yildirim, F.; Bilaloglu, B.; Karamanlioglu, D.; Kayacan, E.; Dizbay, M.; Turkoglu, M.; Aygencel, G. Comparison of the Efficacy of Colistin Monotherapy and Colistin Combination Therapies in the Treatment of Nosocomial Pneumonia and Ventilator-Associated Pneumonia Caused by *Acinetobacter baumannii*. *South Afr. J. Crit. Care* **2015**, *31*, 51–58. [CrossRef]
46. Qin, Y.; Zhang, J.; Wu, L.; Zhang, D.; Fu, L.; Xue, X. Comparison of the Treatment Efficacy between Tigecycline plus High-Dose Cefoperazone-Sulbactam and Tigecycline Monotherapy against Ventilator-Associated Pneumonia Caused by Extensively Drug-Resistant *Acinetobacter baumannii*. *Int. J. Clin. Pharmacol. Ther.* **2018**, *56*, 120–129. [CrossRef]
47. Kanchanasuwan, S.; Kositpantawong, N.; Singkhamanan, K.; Hortiwakul, T.; Charoenmak, B.; Ozioma, F.N.; Doi, Y.; Chusri, S. Outcomes of Adjunctive Therapy with Intravenous Cefoperazone-Sulbactam for Ventilator-Associated Pneumonia Due to Carbapenem-Resistant *Acinetobacter baumannii*. *Infect. Drug Resist.* **2021**, *14*, 1255–1264. [CrossRef]
48. Lv, Q.; Deng, Y.; Zhu, X.; Ruan, J.; Chen, L. Effectiveness of Cefoperazone-Sulbactam Alone and Combined with Tigecycline in the Treatment of Multi-Drug Resistant *Acinetobacter baumannii* Pulmonary Infection. *J. Coll. Physicians Surg. Pak.* **2020**, *30*, 332–334. [CrossRef] [PubMed]
49. Giacobbe, D.R.; Bassetti, M.; De Rosa, F.G.; Del Bono, V.; Grossi, P.A.; Menichetti, F.; Pea, F.; Rossolini, G.M.; Tumbarello, M.; Viale, P.; et al. Ceftolozane/Tazobactam: Place in Therapy. *Expert Rev. Anti. Infect. Ther.* **2018**, *16*, 307–320. [CrossRef]
50. Lizza, B.D.; Betthauser, K.D.; Ritchie, D.J.; Micek, S.T.; Kollef, M.H. New Perspectives on Antimicrobial Agents: Ceftolozane-Tazobactam. *Antimicrob. Agents Chemother.* **2021**, *65*, e0231820. [CrossRef]
51. Cluck, D.; Lewis, P.; Stayer, B.; Spivey, J.; Moorman, J. Ceftolozane-Tazobactam: A New-Generation Cephalosporin. *Am. J. Health Pharm.* **2015**, *72*, 2135–2146. [CrossRef] [PubMed]
52. van Duin, D.; Bonomo, R.A. Ceftazidime/Avibactam and Ceftolozane/Tazobactam: Second-Generation β-Lactam/β-Lactamase Inhibitor Combinations. *Clin. Infect. Dis.* **2016**, *63*, 234–241. [CrossRef] [PubMed]
53. Sheffield, M.; Nelson, D.; O'Neal, M.; Gould, A.P.; Bouchard, J.; Nicolau, D.; Justo, J.A.; Hucks, J.; Bookstaver, P.B. Use of Continuous-Infusion Ceftolozane/Tazobactam for Resistant Gram-Negative Bacterial Infections: A Retrospective Analysis and Brief Review of the Literature. *Int. J. Antimicrob. Agents* **2020**, *56*, 106158. [CrossRef] [PubMed]
54. Rusu, A.; Lungu, I.-A. The New Fifth-Generation Cephalosporins—A Balance between Safety and Efficacy. *Rom. J. Pharm. Pract.* **2020**, *13*, 121–126. [CrossRef]
55. Craig, W.A.; Andes, D.R. In Vivo Activities of Ceftolozane, a New Cephalosporin, with and without Tazobactam against *Pseudomonas aeruginosa* and Enterobacteriaceae, Including Strains with Extended-Spectrum β-Lactamases, in the Thighs of Neutropenic Mice. *Antimicrob. Agents Chemother.* **2013**, *57*, 1577–1582. [CrossRef] [PubMed]

56. Haidar, G.; Philips, N.J.; Shields, R.K.; Snyder, D.; Cheng, S.; Potoski, B.A.; Doi, Y.; Hao, B.; Press, E.G.; Cooper, V.S.; et al. Ceftolozane-Tazobactam for the Treatment of Multidrug-Resistant *Pseudomonas aeruginosa* Infections: Clinical Effectiveness and Evolution of Resistance. *Clin. Infect. Dis.* **2017**, *65*, 110–120. [CrossRef] [PubMed]
57. Pérez, A.; Gato, E.; Pérez-Llarena, J.; Fernández-Cuenca, F.; Gude, M.J.; Oviaño, M.; Pachón, M.E.; Garnacho, J.; González, V.; Pascual, Á.; et al. High Incidence of MDR and XDR *Pseudomonas aeruginosa* Isolates Obtained from Patients with Ventilator-Associated Pneumonia in Greece, Italy and Spain as Part of the MagicBullet Clinical Trial. *J. Antimicrob. Chemother.* **2019**, *74*, 1244–1252. [CrossRef] [PubMed]
58. Sader, H.S.; Flamm, R.K.; Dale, G.E.; Rhomberg, P.R.; Castanheira, M. Murepavadin Activity Tested against Contemporary (2016–17) Clinical Isolates of XDR *Pseudomonas aeruginosa*. *J. Antimicrob. Chemother.* **2018**, *73*, 2400–2404. [CrossRef] [PubMed]
59. Pfaller, M.A.; Shortridge, D.; Arends, S.J.R.; Duncan, L.R.; Streit, J.M.; Flamm, R.K. Activity of Ceftolozane-Tazobactam and Comparators When Tested against Bacterial Surveillance Isolates Collected from Patients at Risk of Infections Caused by Resistant Gram-Negative Pathogens. *Diagn. Microbiol. Infect. Dis.* **2020**, *98*, 115101. [CrossRef]
60. Idowu, T.; Zhanel, G.G.; Schweizer, F. A Dimer, but Not Monomer, of Tobramycin Potentiates Ceftolozane against Multidrug-Resistant and Extensively Drug-Resistant *Pseudomonas aeruginosa* and Delays Resistance Development. *Antimicrob. Agents Chemother.* **2020**, *64*, e02055-19. [CrossRef]
61. Karlowsky, J.A.; Lob, S.H.; Young, K.; Motyl, M.R.; Sahm, D.F. Activity of Ceftolozane/Tazobactam against Gram-Negative Isolates from Patients with Lower Respiratory Tract Infections—SMART United States 2018–2019. *BMC Microbiol.* **2021**, *21*, 74. [CrossRef] [PubMed]
62. Shortridge, D.; Pfaller, M.A.; Arends, S.J.R.; Raddatz, J.; DePestel, D.D.; Flamm, R.K. Comparison of the In Vitro Susceptibility of Ceftolozane-Tazobactam with the Cumulative Susceptibility Rates of Standard Antibiotic Combinations When Tested Against *Pseudomonas aeruginosa* from ICU Patients with Bloodstream Infections or Pneumonia. *Open Forum Infect. Dis.* **2019**, *6*, ofz240. [CrossRef] [PubMed]
63. Shortridge, D.; Carvalhaes, C.G.; Streit, J.M.; Flamm, R.K. Susceptibility Trends of Ceftolozane/Tazobactam and Comparators When Tested against U.S. Gram-Negative Bacterial Surveillance Isolates (2012–2018). *Diagn. Microbiol. Infect. Dis.* **2021**, *100*, 115302. [CrossRef] [PubMed]
64. Lob, S.H.; DePestel, D.D.; DeRyke, C.A.; Kazmierczak, K.M.; Young, K.; Motyl, M.R.; Sahm, D.F. Ceftolozane/Tazobactam and Imipenem/Relebactam Cross-Susceptibility Among Clinical Isolates of *Pseudomonas aeruginosa* from Patients with Respiratory Tract Infections in ICU and Non-ICU Wards—SMART United States 2017–2019. *Open Forum Infect. Dis.* **2021**, *8*, ofab320. [CrossRef] [PubMed]
65. Candel, F.J.; Santerre Henriksen, A.; Longshaw, C.; Yamano, Y.; Oliver, A. In Vitro Activity of the Novel Siderophore Cephalosporin, Cefiderocol, in Gram-Negative Pathogens in Europe by Site of Infection. *Clin. Microbiol. Infect.* **2022**, *28*, 447.e1–447.e6. [CrossRef] [PubMed]
66. Kollef, M.H.; Nováček, M.; Kivistik, Ü.; Réa-Neto, Á.; Shime, N.; Martin-Loeches, I.; Timsit, J.-F.; Wunderink, R.G.; Bruno, C.J.; Huntington, J.A.; et al. Ceftolozane-Tazobactam versus Meropenem for Treatment of Nosocomial Pneumonia (ASPECT-NP): A Randomised, Controlled, Double-Blind, Phase 3, Non-Inferiority Trial. *Lancet Infect. Dis.* **2019**, *19*, 1299–1311. [CrossRef] [PubMed]
67. Timsit, J.-F.; Huntington, J.A.; Wunderink, R.G.; Shime, N.; Kollef, M.H.; Kivistik, Ü.; Nováček, M.; Réa-Neto, Á.; Martin-Loeches, I.; Yu, B.; et al. Ceftolozane/Tazobactam versus Meropenem in Patients with Ventilated Hospital-Acquired Bacterial Pneumonia: Subset Analysis of the ASPECT-NP Randomized, Controlled Phase 3 Trial. *Crit. Care* **2021**, *25*, 290. [CrossRef] [PubMed]
68. Johnson, M.G.; Bruno, C.; Castanheira, M.; Yu, B.; Huntington, J.A.; Carmelitano, P.; Rhee, E.G.; De Anda, C.; Motyl, M. Evaluating the Emergence of Nonsusceptibility among *Pseudomonas aeruginosa* Respiratory Isolates from a Phase-3 Clinical Trial for Treatment of Nosocomial Pneumonia (ASPECT-NP). *Int. J. Antimicrob. Agents* **2021**, *57*, 106278. [CrossRef] [PubMed]
69. Pogue, J.M.; Kaye, K.S.; Veve, M.P.; Patel, T.S.; Gerlach, A.T.; Davis, S.L.; Puzniak, L.A.; File, T.M.; Olson, S.; Dhar, S.; et al. Ceftolozane/Tazobactam vs Polymyxin or Aminoglycoside-Based Regimens for the Treatment of Drug-Resistant *Pseudomonas aeruginosa*. *Clin. Infect. Dis.* **2020**, *71*, 304–310. [CrossRef]
70. Mogyoródi, B.; Csékó, A.B.; Hermann, C.; Gál, J.; Iványi, Z.D. Ceftolozane/Tazobactam versus Colistin in the Treatment of Ventilator-Associated Pneumonia Due to Extensively Drug-Resistant *Pseudomonas aeruginosa*. *Sci. Rep.* **2022**, *12*, 4455. [CrossRef]
71. Bassetti, M.; Castaldo, N.; Cattelan, A.; Mussini, C.; Righi, E.; Tascini, C.; Menichetti, F.; Mastroianni, C.M.; Tumbarello, M.; Grossi, P.; et al. Ceftolozane/Tazobactam for the Treatment of Serious *Pseudomonas aeruginosa* Infections: A Multicentre Nationwide Clinical Experience. *Int. J. Antimicrob. Agents* **2019**, *53*, 408–415. [CrossRef]
72. Zhanel, G.G.; Dhami, R.; Baxter, M.; Kosar, J.; Cervera, C.; Irfan, N.; Zvonar, R.; Borgia, S.; Tessier, J.-F.; Dow, G.; et al. Real-Life Experience with Ceftolozane/Tazobactam in Canada: Results from the CLEAR (Canadian LEadership on Antimicrobial Real-Life Usage) Registry. *J. Glob. Antimicrob. Resist.* **2021**, *25*, 346–350. [CrossRef] [PubMed]
73. Gallagher, J.C.; Satlin, M.J.; Elabor, A.; Saraiya, N.; McCreary, E.K.; Molnar, E.; El-Beyrouty, C.; Jones, B.M.; Dixit, D.; Heil, E.L.; et al. Ceftolozane-Tazobactam for the Treatment of Multidrug-Resistant *Pseudomonas aeruginosa* Infections: A Multicenter Study. *Open Forum Infect. Dis.* **2018**, *5*, ofy280. [CrossRef] [PubMed]

74. Holger, D.J.; Rebold, N.S.; Alosaimy, S.; Morrisette, T.; Lagnf, A.; Belza, A.C.; Coyne, A.J.K.; El Ghali, A.; Veve, M.P.; Rybak, M.J. Impact of Ceftolozane-Tazobactam vs. Best Alternative Therapy on Clinical Outcomes in Patients with Multidrug-Resistant and Extensively Drug-Resistant *Pseudomonas aeruginosa* Lower Respiratory Tract Infections. *Infect. Dis. Ther.* **2022**, *11*, 1965–1980. [CrossRef]
75. Wang, Y.; Wang, J.; Wang, R.; Cai, Y. Resistance to Ceftazidime-Avibactam and Underlying Mechanisms. *J. Glob. Antimicrob. Resist.* **2020**, *22*, 18–27. [CrossRef]
76. Zhanel, G.G.; Lawson, C.D.; Adam, H.; Schweizer, F.; Zelenitsky, S.; Lagacé-Wiens, P.R.S.; Denisuik, A.; Rubinstein, E.; Gin, A.S.; Hoban, D.J.; et al. Ceftazidime-Avibactam: A Novel Cephalosporin/β-Lactamase Inhibitor Combination. *Drugs* **2013**, *73*, 159–177. [CrossRef] [PubMed]
77. Matesanz, M.; Mensa, J. Ceftazidime-Avibactam. *Rev. Esp. Quimioter.* **2021**, *34* (Suppl. 1), 38–40. [CrossRef]
78. Falcone, M.; Paterson, D. Spotlight on Ceftazidime/Avibactam: A New Option for MDR Gram-Negative Infections. *J. Antimicrob. Chemother.* **2016**, *71*, 2713–2722. [CrossRef]
79. Carmeli, Y.; Armstrong, J.; Laud, P.J.; Newell, P.; Stone, G.; Wardman, A.; Gasink, L.B. Ceftazidime-Avibactam or Best Available Therapy in Patients with Ceftazidime-Resistant Enterobacteriaceae and *Pseudomonas aeruginosa* Complicated Urinary Tract Infections or Complicated Intra-Abdominal Infections (REPRISE): A Randomised, Pathogen-Directed, Phase 3 Study. *Lancet Infect. Dis.* **2016**, *16*, 661–673. [CrossRef]
80. Drwiega, E.N.; Rodvold, K.A. Penetration of Antibacterial Agents into Pulmonary Epithelial Lining Fluid: An Update. *Clin. Pharmacokinet.* **2022**, *61*, 17–46. [CrossRef]
81. Ehmann, D.E.; Jahić, H.; Ross, P.L.; Gu, R.F.; Hu, J.; Kern, G.; Walkup, G.K.; Fisher, S.L. Avibactam Is a Covalent, Reversible, Non-β-Lactam β-Lactamase Inhibitor. *Proc. Natl. Acad. Sci. USA* **2012**, *109*, 11663–11668. [CrossRef]
82. Sader, H.S.; Castanheira, M.; Flamm, R.K.; Mendes, R.E.; Farrell, D.J.; Jones, R.N. Ceftazidime/Avibactam Tested against Gram-Negative Bacteria from Intensive Care Unit (ICU) and Non-ICU Patients, Including Those with Ventilator-Associated Pneumonia. *Int. J. Antimicrob. Agents* **2015**, *46*, 53–59. [CrossRef]
83. Zhang, W.; Guo, Y.; Li, J.; Zhang, Y.; Yang, Y.; Dong, D.; Zhu, D.; He, P.; Hu, F. In Vitro and in Vivo Bactericidal Activity of Ceftazidime-Avibactam against Carbapenemase-Producing *Klebsiella pneumoniae*. *Antimicrob. Resist. Infect. Control* **2018**, *7*, 142. [CrossRef]
84. Stone, G.G.; Newell, P.; Gasink, L.B.; Broadhurst, H.; Wardman, A.; Yates, K.; Chen, Z.; Song, J.; Chow, J.W. Clinical Activity of Ceftazidime/Avibactam against MDR Enterobacteriaceae and *Pseudomonas aeruginosa*: Pooled Data from the Ceftazidime/Avibactam Phase III Clinical Trial Programme. *J. Antimicrob. Chemother.* **2018**, *73*, 2519–2523. [CrossRef]
85. Stone, G.G.; Bradford, P.A.; Tawadrous, M.; Taylor, D.; Cadatal, M.J.; Chen, Z.; Chow, J.W. In Vitro Activity of Ceftazidime-Avibactam against Isolates from Respiratory and Blood Specimens from Patients with Nosocomial Pneumonia, Including Ventilator-Associated Pneumonia, in a Phase 3 Clinical Trial. *Antimicrob. Agents Chemother.* **2020**, *64*, e02356-19. [CrossRef]
86. Sader, H.S.; Castanheira, M.; Flamm, R.K. Antimicrobial Activity of Ceftazidime-Avibactam against Gram-Negative Bacteria Isolated from Patients Hospitalized with Pneumonia in U.S. Medical Centers, 2011 to 2015. *Antimicrob. Agents Chemother.* **2017**, *61*, e02083-16. [CrossRef] [PubMed]
87. Torres, A.; Zhong, N.; Pachl, J.; Timsit, J.-F.; Kollef, M.; Chen, Z.; Song, J.; Taylor, D.; Laud, P.J.; Stone, G.G.; et al. Ceftazidime-Avibactam versus Meropenem in Nosocomial Pneumonia, Including Ventilator-Associated Pneumonia (REPROVE): A Randomised, Double-Blind, Phase 3 Non-Inferiority Trial. *Lancet Infect. Dis.* **2018**, *18*, 285–295. [CrossRef] [PubMed]
88. Torres, A.; Rank, D.; Melnick, D.; Rekeda, L.; Chen, X.; Riccobene, T.; Critchley, I.A.; Lakkis, H.D.; Taylor, D.; Talley, A.K. Randomized Trial of Ceftazidime-Avibactam vs Meropenem for Treatment of Hospital-Acquired and Ventilator-Associated Bacterial Pneumonia (REPROVE): Analyses per US FDA–Specified End Points. *Open Forum Infect. Dis.* **2019**, *6*, ofz149. [CrossRef] [PubMed]
89. Shi, Y.; Hu, J.; Liu, P.; Wang, T.; Wang, H.; Liu, Y.; Cao, Q.; Zuo, X. Ceftazidime-Avibactam-Based Versus Tigecycline-Based Regimen for the Treatment of Carbapenem-Resistant *Klebsiella pneumoniae*-Induced Pneumonia in Critically Ill Patients. *Infect. Dis. Ther.* **2021**, *10*, 2721–2734. [CrossRef]
90. Tsolaki, V.; Mantzarlis, K.; Mpakalis, A.; Malli, E.; Tsimpoukas, F.; Tsirogianni, A.; Papagiannitsis, C.; Zygoulis, P.; Papadonta, M.-E.; Petinaki, E.; et al. Ceftazidime-Avibactam To Treat Life-Threatening Infections by Carbapenem-Resistant Pathogens in Critically Ill Mechanically Ventilated Patients. *Antimicrob. Agents Chemother.* **2020**, *64*, e02320-19. [CrossRef]
91. van Duin, D.; Lok, J.J.; Earley, M.; Cober, E.; Richter, S.S.; Perez, F.; Salata, R.A.; Kalayjian, R.C.; Watkins, R.R.; Doi, Y.; et al. Colistin Versus Ceftazidime-Avibactam in the Treatment of Infections Due to Carbapenem-Resistant Enterobacteriaceae. *Clin. Infect. Dis.* **2018**, *66*, 163–171. [CrossRef] [PubMed]
92. Zheng, G.; Zhang, J.; Wang, B.; Cai, J.; Wang, L.; Hou, K.; Zhang, Y.; Zhang, L.; Yang, Z.; He, J.; et al. Ceftazidime-Avibactam in Combination with In Vitro Non-Susceptible Antimicrobials Versus Ceftazidime-Avibactam in Monotherapy in Critically Ill Patients with Carbapenem-Resistant *Klebsiella pneumoniae* Infection: A Retrospective Cohort Study. *Infect. Dis. Ther.* **2021**, *10*, 1699–1713. [CrossRef] [PubMed]
93. Mantzarlis, K.; Manoulakas, E.; Parisi, K.; Sdroulia, E.; Zapaniotis, N.; Tsolaki, V.; Zakynthinos, E.; Makris, D. Meropenem plus Ertapenem and Ceftazidime-Avibactam plus Aztreonam for the Treatment of Ventilator Associated Pneumonia Caused by Pan-Drug Resistant *Klebsiella pneumonia*. *Antibiotics* **2024**, *13*, 141. [CrossRef] [PubMed]

94. Burastero, G.J.; Orlando, G.; Santoro, A.; Menozzi, M.; Franceschini, E.; Bedini, A.; Cervo, A.; Faltoni, M.; Bacca, E.; Biagioni, E.; et al. Ceftazidime/Avibactam in Ventilator-Associated Pneumonia Due to Difficult-to-Treat Non-Fermenter Gram-Negative Bacteria in COVID-19 Patients: A Case Series and Review of the Literature. *Antibiotics* **2022**, *11*, 1007. [CrossRef] [PubMed]
95. Corbella, L.; Boán, J.; San-Juan, R.; Fernández-Ruiz, M.; Carretero, O.; Lora, D.; Hernández-Jiménez, P.; Ruiz-Ruigómez, M.; Rodríguez-Goncer, I.; Silva, J.T.; et al. Effectiveness of Ceftazidime-Avibactam for the Treatment of Infections Due to *Pseudomonas aeruginosa*. *Int. J. Antimicrob. Agents* **2022**, *59*, 106517. [CrossRef] [PubMed]
96. Jorgensen, S.C.J.; Trinh, T.D.; Zasowski, E.J.; Lagnf, A.M.; Bhatia, S.; Melvin, S.M.; Steed, M.E.; Simon, S.P.; Estrada, S.J.; Morrisette, T.; et al. Real-World Experience With Ceftazidime-Avibactam for Multidrug-Resistant Gram-Negative Bacterial Infections. *Open Forum Infect. Dis.* **2019**, *6*, ofz522. [CrossRef] [PubMed]
97. Domingues, S.; Lima, T.; Saavedra, M.J.; Da Silva, G.J. An Overview of Cefiderocol's Therapeutic Potential and Underlying Resistance Mechanisms. *Life* **2023**, *13*, 1427. [CrossRef] [PubMed]
98. El-Lababidi, R.M.; Rizk, J.G. Cefiderocol: A Siderophore Cephalosporin. *Ann. Pharmacother.* **2020**, *54*, 1215–1231. [CrossRef] [PubMed]
99. McCreary, E.K.; Heil, E.L.; Tamma, P.D. New Perspectives on Antimicrobial Agents: Cefiderocol. *Antimicrob. Agents Chemother.* **2021**, *65*, e0217120. [CrossRef]
100. Sato, T.; Yamawaki, K. Cefiderocol: Discovery, Chemistry, and In Vivo Profiles of a Novel Siderophore Cephalosporin. *Clin. Infect. Dis.* **2019**, *69*, S538–S543. [CrossRef]
101. Katsube, T.; Nicolau, D.P.; Rodvold, K.A.; Wunderink, R.G.; Echols, R.; Matsunaga, Y.; Menon, A.; Portsmouth, S.; Wajima, T. Intrapulmonary Pharmacokinetic Profile of Cefiderocol in Mechanically Ventilated Patients with Pneumonia. *J. Antimicrob. Chemother.* **2021**, *76*, 2902–2905. [CrossRef] [PubMed]
102. Zhanel, G.G.; Golden, A.R.; Zelenitsky, S.; Wiebe, K.; Lawrence, C.K.; Adam, H.J.; Idowu, T.; Domalaon, R.; Schweizer, F.; Zhanel, M.A.; et al. Cefiderocol: A Siderophore Cephalosporin with Activity Against Carbapenem-Resistant and Multidrug-Resistant Gram-Negative Bacilli. *Drugs* **2019**, *79*, 271–289. [CrossRef] [PubMed]
103. Rahman, M.S.; Koh, Y.-S. A Novel Antibiotic Agent, Cefiderocol, for Multidrug-Resistant Gram-Negative Bacteria. *J. Bacteriol. Virol.* **2020**, *50*, 218–226. [CrossRef]
104. Yamano, Y. In Vitro Activity of Cefiderocol Against a Broad Range of Clinically Important Gram-Negative Bacteria. *Clin. Infect. Dis.* **2019**, *69*, S544–S551. [CrossRef] [PubMed]
105. Tsuji, M.; Hackel, M.; Echols, R.; Yamano, Y.; Sahm, D. In Vitro Activity of Cefiderocol against Globally Collected Carbapenem-Resistant Gram-Negative Bacteria Isolated from Urinary Track Source: SIDERO-CR-2014/2016. *Open Forum Infect. Dis.* **2017**, *4*, S366. [CrossRef]
106. Shortridge, D.; Streit, J.M.; Mendes, R.; Castanheira, M. In Vitro Activity of Cefiderocol against U.S. and European Gram-Negative Clinical Isolates Collected in 2020 as Part of the SENTRY Antimicrobial Surveillance Program. *Microbiol. Spectr.* **2022**, *10*, e0271221. [CrossRef] [PubMed]
107. Wunderink, R.G.; Matsunaga, Y.; Ariyasu, M.; Clevenbergh, P.; Echols, R.; Kaye, K.S.; Kollef, M.; Menon, A.; Pogue, J.M.; Shorr, A.F.; et al. Cefiderocol versus High-Dose, Extended-Infusion Meropenem for the Treatment of Gram-Negative Nosocomial Pneumonia (APEKS-NP): A Randomised, Double-Blind, Phase 3, Non-Inferiority Trial. *Lancet Infect. Dis.* **2021**, *21*, 213–225. [CrossRef]
108. Bassetti, M.; Echols, R.; Matsunaga, Y.; Ariyasu, M.; Doi, Y.; Ferrer, R.; Lodise, T.P.; Naas, T.; Niki, Y.; Paterson, D.L.; et al. Efficacy and Safety of Cefiderocol or Best Available Therapy for the Treatment of Serious Infections Caused by Carbapenem-Resistant Gram-Negative Bacteria (CREDIBLE-CR): A Randomised, Open-Label, Multicentre, Pathogen-Focused, Descriptive, Phase 3 Trial. *Lancet Infect. Dis.* **2021**, *21*, 226–240. [CrossRef]
109. Falcone, M.; Tiseo, G.; Nicastro, M.; Leonildi, A.; Vecchione, A.; Casella, C.; Forfori, F.; Malacarne, P.; Guarracino, F.; Barnini, S.; et al. Cefiderocol as Rescue Therapy for *Acinetobacter baumannii* and Other Carbapenem-Resistant Gram-Negative Infections in Intensive Care Unit Patients. *Clin. Infect. Dis.* **2021**, *72*, 2021–2024. [CrossRef]
110. Falcone, M.; Tiseo, G.; Leonildi, A.; Della Sala, L.; Vecchione, A.; Barnini, S.; Farcomeni, A.; Menichetti, F. Cefiderocol- Compared to Colistin-Based Regimens for the Treatment of Severe Infections Caused by Carbapenem-Resistant *Acinetobacter baumannii*. *Antimicrob. Agents Chemother.* **2022**, *66*, e0214221. [CrossRef]
111. Rando, E.; Segala, F.V.; Vargas, J.; Seguiti, C.; De Pascale, G.; Murri, R.; Fantoni, M. Cefiderocol for Severe Carbapenem-Resistant *A. baumannii* Pneumonia: Towards the Comprehension of Its Place in Therapy. *Antibiotics* **2021**, *11*, 3. [CrossRef] [PubMed]
112. Russo, A.; Bruni, A.; Gullì, S.; Borrazzo, C.; Quirino, A.; Lionello, R.; Serapide, F.; Garofalo, E.; Serraino, R.; Romeo, F.; et al. Efficacy of Cefiderocol- vs Colistin-Containing Regimen for Treatment of Bacteraemic Ventilator-Associated Pneumonia Caused by Carbapenem-Resistant *Acinetobacter baumannii* in Patients with COVID-19. *Int. J. Antimicrob. Agents* **2023**, *62*, 106825. [CrossRef] [PubMed]
113. Rando, E.; Cutuli, S.L.; Sangiorgi, F.; Tanzarella, E.S.; Giovannenze, F.; De Angelis, G.; Murri, R.; Antonelli, M.; Fantoni, M.; De Pascale, G. Cefiderocol-Containing Regimens for the Treatment of Carbapenem-Resistant *A. baumannii* Ventilator-Associated Pneumonia: A Propensity-Weighted Cohort Study. *JAC-Antimicrobial Resist.* **2023**, *5*, dlad085. [CrossRef] [PubMed]
114. Meschiari, M.; Volpi, S.; Faltoni, M.; Dolci, G.; Orlando, G.; Franceschini, E.; Menozzi, M.; Sarti, M.; Del Fabro, G.; Fumarola, B.; et al. Real-Life Experience with Compassionate Use of Cefiderocol for Difficult-to-Treat Resistant *Pseudomonas aeruginosa* (DTR-P) Infections. *JAC-Antimicrobial Resist.* **2021**, *3*, dlab188. [CrossRef]

115. Dalfino, L.; Stufano, M.; Bavaro, D.F.; Diella, L.; Belati, A.; Stolfa, S.; Romanelli, F.; Ronga, L.; Di Mussi, R.; Murgolo, F.; et al. Effectiveness of First-Line Therapy with Old and Novel Antibiotics in Ventilator-Associated Pneumonia Caused by Carbapenem-Resistant *Acinetobacter baumannii*: A Real Life, Prospective, Observational, Single-Center Study. *Antibiotics* **2023**, *12*, 1048. [CrossRef] [PubMed]
116. de la Fuente, C.; Rodríguez, M.; Merino, N.; Carmona, P.; Machuca, I.; Córdoba-Fernández, M.; Guzmán-Puche, J.; Domínguez, A.; López-Viñau, T.; García, L.; et al. Real-Life Use of Cefiderocol for Salvage Therapy of Severe Infections Due to Carbapenem-Resistant Gram-Negative Bacteria. *Int. J. Antimicrob. Agents* **2023**, *62*, 106818. [CrossRef] [PubMed]
117. Mercadante, S.; Tripiciano, C.; Romani, L.; Di Nardo, M.; Bottari, G.; Goffredo, B.M.; Simeoli, R.; Guzzo, I.; Lancella, L.; Antachopoulos, C.; et al. The Use of Cefiderocol as Salvage Therapy in an Infant Receiving ECMO and Continuous Renal Replacement Therapy. *Antibiotics* **2023**, *13*, 37. [CrossRef]
118. Roy, A.M.; Hodgson, M.; Brickman, C.; MacDougall, C. Cefiderocol as Adjunctive Treatment of Necrotizing Ventilator-Associated Pneumonia Due to Extensively Drug-Resistant *Pseudomonas aeruginosa*. *Infect. Dis. Clin. Pract.* **2021**, *29*, e111–e113. [CrossRef]

antibiotics

Article

Chemical Characterization and Biological Properties Assessment of *Euphorbia resinifera* and *Euphorbia officinarum* Moroccan Propolis

Oumaima Boutoub [1,2], Soukaina El-Guendouz [2], Isabel Matos [1,3], Lahsen El Ghadraoui [2], Maria Clara Costa [1,4], Jorge Dias Carlier [1,4], Maria Leonor Faleiro [1,3,5,*], Ana Cristina Figueiredo [6], Letícia M. Estevinho [7,8] and Maria Graça Miguel [1,9,*]

[1] Faculdade de Ciências e Tecnologia, Universidade do Algarve, Campus de Gambelas, 8005-139 Faro, Portugal; boutoub.oumaima@gmail.com (O.B.); icmatos@ualg.pt (I.M.); mcorada@ualg.pt (M.C.C.); jcarlier@ualg.pt (J.D.C.)

[2] Laboratory of Functional Ecology and Environmental Engineering, Faculty of Science and Technology, Sidi Mohamed Ben Abdellah University, P.O. Box 2202, Fez 30050, Morocco; soukainaelguendouz@gmail.com (S.E.-G.); lahsen.elghadraoui@usmba.ac.ma (L.E.G.)

[3] Algarve Biomedical Center-Research Institute, 8005-139 Faro, Portugal

[4] Centro de Ciências do Mar (CCMAR), Faculdade de Ciências e Tecnologia, Universidade do Algarve, Campus de Gambelas, 8005-139 Faro, Portugal

[5] Champalimaud Researh Program, Chaupalimaud Centre for the Unknown, 1400-038 Lisbon, Portugal

[6] Centro de Estudos do Ambiente e do Mar (CESAM Ciências), Faculdade de Ciências da Universidade de Lisboa (FCUL), Biotecnologia Vegetal, DBV, C2, Campo Grande, 1749-016 Lisboa, Portugal; acfigueiredo@fc.ul.pt

[7] Centro de Investigação de Montanha (CIMO), Instituto Politécnico de Bragança, Campus de Santa Apolónia, 5300-253 Bragança, Portugal; leticia@ipb.pt

[8] Laboratório Associado para a Sustentabilidade e Tecnologia em Regiões de Montanha (SusTEC), Instituto Politécnico de Bragança, Campus de Santa Apolónia, 5300-253 Bragança, Portugal

[9] Instituto Mediterrâneo para a Agricultura, Ambiente e Desenvolvimento (MED), Faculdade de Ciências e Tecnologia, Universidade do Algarve, 8005-139 Faro, Portugal

* Correspondence: mfaleiro@ualg.pt (M.L.F.); mgmiguel@ualg.pt (M.G.M.)

Citation: Boutoub, O.; El-Guendouz, S.; Matos, I.; El Ghadraoui, L.; Costa, M.C.; Carlier, J.D.; Faleiro, M.L.; Figueiredo, A.C.; Estevinho, L.M.; Miguel, M.G. Chemical Characterization and Biological Properties Assessment of *Euphorbia resinifera* and *Euphorbia officinarum* Moroccan Propolis. *Antibiotics* **2024**, *13*, 230. https://doi.org/10.3390/antibiotics13030230

Academic Editor: Marc Maresca

Received: 18 January 2024
Revised: 20 February 2024
Accepted: 27 February 2024
Published: 29 February 2024

Abstract: Although the plants of the genus *Euphorbia* are largely exploited by therapists in Morocco, the composition and antibacterial activities of propolis from these plants are still unknown. To address this gap, this study aimed to characterize the pollen type, the volatile compounds, and the phenolic and mineral profiles of three *Euphorbia* propolis samples collected in Morocco and evaluate their antimicrobial activities. The minimum inhibitory concentration of the propolis samples was determined by the microdilution method, and the anti-adherence activity was evaluated by the crystal violet assay. The examination of anti-quorum-sensing proprieties was performed using the biosensor *Chromobacterium violaceum* CV026. Pollen analysis revealed that *Euphorbia resinifera* pollen dominated in the P1 sample (58%), while *E. officinarum* pollen dominated in the P2 and P3 samples (44%). The volatile compounds were primarily composed of monoterpene hydrocarbons, constituting 35% in P1 and 31% in P2, with α-pinene being the major component in both cases, at 16% in P1 and 15% in P2. Calcium (Ca) was the predominant mineral element in both *E. resinifera* (P1) and *E. officinarum* (P2 and P3) propolis samples. Higher levels of phenols, flavonoids and dihydroflavonoids were detected in the *E. officinarum* P2 sample. The minimum inhibitory concentration (MIC) value ranged from 50 to 450 μL/mL against Gram-positive and Gram-negative bacteria. *Euphorbia* propolis displayed the ability to inhibit quorum sensing in the biosensor *C. violaceum* CV026 and disrupted bacterial biofilm formation, including that of resistant bacterial pathogens. In summary, the current study evidences the potential use of *E. officinarum* propolis (P2 and P3) to combat important features of resistant pathogenic bacteria, such as quorum sensing and biofilm formation.

Keywords: volatiles; *Euphorbia* propolis; pollen analysis; quorum sensing; anti-biofilm; methicillin-resistant *Staphylococcus aureus*

1. Introduction

Propolis, also called bee glue, is a natural substance produced by bees from plant resins, sap and other botanical sources. It is used by bees to seal small gaps and cracks in their hives, as well as to sterilize and protect the hive from bacterial and fungal infections [1]. The biological activities of propolis are thought to be due to the presence of a wide range of bioactive compounds, including flavonoids, phenolic acids, terpenes and volatiles [2]. The composition of propolis drives its activity, and the content of flavonoids, the main polyphenols in propolis, is influenced by the botanical sources available in the bee environment, as well as the honey bee species present [3–5], which in turn will impact its antibacterial action. The proposed mechanism of action of propolis against bacteria is based on the attachment of active components of propolis to the bacterial cellular membrane, which affects its stability and integrity and induces the loss of cellular content, causing cell death. Moreover, the flavonoids may act at the molecular level by inhibiting the activity of DNA gyrase, affecting bacterial replication [6]. Besides the antimicrobial activity of propolis, other important features of this bee product have been reported, particularly its anticancer activity [7,8], anti-inflammatory activity [9,10] and anti-diabetic activity [11–13]. Altogether, the reported studies evidence the great therapeutic potential of propolis in the medical field.

Presently, the increase in antibiotic resistance exhibited by several human pathogen bacteria is a huge challenge for healthcare institutions. A particular group of bacteria is of special concern due to its extraordinary ability to develop a multiresistant profile; it includes vancomycin-resistant *Enterococcus faecium* (VRE); *Staphylococcus aureus*, methicillin-resistant, vancomycin-intermediate and -resistant (MRSA/VRSA); and carbapenem-resistant and third-generation cephalosporin-resistant *Klebsiella pneumoniae, Acinetobacter baumanii, Pseudomonas aeruginosa* and Enterobacteriaceae, abbreviated by the designation ESKAPE. These bacteria master antibiotic resistance mechanisms, namely (i) enzymatic destruction or alteration of the agent, (ii) modification or protection of the molecular target, (iii) obstruction of the entry of the antibiotic into the cell, (iv) use of efflux pumps [14]. In Morocco, a study carried out in 2020 at the Mohamed V hospital center, in the Meknes region, showed that *S. aureus* alone is responsible for 86.6% of bacterial infections in burns, and all of them were identified as methicillin-resistant *S. aureus* (MRSA) [15]. The problem with healthcare-associated infections is mainly associated with the ability of bacteria to form biofilms (adhere to biotic and abiotic surfaces and establish a cell–cell communication community) [16]. A biofilm develops along determined phases: the bacterial cells adhere to a surface (the initial phase can be reversible), followed by a permanent adhesion (irreversible), the production of exopolysaccharides that allow the aggregation of the bacterial cells, biofilm maturation, and finally disaggregation of the sessile cells (adherent cells) by the action of fluids and natural processes inside the aggregate [17]. The bacterial colonization of medical devices (e.g., urinary catheters, central venous catheters, and prosthetic joints) in the form of biofilms constitutes a serious risk to the health of patients [18].

The present work aimed to research the chemical composition and the potential antibacterial activity of *Euphorbia resinifera* and *E. officinarum* propolis due to the lack of information on the important propolis obtained from these plants, which are largely used by therapists in Morocco. The antibacterial activity of the propolis samples was evaluated against susceptible *S. aureus*, methicillin-resistant *S. aureus* (MRSA) and susceptible *Escherichia coli*. Their capacity to prevent bacterial biofilm formation and also their ability to disrupt mature biofilm were tested against susceptible and multiresistant Gram-positive and Gram-negative bacteria.

2. Results

2.1. Pollen Grains

Propolis sample P1, collected in Beni Mellal-Khénifra, showed predominantly pollen grains of *E. resinifera* (59%) (Table 1), followed by *Genista hirsuta* (9%). P2 and P3 propolis samples were dominated by *E. officinarum* pollen grains (47% and 45%, respectively), followed by *Hypericum elodes* (14%) and *Smilax aspera* (21%), respectively.

Table 1. Propolis samples, place, year of production and the most predominant pollen of three *Euphorbia* propolis samples from Morocco.

Sample	Propolis Type	Pollen Species	(%)	Region/Year	Coordinates
P1	*Euphorbia resinifera*	*Euphorbia resinifera*	58.6 ± 1.1	Beni Mellal-Khénifra (2019)	32°22′06″ N, 6°22′09″ W
		Genista hirsuta	9.0 ± 0.4		
		Asparagus albus	8.0 ± 1.4		
		Populus nigra	7.1 ± 1.0		
		Kleinia anteuphorbium	5.9 ± 0.4		
		Caesalpinia spinosa	4.0 ± 0.7		
		Pinus pinaster	3.9 ± 0.3		
		Eucalyptus cinerea	1.6 ± 0.4		
		Convolvulus arvensis	1.6 ± 0.4		
P2	*Euphorbia officinarum*	*Euphorbia officinarum*	46.6 ± 0.5	Souss-Massa-Tiznit (2019)	29°43′ N, 8°58′ W
		Hypericum elodes	13.9 ± 1.5		
		Quercus rotindifolia	10.3 ± 1.0		
		Populus nigra	7.1 ± 1.0		
		Eucalyptus cinerea	6.7 ± 0.5		
		Pinus pinaster	6.4 ± 0.2		
		Kleinia anteuphorbium	3.4 ± 0.8		
		Ilex aquifolium	3.3 ± 0.3		
		Asparagus albus	1.9 ± 0.6		
P3	*Euphorbia officinarum*	*Euphorbia officinarum*	44.8 ± 0.5	Guelmim-Oued noun (2019)	28°27′ N, 10°07′ W
		Smilax aspera	21.0 ± 1.4		
		Cistus crepis	14.6 ± 1.0		
		Caesalpinia spinosa	9.9 ± 0.8		
		Campanula rotundifolia	5.7 ± 0.3		
		Hypericum elodes	3.9 ± 0.6		
		Populus nigra	3.8 ± 0.8		
		Convolvulus arvensis	3.0 ± 0.17		
		Ilex aquifolium	2.9 ± 0.2		
		Quercus rotindifolia	1.8 ± 0.3		

As far as we know, this is the first time that palynological data have been provided for *Euphorbia* propolis. These differences can be attributed to the different regions where the samples were collected (Souss-Massa-Tiznit and Guelmim-Oued noun).

2.2. Propolis Volatile Profile

As stated in the materials and methods section, only *E. resinifera* P1 and *E. officinarum* P2 propolis samples were available for volatile determination, and their profile is depicted in Table 2, in accordance with their elution order on the DB-1 column. In total, ninety-nine components were identified, accounting for 85% and 88%, respectively, of the total volatiles. Monoterpenes constituted the major fraction, with hydrocarbons (35% in *E. resinifera* P1 and 31% in *E. officinarum* P2) and oxygen-containing compounds in high percentages (15% in both cases).

Table 2. Percentage composition of the essential oils isolated by hydrodistillation from P1 and P2.

Components	RI	P1	P2
3-Methyl-2-butenol	726	t	0.3
Hexanal	739	t	
n-Octane	800	t	
Hexanol	883	t	
Heptanal	897	t	

Table 2. *Cont.*

Components	RI	P1	P2
n-Nonane	900	0.3	0.6
Tricyclene	921	0.4	0.2
α-Thujene	924	0.7	0.3
α-Pinene	930	15.9	14.7
Camphene	938	1.4	1.0
Thuja-2,4(10)-diene *	940	1.5	1.0
Sabinene	958	3.0	2.1
β-Pinene	963	6.8	6.5
n-Octanal	973	1.0	1.2
1,2,4-Trimethyl benzene	975		t
β-Myrcene	975	0.5	t
Hexyl acetate	995	0.5	0.8
α-Terpinene	1002	0.7	0.3
p-Cymene	1003	1.3	1.3
1,8-Cineole	1005	0.2	1.2
β-Phellandrene	1005	t	t
Limonene	1009	0.7	1.3
2-Methyl butyric acid butyl ester	1017		t
trans-β-Ocimene	1027	t	t
γ-Terpinene	1035	1.6	1.2
2,5-Dimethyl styrene	1059	0.4	0.4
Terpinolene	1064	0.4	0.4
n-Nonanal	1073	2.4	3.8
α-Campholenal	1092	1.9	2.0
n-Undecane	1100	t	t
trans-Pinocarveol	1106	1.9	1.4
cis-Verbenol	1114	t	t
trans-Verbenol	1114	0.5	1.1
Pinocarvone	1121	1.2	0.9
Terpinen-4-ol	1148	0.9	0.4
Myrtenal	1153	1.3	1.0
Verbenone	1164	0.4	0.3
Myrtenol	1168	1.9	2.3
Hexyl butanoate (=Hexyl butyrate)	1173	t	t
n-Decanal	1180	4.3	7.8
trans-Carveol	1189	0.2	0.1
n-Dodecane (C12)	1200		t
Cuminaldehyde	1200		t
Carvone	1210		t
Hexyl 2-methyl butyrate	1220	0.6	1.1
2-*trans*-Decenal	1236	0.7	0.4
Nonanoic acid	1263	t	
Bornyl acetate (=Borneol acetate)	1265	3.3	3.0
n-Undecanal	1288	0.2	0.2
trans-Theaspirane	1300		
n-Tridecane	1300	t	0.2
α-Terpenyl acetate	1334	1.3	1.2
trans-2-Undecenal	1334	0.3	0.2
α-Cubebene	1345	0.3	0.2
α-Copaene	1375	t	0.2
Hexyl hexanoate	1375	0.3	t
β-Bourbonene	1379	2.3	2.0
n-Dodecanal	1397	0.7	1.0
n-Tetradecane	1400	t	0.1

Table 2. *Cont.*

Components	RI	P1	P2
β-Caryophyllene	1414	0.8	0.6
β-Copaene	1426	0.2	0.1
Aromadendrene	1428		0.2
α-Humulene	1447	0.4	0.2
allo-Aromadendrene	1456	0.7	1.5
Germacrene D	1474	0.4	0.2
Valencene	1484		0.3
α-Muurolene	1494	0.2	
trans,trans-α-Farnesene	1500	0.9	1.1
trans-Calamenene	1505	t	t
δ-Cadinene	1505	0.6	0.4
α-Calacorene	1525	0.2	0.2
Elemol	1530	t	0.1
Spathulenol	1551	6.4	3.0
β-Caryophyllene oxide	1561	0.3	0.5
Cedrol	1574	1.5	1.0
γ-Eudesmol	1609		0.2
T-Cadinol	1616	0.3	
δ-Cadinol	1621	0.4	
β-Eudesmol	1622	t	0.4
α-Eudesmol	1634	0.3	0.7
Cadalene	1640		0.5
α-Bisabolol	1656	2.0	0.7
n-Heptadecane	1700	0.3	0.6
n-Octadecane	1800		t
n-Nonadecane	1900	0.7	1.7
Hexadecanoic acid (=Palmitic acid)	1908		0.1
n-Eicosane	2000	t	t
Abietatriene	2045	0.1	0.2
n-Heneicosane	2100	0.7	1.7
n-Docosane	2200		0.2
n-Tricosane	2300	1.0	2.0
n-Tetracosane	2400	t	t
n-Pentacosane	2500	0.7	1.4
n-Hexacosane	2600	t	0.1
n-Octacosane	2800	t	0.1
n-Heptacosane	2700	0.8	t
n-Nonacosane	2900	0.4	0.8
n-Triacontane	3000	t	0.1
n-Hentriacontane	3100	0.4	1.0
% Identification		84.9	87.6
Grouped components			
Monoterpene hydrocarbons		35.3	30.7
Oxygen-containing monoterpenes		15.0	14.9
Sesquiterpene hydrocarbons		7.0	7.7
Oxygen-containing sesquiterpenes		11.2	6.6
Diterpene hydrocarbons		0.1	0.2
Oxygen-containing diterpenes		t	t
Phenylpropanoids		t	t
Fatty acids		t	0.1
Alkanes		5.0	10.0
Others		11.3	17.4

RI: Lab-calculated retention index relative to C7-C31 *n*-alkanes on the DB-1 column. * Identification based on mass spectra only. t: trace (<0.1%).

The main difference between P1 and P2 was the percentages of sesquiterpenes, alkanes and other compounds. P2 had a higher percentage of alkanes (10%) and other compounds (non-terpene aldehydes, alcohols, and esters) (17%) and lower percentages of oxygen-containing sesquiterpenes (7%) than P1 (5% and 11%, respectively). The percentages of spathulenol, α-bisabolol and cedrol were relatively higher in P1 than in P2, which may partially explain the highest percentage of the oxygen-containing sesquiterpenes in the P1 sample. Higher percentages of the alkanes heneicosane, tricosane, pentacosane, heptacosane, nonacosane and hentriacontane in P2 contributed to P2 having a higher percentage of alkanes than P1. Nonanal (4%), decanal (8%) and hexyl 2-methyl butyrate (1%) were present in higher percentages in P2 samples than in P1 ones (2%, 4%, and 1%, respectively) (Table 2).

2.3. Mineral Element Compounds

The P1 sample (*E. resinifera*) revealed higher contents of the analyzed minerals except for zinc compared with samples P2 and P3 (both *E. officinarum*); calcium (Ca) was the most abundant element in the three propolis samples (Table 3).

Table 3. Element content (mg/g) in Moroccan *Euphorbia* propolis.

	Element Content (mg/g)		
Sample	**P1**	**P2**	**P3**
Ca	16.61 ± 0.20 ^{a}	1.35 ± 0.16 ^{a}	1.13 ± 0.2 ^{a}
Co	<LOD^{1}	<LOD^{1}	<LOD^{1}
Cr	<LOD^{2}	<LOD^{2}	<LOD^{2}
Cu	<LOD^{3}	<LOD^{3}	<LOD^{3}
Fe	1.20 ± 0.02 ^{d}	0.75 ± 0.12 ^{b}	0.19 ± 0.009 ^{c}
K	1.46 ± 0.03 ^{c}	0.65 ± 0.005 ^{c}	0.43 ± 0.01 ^{b}
Mg	2.39 ± 0.02 ^{b}	0.53 ± 0.05 ^{d}	0.19 ± 0.02 ^{c}
Mn	0.04 ± 0.01 ^{f}	0.017 ± 0.01 ^{g}	0.0047 ± 0.0005 ^{g}
Na	0.49 ± 0.01 ^{e}	0.09 ± 0.001 ^{e}	0.08 ± 0.01 ^{e}
Ni	0.0037 ± 0.0006 ^{h}	0.0026 ± 0.0001 ^{h}	0.0042 ± 0.0004 ^{g}
Zn	0.033 ± 0.001 ^{g}	0.04 ± 0.03 ^{f}	0.005 ± 0.001 ^{f}

LOD^{1}: 0.0089 mg/g. LOD^{2}: 0.0081 mg/g. LOD^{3}: 0.0126 mg/g. The values in the same column followed by the same letter are not significantly different ($p < 0.05$) by Tukey's multiple range test.

2.4. Total Phenol, Flavone, Flavonol, Flavanone and Dihydroflavonol Contents

The hydro-alcoholic extracts of *E. resinifera* P1 and *E. officinarum* P2 propolis had the highest levels of total phenolic compounds (39.7 and 21.7 mg GAE/g, respectively) (Table 4). They also had the highest levels of total flavonoids (flavonols/flavones and dihydroflavonols). Nevertheless, P2 always showed the highest amounts of all groups of phenols, followed by P1 and P3, despite P2 and P3 having predominantly *E. officinarum* pollen grains but being collected in different regions. Therefore, the amounts of total phenols or flavonoids did not provide information about the collection region or the type of pollen present in the samples.

Table 4. Phenol, flavonol/flavone and dihydroflavonol contents of the three analyzed hydro-alcoholic propolis extracts.

Sample	Phenol (mg GAE/g Propolis)	Flavonol/Flavones (mg QE/g Propolis)	Dihydroflavonol (mg Naringenin Eq/g Propolis)
P1	21.7 ± 1.2 ^{b}	0.4 ± 0.0 ^{b}	6.1 ± 0.1 ^{b}
P2	39.7 ± 1.0 ^{a}	0.7 ± 0.1 ^{a}	7.6 ± 0.2 ^{a}
P3	1.3 ± 0.1 ^{c}	0.1 ± 0.0 ^{c}	3.1 ± 0.1 ^{c}

The results are shown as the mean ± standard error ($n = 3$). The values in the same column followed by the same letter are not significantly different ($p > 0.05$) by Tukey's multiple range test. GAE—gallic acid equivalents, QE—quercetin equivalents, Eq—equivalents.

2.5. Antimicrobial Activity

2.5.1. Antimicrobial Properties

The antibacterial activity of the hydro-alcoholic extract of two types of *Euphorbia* propolis, *Euphorbia resinifera* (P1) and *Euphorbia officinarum* (P2 and P3), was examined by determining the minimum inhibitory concentration (MIC) and minimum bactericidal concentration (MBC) (Table 5).

Table 5. Minimum inhibitory concentration (MIC) and minimum bactericidal concentration (MBC) of *Euphorbia* propolis extracts *.

	Bacteria	MIC (µL/mL) ‡	MBC (µL/mL) ‡
Euphorbia resinifera (P1)	*S. aureus* ATCC 6538	50 ^{a}	150 ^{a}
	MRSA12	100 ^{b}	300 ^{c}
	MRSA15	120 ^{c}	250 ^{b}
	E. coli DSM 1077	150 ^{d}	250 ^{b}
Euphorbia officinarum (P2)	*S. aureus* ATCC 6538 (C48)	150 ^{a}	200 ^{a}
	MRSA12	100 ^{b}	200 ^{a}
	MRSA15	250 ^{d}	300 ^{b}
	E. coli DSM 1077	200 ^{c}	350 ^{c}
Euphorbia officinarum (P3)	*S. aureus* ATCC 6538 (C48)	250 ^{a}	300 ^{a}
	MRSA12	200 ^{b}	400 ^{b}
	MRSA15	300 ^{c}	450 ^{c}
	E. coli DSM 1077	350 ^{d}	400 ^{b}

* The antibiotic chloramphenicol (CLO) (30 µg/mL) was used as control, and no bacterial growth was observed (the MIC value of CLO for all strains tested is <8 µg/mL). ‡ The superscript letters indicate significantly different values ($p < 0.05$) for each propolis sample across the tested bacteria.

Regarding the MIC values, *E. resinifera* propolis (P1) showed the lowest values, followed by *E. officinarum* (P2), evidencing the highest susceptibility of the tested bacteria to P1 propolis; in particular, *S. aureus* ATCC 6530 showed an MIC value of 50 µL/mL, in contrast to P2 and P3, the MIC values of which reached 150 and 250 µL/mL, respectively. *S. aureus* MRSA 12 showed similar MIC values for P1 and P2 (100 µL/mL), in contrast to *S. aureus* MRSA 15, which showed increased MIC values; specifically, the lowest value, 120 µL/mL, was achieved using P1 propolis, followed by 250 µL/mL for P2 and 300 µL/mL for P3. A similar behavior was observed in *E. coli* DSM 1077, with an MIC value of 150 µL/mL for P1 propolis, 200 µL/mL for P2 and 350 µL/mL for P3. All tested bacterial strains were less susceptible to P3. Regarding the MBC value, which is the lowest concentration of an agent that kills the target bacterium, it was observed that the strain *S. aureus* ATCC 6530 showed the lowest MBC value, 150 µL/mL for P1 propolis, whereas the highest was reached by *S. aureus* MRSA 15, 450 µL/mL with P3 propolis (Table 5). These results evidence the best antibacterial activity of P1 and P2 propolis.

2.5.2. Anti-Quorum-Sensing Activity

In the current study, the anti-quorum-sensing activity of *E. resinifera* and *E. officinarum* propolis extract was determined using the *C. violaceum* CV026 biosensor, in which the production of the violacein is regulated by the QS system [19]. The results of the anti-QS ability are illustrated in Figure 1. The *Euphorbia* propolis samples (P1, P2 and P3) were tested using five different concentrations (0.3, 0.6, 0.75, 1 and 1.5 mg/mL). The highest inhibition of violacein production was observed at the lowest concentrations, namely 0.3 and 0.6 mg/mL for *E. resinifera* (P1) and *E. officinarum* (P2). In contrast, the (P3) sample did not record a remarkable inhibition at all tested concentrations. The growth of the *C. violaceum* CV026 biosensor (control) was not affected by the exposure to the different concentrations of the tested propolis samples (Figure 1).

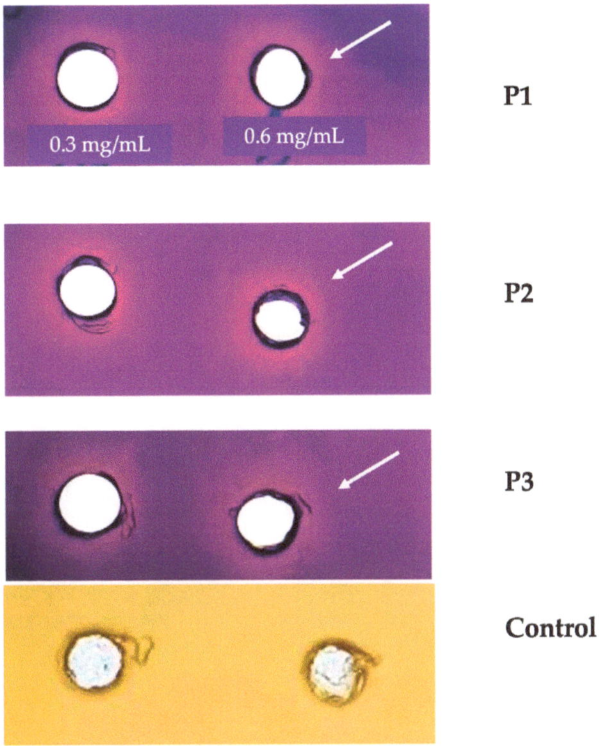

Figure 1. Anti-QS properties of propolis. P1: *Euphorbia resinifera* propolis. P2 and P3: *Euphorbia officinarum* propolis. N-hexanoylhomoserine lactone (C6-HSL) at 0.12 µg/mL was added to the culture medium. Control: No addition of C6-HSL to the culture medium. The white arrow highlights the concentration (0.6 mg/mL) at which the zone of inhibition of the production of the pigment violacein is observed. The assay was conducted using three independent triplicates.

2.5.3. Anti-Adherence and Anti-Biofilm Activity

The results of the impact of *E. resinifera* and *E. officinarum* propolis on the bacterial adherence ability are illustrated in Figure 2. Our results evidence that the *E. resinifera* (P1) and *E. officinarum* samples at the concentration of 250 µL/mL were able to reduce the adherence of *S. aureus* ATCC 6538 significantly ($p < 0.0001$) in comparison with chlorohexidine (0.2%, v/v). In contrast, P1 stimulated the adherence of both *S. aureus* MRSA 12 and MRSA 15 in a concentration-dependent manner, unlike the P2 and P3 samples, which were able to efficiently inhibit the adherence of the two MRSA strains. Interestingly, the P1 sample also induced the adherence of *E. coli* DSM 1077 in a concentration-dependent manner, but such an effect was not observed on the multiresistant strain of *E. coli* I73194. The adherence of *E. coli* DSM 1077 and the multiresistant strain of *E. coli* I73194 was better inhibited by the propolis sample P3 (Figure 2). The propolis sample P3, as mentioned above, showed lower phenol, flavonol/flavones and dihydroflavonol contents in comparison with the propolis samples P1 and P2 (Table 4), and we can anticipate that this lower content is not beneficial for the control of the bacterial growth but can act to promote the attachment of the bacterial cells. Contrarily, P1 and P2 showed better antibacterial activity and higher contents of those components, and the adherence of *S. aureus* ATCC 6538, MRSA and *E. coli* DSM 1077 cells in the presence of P1 propolis may have been induced as a response to a stress condition.

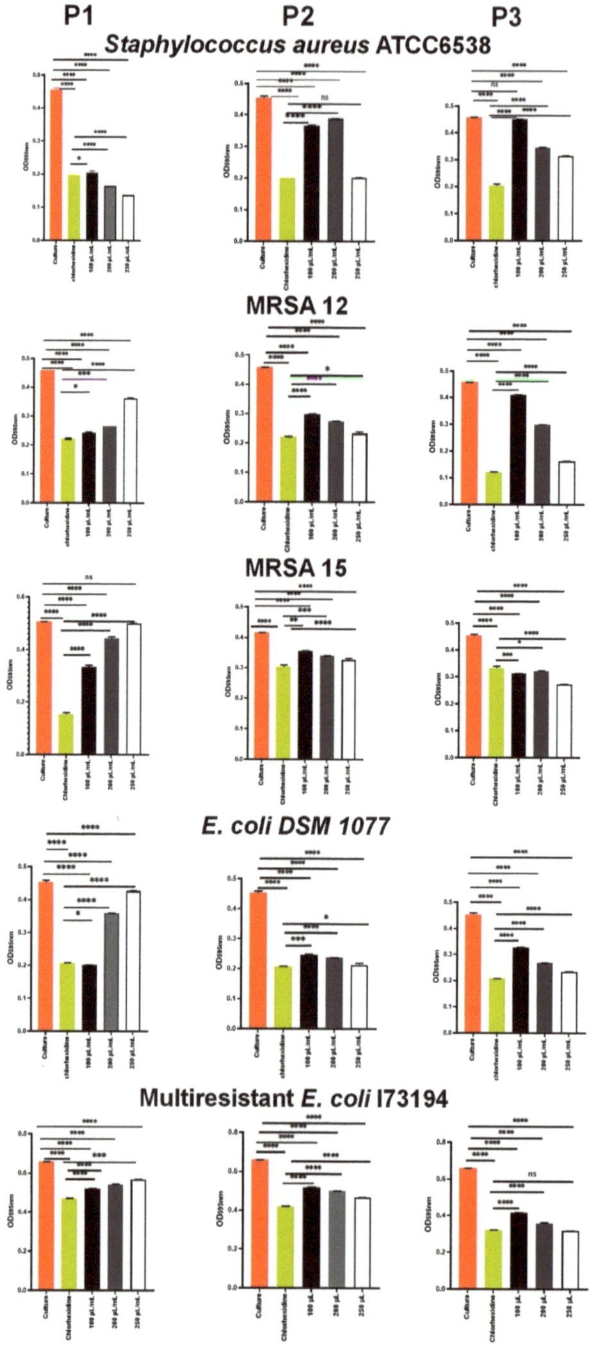

Figure 2. Inhibition of the bacterial adherence by the propolis samples P1, P2 and P3. Data represent the mean of three biological replicates. Error bars represent the standard deviation. * $p < 0.05$. ** $p < 0.01$. *** $p < 0.001$. **** $p < 0.0001$. ns: not significant.

3. Discussion

3.1. Pollen Grains

Pollen grains from anemophilous or entomophilous flowers can adhere to resins when collected by honeybees or may come from harvested pollen inside the hives [20]. This fact may provide an indication of the vegetation around the beehives as well as the geographical origin of propolis [21,22]. In this context, some decades ago, d'Albore [23] examined 56 propolis samples from several countries to establish the geographical origin of propolis based on palynological studies. For example, the author reported that a very high percentage of pollen grains of *Eucalyptus* and a considerable percentage of *Daphne* pollen grains characterized Moroccan propolis. To the best of our knowledge, this is the first time that palynological data have been provided for *Euphorbia* propolis from Morocco, and according to the results obtained, it should be considered that other types of pollen grains can be found in Moroccan propolis. Overall, many more palynological studies are needed to better respond to the need for an acceptable medicinal quality of propolis, which is more and more required by consumers. For example, for honey, monofloral honey is more appreciated, not only for its characteristic flavor and aroma but also for its intrinsic biological properties [24]; the same can be considered for propolis.

3.2. Propolis Volatile Profile

Heneicosane, tricosane, pentacosane, heptacosane, nonacosane, hentriacontane, nonanal and decanal, present in the P2 sample in relatively high amounts, were also reported as constituents of several Moroccan propolis volatiles. Their percentages also varied, which allowed distinguishing two clusters [25]; nevertheless, hexyl 2-methyl butyrate, present in the P2 sample, was not previously found in Moroccan propolis volatiles.

Seemingly, this is the first time that the chemical composition of the volatiles isolated from propolis where *E. resinifera* and *E. officinarum* pollen grains dominate, from Morocco, was studied. Keeping in mind that the volatile chemical composition of propolis is strongly dependent on the local flora at the harvesting location [26], differences are expected to occur in the volatile profiles of propolis samples.

3.3. Mineral Element Compounds

The samples presented a similar profile to other previously reported Moroccan propolis samples in which Ca or Na generally dominated [27–29], although Laaroussi et al. [28] had also reported K and Mg as main elements. The levels of Fe, Zn and Ni in the propolis of the present work were generally higher than those reported by Menyiy et al. [27] for Moroccan propolis, or those reported for Polish propolis [30]. The mineral content in propolis samples can be informative of the geographical region where they are gathered, but in this case, they can also be an indicator of environmental pollution [31] due to relatively high amounts of Fe, Zn and Ni. We do not consider that pollution can be present in beehive surroundings; nevertheless, it is something that must be considered because safety and quality must be guaranteed for the utilization of this bee product as a food supplement [32].

3.4. Total Phenol, Flavone, Flavonol, Flavanone and Dihydroflavonol Contents

The values of total phenols are within the range of samples reported by other authors for propolis collected in Morocco and expressed as mg GAE/g: 0.74–92.22 mg/g GAE [33], 12.02–134.04 mg GAE/g [34], 19.91 mg GAE/g [35]. For flavonols/flavones, the amounts were generally lower than those previously reported and expressed as mg QE/g: 0.16–129.60 mg QE/g [19], 0.20–34.27 mg QE/g [33], 3.28 mg QE/g [35], 1.19–108.11 mg QE/g [27]. Regarding the dihydroflavonol amounts, they are within the range of those previously reported for Moroccan samples (0.92–15.95 mg eriodictyol/g) [19], although the reference used was different in both cases (naringenin in the present work and eriodictyol by those authors).

3.5. Antimicrobial Activity

3.5.1. MIC and MBC Values

The MIC and MBC values of propolis can vary depending on the type and origin of propolis, as well as the strain of the tested bacteria. From several studies, it has been observed that propolis is more effective against Gram-positive bacteria than Gram-negative bacteria [19,36,37], which differ in cell wall composition; in particular, Gram-positive bacteria have a thick layer of peptidoglycan in their cell wall, while Gram-negative bacteria have an outer membrane rich in lipopolysaccharides and a thinner layer of peptidoglycan in their cell wall. Therefore, the powerful effect of propolis against Gram-positive bacteria in comparison with Gram-negative bacteria could be explained by the protection given by the outer membrane structure of Gram-negative bacteria and the production of hydrolytic enzymes that block and break down the active ingredients of propolis [37]. However, besides the accumulated knowledge about the composition and content of volatiles, phenols and flavonoids of propolis, the understanding of the exact contribution of these compounds to the antibacterial capacity of propolis is still very limited [38]. In a study conducted by Abdullah et al. [39], it was reported that the antibacterial capacity may be due to the nature and grouping of those chemical compounds; in other words, chemical compounds having electronegative carbonyl, amine, imine, sulfide, thiol, methoxyl and hydroxyl groups are highly polar and lipophilic, and having these features, when they are in contact with bacterial cells, they can injure the cellular membrane structure, allowing the escape of the cellular contents and therefore arresting the bacterial growth and ultimately causing cell death. The characterization of our propolis samples evidenced that the propolis samples of *E. resinifera* (P1) and *E. officinarum* (P2) were the two samples enriched in phenols, flavonoids and dihydroflavonols (Table 4). α-Pinene has been shown to have antimicrobial activity against a variety of bacteria, including *S. aureus*, *E. coli* and *Pseudomonas aeruginosa* [40]. Sesquiterpenes have been responsible for the growth inhibition of both Gram-positive and Gram-negative bacteria [41,42]. These terpenes may be present in the extracts and contribute in some manner to the activities detected, even if they were not identified or quantified by gas chromatography coupled to mass spectrometry (GC-MS) in the current research.

3.5.2. Anti-Quorum-Sensing Activity

Quorum sensing (QS) is a process by which bacteria communicate with each other and coordinate their behavior through the production and sensing of small signaling molecules called auto-inducers [43]. The inhibition of quorum sensing (quorum quenching, QQ) has emerged as a promising strategy for combating bacterial infections, as it can disrupt bacterial biofilm formation and impair other bacterial virulence traits.

The inhibition of QS by Moroccan propolis was reported previously [19], where the tested Moroccan propolis showed the ability to inhibit the QS system of *C. violaceum* CV026 at 1.22 mg/mL. The *Euphorbia* propolis showed higher anti-QS activity since the QS system was inhibited at a much lower concentration (0.3 mg/mL). The difference in results between the two studies may be explained by the chemical composition and the geographical origin of the tested propolis samples.

3.5.3. Anti-Adherence and Anti-Biofilm Activity

The inhibition of adherence to surfaces (biotic or abiotic) can prevent the establishment of bacterial communities that will form a mature biofilm (sessile cells) that is very difficult to eliminate. The propolis sample P1 was selected for testing its ability to disrupt the biofilm formed by *S. aureus* MRSA 12 and the multiresistant strain *E. coli* I73194 since it showed the lowest MIC and MBC values for the tested bacteria (Table 5). The results evidenced the ability of this propolis sample P1 to disturb the biofilm produced by these two resistant bacteria (Figure 3A,B,D,E). In contrast, the exposure of the bacterial biofilm to 70% ethanol (control) did not cause any injury to the sessile cells (Figure 3C,F). Our findings are in accordance with other studies that reported the ability of propolis to damage bacterial biofilms, and this capacity is particularly beneficial in cases of chronic infections that are difficult to treat [44–46].

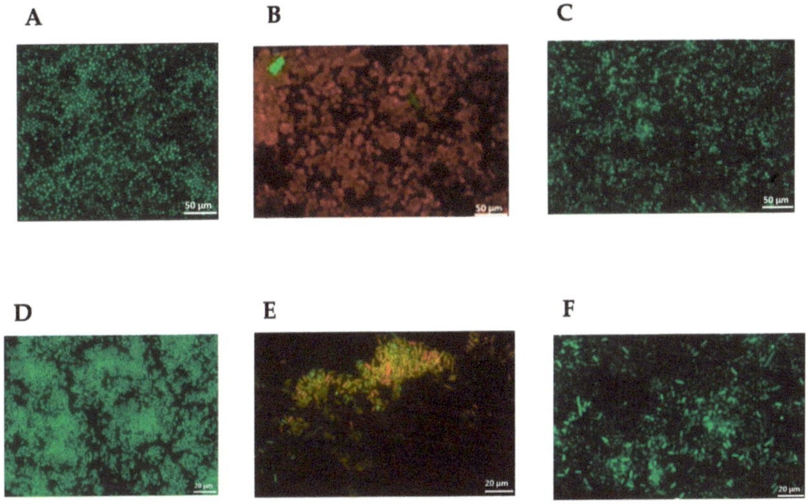

Figure 3. The impact of the *E. resinifera* (P1) propolis hydro-alcoholic extract on the disruption of biofilm formed by MRSA12 ((**A**) control, no exposure to anti-biofilm agent; (**B**) after exposure to propolis P1; (**C**) after exposure to 70% ethanol) and multiresistant *E. coli* I73194 ((**D**) control, no exposure to anti-biofilm agent; (**E**) after exposure to propolis P1; (**F**) after exposure to 70% ethanol). The bacterial cells formed a biofilm over 24 h and were visualized after staining with LIVE/DEAD Baclight.

4. Material and Methods

4.1. Propolis Collection

Euphorbia resinifera (P1) and *E. officinarum* propolis samples (P2 and P3) were obtained in three apiaries located in the Morocco regions of Beni Mellal-Khénifra (P1), Souss-Massa-Tiznit (P2) and Guelmim-Oued noun (P3) (Figure 4).

Figure 4. From right to left, the geographical location of Morocco and of the apiaries where the samples of *Euphorbia resinifera* (P1) and *E. officinarum* propolis (P2 and P3) were obtained. On the left are the flowers and propolis of *E. resinifera* (P1) and *E. officinarum* (P2 and P3).

To harvest propolis, propolis traps (special grids) were placed inside the hives at the beginning of June 2019. As the bees try to seal the grid holes, the trap becomes filled with propolis. After the traps were collected by the end of July, the propolis was recovered and kept in the dark at −4 °C.

4.2. Evaluation of Pollen Grains

The analysis of the qualitative and quantitative spectrum of the propolis samples' pollen was performed according to the International Commission for Bee Botany (ICBB), as previously detailed [47]. Pollen identification and count were carried out using a light microscope (Leitz Messtechnik GmbH; Wetzlar, Germany) at 400× and 1000×.

4.3. Volatile Organic Compound Extraction, Analysis and Identification

Due to a shortage of P3 propolis, the volatile profile was determined only in the P1 and P2 samples. The volatile organic compounds (VOCs) were isolated from 10 g of propolis by hydrodistillation in a Clevenger-type apparatus for 3 h [48]. At the end of the distillation procedure, the apparatus was cooled down for approximately 10 to 15 min, and the VOCs were recovered from the graduated tube by rinsing it with *n*-pentane distilled in the laboratory. The yields were 0.06 and 0.09% for *E. resinifera* and *E. officinarum*, respectively. The mixture of distilled *n*-pentane and volatiles was transferred to a clean glass vial and concentrated to approximately 10 μL using a blow-down evaporator system under a flux of nitrogen at room temperature. The concentrated samples were stored at −20 °C in the dark until further analysis. VOCs were identified and quantified as reported by Elamine et al. [49].

4.4. Quantification of Mineral Elements

Eleven mineral elements were quantified (Ca, Co, Cr, Cu, Fe, K, Mg, Mn, Na, Ni and Zn). The mineral analysis was performed as described by Boutoub et al. [50]. Half a gram of raw propolis per sample was added to an acid mixture of 15 mL nitric acid (65%) + 5 mL hydrogen peroxide (30%) and digested in a pressurized system heated by microwave (Discover SP-D 80; CEM), using special 80 mL quartz tubes suitable for the equipment, following a gradual digestion program (Table 6). Afterward, still in the quartz tube, about 20 mL of water was added (exothermic reaction), and then the sample was transferred to 50 mL volumetric flasks, and the volume was completed with water.

Table 6. Digestion program for mineral analysis.

Step	Temperature (°C)	Slope Time (min:s)	Step Time (min:s)	Pressure (psi)	Magnetic Stirring
1	Room temperature	-	00:15	<50	Mean
2	90	02:00	02:00	250	Mean
3	190	06:00	05:00	300	Mean
4	230	10:00	05:00	300	Mean

Measurements were performed through flame atomic absorption spectroscopy with air–acetylene mixtures according to the manufacturer's programs using a novAA 350 system (Analytik Jena, Jena, Germany). The concentrations were expressed as mg/kg propolis.

4.5. Hydro-Alcoholic Propolis Extraction

The hydro-alcoholic extraction of the three samples of propolis (P1, P2 and P3) was performed by maceration as previously described [9].

4.6. Quantification of Total Phenol, Flavones, Flavonol, Flavanone and Dihydrofavonol Contents

4.6.1. Total Phenol Content

The total phenol content of the samples was determined using the Folin–Ciocalteu method as previously described [19]. The total polyphenol content was expressed as mg per g of gallic acid equivalents (GAE) using a calibration curve.

4.6.2. Total Flavone and Flavonol Contents

The content of flavones and flavonol in the three hydro-alcoholic extracts was evaluated by the Al_2Cl_3 method as previously reported [19]. The total flavone and flavonol contents were expressed as mg per g of quercetin equivalents (QE) using a calibration curve. Tests were carried out in triplicate.

4.6.3. Total Flavanone and Dihydroflavonol Contents

The amounts of flavanones and dihydroflavonol in *Euphorbia* propolis extracts were determined using the 2,4-dinitrophenylhydrazine (DNP) method as described by El-Guendouz et al. [19]. Total flavanone and dihydroflavonol contents are presented as naringenin equivalents (NE) (mg per g) using a calibration curve. Tests were carried out in triplicate.

4.7. Antimicrobial Activity

4.7.1. Determination of the Minimum Inhibitory Concentration

The minimum inhibitory concentration (MIC) of the hydro-alcoholic extract of *E. resinifera* and *E. officinarum* propolis was determined by the microdilution method previously described [19] with slight modifications. Three Gram-positive bacteria, namely *S. aureus* ATCC 6538, methicillin-resistant *S. aureus* 12 (MRSA12) and methicillin-resistant *S. aureus* 15 (MRSA15), and the Gram-negative *E. coli* DSM 1077 were tested. All bacterial strains were maintained at −80 °C and when required were grown in fresh Brain Heart Infusion agar plates (BHI agar). Inoculated plates were incubated at 37 °C for 24 h. The tested concentrations of propolis extracts were 50, 100, 120, 150, 200, 250 and 300 µL/mL and were prepared in BHI broth. The bacterial strains were previously grown in 15 mL of BHI and incubated overnight in a water shaking bath at 37 °C. Afterward, 100 µL of the BHI culture medium with double the appropriate concentrations of propolis was distributed in the wells of a microplate containing 100 µL of the bacterial culture prepared the previous night. The inoculated microplates were incubated at 37 °C. Optical density readings were performed at 600 nm (OD_{600nm}) using a microplate reader (Tecan Infinite, M200, Männedorf, Switzerland). The antibiotic chloramphenicol (30 µg/mL) was used as a control. As a negative control, the solvent (ethanol at 70%) was used. The concentration of propolis that inhibited 95–100% of growth was considered the MIC value, and the lowest concentration of propolis that did not allow the recovery of the bacterial cells in culture plates was considered the minimum bactericidal concentration (MBC).

4.7.2. Evaluation of Anti-Adherence Activity

The anti-adherence activity of *Euphorbia* propolis extracts was evaluated as described by Apolónio et al. [51] with slight modifications; all of the bacterial suspensions, *S. aureus* ATCC 6538, MRSA12, *E. coli* DSM 1077 and the multiresistant strain *E. coli* I73194, were exposed to propolis extracts at concentrations of 100, 200 and 250 µL/mL. For this, 200 µL of the culture was distributed across a flat-bottom 96-well microplate and maintained at room temperature inside a flow cabinet for 30 min. Afterward, the bacterial suspension was collected, and the wells were washed with phosphate-buffered saline (PBS). Then, the microplate was dried at 80 °C for 30 min for heat fixation of the bacterial cells. After cooling, the adherent cells were stained for 1 min with 220 µL of crystal violet (0.1%). The stain was removed, and the wells were washed twice with PBS, followed by the dissolution of the stain with 220 µL of ethanol–acetone (80:20), and after 15 min, the OD 595 nm was determined using a microplate reader (Tecan Infinite, M200, Männedorf, Switzerland).

4.7.3. Determination of Anti-Biofilm Activity

The anti-biofilm activity of the propolis samples was evaluated according to the method reported by Walker and Horswill [52] with slight modifications. The bacterial suspensions of MRSA12 and the multiresistant *E. coli* I73194 were previously grown in BHI. Each bacterial culture was diluted 1:10 with fresh BHI and further grown at 37 °C until OD_{600nm} = 0.2 (2.0×10^7 CFU/mL). Six sterile non-breakable coverslips (22 × 22 mm × 0.25 mm) were distributed across a 6-well plate, and each coverslip was covered with 3 mL of a diluted bacterial suspension (1:10). The 6-well plate was incubated for 24 h at 37 °C to allow the production of a mature biofilm. Afterward, the bacterial suspension was removed, and each well was washed thrice with PBS. Then, the formed biofilm was exposed to 2.5 mL of propolis samples. The solvent (ethanol at 70%) was used as a control. The viability of sessile cells was determined after 6 and 24 h of exposure. The experiment was conducted using three biological and two technical replicates.

The count of sessile cells was determined according to the method described by Sakimura et al. [53]. Briefly, the coverslips were transferred into 10 mL of BHI, and the tubes were then sonicated for 7 min. After sonication, each coverslip was rapidly removed, and serial decimal dilutions were prepared and inoculated in BHI agar. The inoculated plates were incubated at 37 °C for 24–48 h, and the colonies were counted.

4.7.4. Visualization of Biofilm Cells by Fluorescence Staining

LIVE/DEAD Baclight (Invitrogen Molecular Probes, Eugene, OR, USA) was used to visualize the biofilm cells. For this, each coverslip was inverted and mounted on a microscope slide with 25 µL LIVE/DEAD fluorescent dye and stained for 15 min before observation. The observation was performed using an Axio Imager Z2 microscope (Zeiss, Oberkochen, Germany).

4.8. Statistical Analysis

The results are reported as mean ± standard deviation (SD) of three independent replicates. Statistical analysis of the data was carried out using SPSS (Version 23.0, Inc., and Chicago, IL, USA). One-way ANOVA and Tukey post hoc multiple comparison tests were used to analyze data, using Graph Pad Prism 9 statistical software. *p*-values less than 0.05 were considered significant.

5. Conclusions

In summary, our study shows for the first time the high antibacterial potential of propolis collected from *E. resinifera* and *E. officinarum* plants. This antibacterial activity is reinforced by anti-quorum-sensing and anti-biofilm properties against dangerous bacterial pathogens, such as MRSA. These bioactivities are believed to be due to the presence of bioactive compounds such as polyphenols, flavonoids, phenolic acids, and some volatiles, including terpenoids, in propolis. This result suggests that propolis is promising as a valuable natural product for the development of therapeutic drugs. However, further studies are needed to elucidate its mechanism of action, optimize its dosage form, and evaluate its safety and efficacy in different populations. Propolis could be used as a natural product or supplement to potentially support the human body's health due to its diverse biological activities.

Author Contributions: M.G.M. conceived the research with inputs from L.M.E., L.E.G., M.C.C., J.D.C., M.L.F., A.C.F. and L.M.E. performed the palynological assays. O.B. and A.C.F. extracted and analyzed metabolites. M.C.C., J.D.C., O.B. and S.E.-G. performed the mineral analyses. O.B. and S.E.-G. performed the antioxidant and inhibitory enzyme assays. O.B., I.M. and M.L.F. performed microbial assays. M.G.M., M.L.F., L.M.E., M.C.C., L.E.G. and A.C.F. provided analytical support and supervision. M.G.M. wrote the manuscript with inputs from O.B., M.L.F. and A.C.F. All authors have read and agreed to the published version of the manuscript.

Funding: This research was supported by funding from the Fundação para a Ciência e a Tecnologia (FCT/MCTES): Grant No. MED UIDB/05183/2020 to M.G.M.; Grant Nos. CESAM UIDP/50017/2020, UIDB/50017/2020 and LA/P/0094/2020 to A.C.F.; and Grant Nos. UIDB/04326/2020, UIDP/04326/2020 and LA/0101/2020 to M.C.C. and J.D.C.

Institutional Review Board Statement: Not applicable.

Informed Consent Statement: Not applicable.

Data Availability Statement: Data are contained within the article.

Acknowledgments: The authors are grateful to the financial support to FCT/MCTES through national funds, under CESAM UIDP/50017/2020+UIDB/50017/2020+LA/P/ 0094/2020, FEDER, PT2020 PACompete 2020. The authors are also grateful for financial support through national funds FCT/MCTES (PIDDAC) to CIMO (UIDB/00690/2020 and UIDP/00690/2020) and SusTEC (LA/P/0007/2020); and MED UIDB/05183/2020; UIDP/04326/2020 and LA/0101/2020. The authors acknowledge Município de Loulé for the support provided during the study. The authors are also thankful for the imaging analysis using the equipment available at the Light Microscopy Unit of ABC-UAlg that was partially supported by National Portuguese funding PPBI-POCI-01-0145-FEDER-22122.

Conflicts of Interest: The authors declare no conflicts of interest.

References

1. Damodaran, T. Chapter 46—Propolis. In *Nutraceuticals: Efficacy, Safety and Toxicity*, 2nd ed.; Ramesh, C., Gupta, L., Srivastava, A., Eds.; Academic Press: Cambridge, MA, USA, 2021; pp. 795–812. [CrossRef]
2. Mohammadzadeh, S.; Sharriatpanahi, M.; Hamedi, M.; Amanzadeh, Y.; Sadat Ebrahimi, S.E.; Ostad, S.N. Antioxidant power of Iranian propolis extract. *Food Chem.* **2007**, *103*, 729–733. [CrossRef]
3. Bankova, V.S.; de Castro, S.L.; Marcucci, M. Propolis: Recent advances in chemistry and plant origin. *Apidologie* **2000**, *31*, 3–15. [CrossRef]
4. Becerra, T.L.; Calla-Poma, R.D.; Requena-Mendizabal, M.F.; Millones-Gómez, P. Antibacterial effect of Peruvian propolis collected during different seasons on the growth of *Streptococcus mutans*. *Open Dentristy J.* **2019**, *13*, 3–14. [CrossRef]
5. Woźniak, M.; Mrówczyńska, L.; Waśkiewicz, A.; Rogoziński, T.; Ratajczak, I. The role of seasonality on the chemical composition, antioxidant activity and cytotoxicity of Polish propolis in human erythrocytes. *Rev. Bras. Farmacogn.* **2019**, *29*, 301–308. [CrossRef]
6. Shamsudin, N.F.; Ahmed, Q.U.; Mahmood, S.; Shah, S.A.A.; Khatib, A.; Mukhtar, S.; Alsharif, M.A.; Parveen, H.; Zakaria, Z.A. Antibacterial effects of flavonoids and their structure-activity relationship study: A comparative interpretation. *Molecules* **2022**, *27*, 1149. [CrossRef]
7. Li, F.; Awale, S.; Tezuka, Y.; Esumi, H.; Kadota, S. Study on the constituents of mexican propolis and their cytotoxic activity against PANC-1 human pancreatic cancer cells. *J. Nat. Prod.* **2010**, *73*, 623–627. [CrossRef]
8. Demir, S.; Aliyazicioglu, Y.; Turan, I.; Misir, S.; Mentese, A.; Yaman, S.O.; Akbulut, K.; Kilinc, K.; Deger, O. Antiproliferative and proapoptotic activity of Turkish propolis on human lung cancer cell line. *Nutr. Cancer* **2016**, *68*, 165–172. [CrossRef]
9. Silva, J.C.; Rodrigues, S.; Feás, X.; Estevinho, L.M. Antimicrobial activity, phenolic profile and role in the inflammation of propolis. *Food Chem. Toxicol.* **2012**, *50*, 1790–1795. [CrossRef] [PubMed]
10. Bueno-Silva, B.; Rosalen, P.L.; Alencar, S.M.; Mayer, M.P.A. Anti-inflammatory mechanisms of neovestitol from Brazilian red propolis in LPS-activated macrophages. *J. Funct. Foods* **2017**, *36*, 440–447. [CrossRef]
11. Liu, Y.; Liang, X.; Zhang, G.; Kong, L.; Peng, W.; Zhang, H. Galangin and pinocembrin from propolis ameliorate insulin resistance in HepG2 cells via regulating Akt/mTOR signaling. *Evid. -Based Complement Altern. Med.* **2018**, *2018*, 7971842. [CrossRef] [PubMed]
12. Pujirahayu, N.; Bhattacharjya, D.K.; Suzuki, T.; Katayama, T. α-Glucosidase inhibitory activity of cycloartane-type triterpenes isolated from indonesian stingless bee propolis and their structure-activity relationship. *Pharmaceuticals* **2019**, *12*, 102. [CrossRef]
13. Laaroussi, H.; Bakour, M.; Ousaaid, D.; Aboulghazi, A.; Ferreira-Santos, P.; Genisheva, Z.; Lyoussi, B. Effect of antioxidant-rich propolis and bee pollen extracts against D-glucose induced type 2 diabetes in rats. *Food Res. Int.* **2020**, *138*, 109802. [CrossRef]
14. Mancuso, G.; Midiri, A.; Gerace, E.; Biondo, C. Bacterial antibiotic resistance: The most critical pathogens. *Pathogens* **2021**, *10*, 1310. [CrossRef] [PubMed]
15. El Hamzaoui, N.; Barguigua, A.; Larouz, S.; Maouloua, M. Epidemiology of burn wound bacterial infections at a Meknes hospital, Morocco. *New Microbes New Infect.* **2020**, *38*, 100764. [CrossRef]
16. Rode, D.K.H.; Singh, P.K.; Drescher, K. Multicellular and unicellular responses of microbial biofilms to stress. *Biol. Chem.* **2020**, *401*, 1365–1374. [CrossRef]
17. Azevedo, N.F.; Allkja, J.; Goeres, D.M. Biofilms vs. cities and humans vs. aliens—A tale of reproducibility in biofilms. *Trends Microbiol.* **2021**, *29*, 1062–1071. [CrossRef] [PubMed]

18. Caldara, M.; Belgiovine, C.; Secchi, E.; Rusconi, R. Environmental, microbiological, and immunological features of bacterial biofilms associated with implanted medical devices. *Clin. Microbiol. Rev.* **2022**, *35*, e0022120. [CrossRef]
19. El-Guendouz, S.; Aazza, S.; Lyoussi, B.; Bankova, V.; Popova, M.; Neto, L.; Faleiro, M.L.; Da Graça Miguel, M. Moroccan Propolis: A natural antioxidant, antibacterial, and antibiofilm against *Staphylococcus aureus* with no induction of resistance after continuous exposure. *Evid. -Based Complement. Altern. Med.* **2018**, *2018*, 9759240. [CrossRef]
20. Barth, O.M.; Da Luz, C.F.P. Palynological analysis of Brazilian geopropolis sediments. *Grana* **2003**, *42*, 121–127. [CrossRef]
21. Moreira, L.; Dias, L.G.; Pereira, J.A.; Estevinho, L. Antioxidant properties, total phenols and pollen analysis of propolis samples from Portugal. *Food Chem. Toxicol.* **2008**, *46*, 3482–3485. [CrossRef]
22. Guzelmeric, E.; Ristivojević, P.; Trifković, J.; Dastan, T.; Yilmaz, O.; Cengiz, O.; Yesilada, E. Authentication of Turkish propolis through HPTLC fingerprints combined with multivariate analysis and palynological data and their comparative antioxidant activity. *LWT-Food Sci. Technol.* **2018**, *87*, 23–32. [CrossRef]
23. d'Albore, G.R. L'origine geographique de la propolis. *Apidologie* **1979**, *10*, 241–267.
24. Villanueva-Gutiérrez, R.; Moguel-Ordóñez, Y.B.; Echazarreta-González, C.M.; Arana-López, G. Monofloral honeys in the Yucatán Peninsula, Mexico. *Grana* **2009**, *48*, 214–223. [CrossRef]
25. El-Guendouz, S.; Lyoussi, B.; Miguel, M.G.; Figueiredo, A.C. Characterization of volatiles from Moroccan propolis samples. *J. Essent. Oil Res.* **2019**, *31*, 27–33. [CrossRef]
26. González, M.; García, M.E.; Slanis, A.; Bonini, A.; Fiedler, S.; Fariña, L.; Dellacassa, E.; Condurso, C.; Lorenzo, D.; Russo, M.; et al. Phytochemical findings evidencing botanical origin of new propolis type from North-West Argentina. *Chem. Biodivers.* **2019**, *16*, e1800442. [CrossRef] [PubMed]
27. El Menyiy, N.; Bakour, M.; El Ghouizi, A.; El Guendouz, S.; Lyoussi, B. Influence of geographic origin and plant source on physicochemical properties, mineral content, and antioxidant and antibacterial activities of Moroccan propolis. *Int. J. Food Sci.* **2021**, *2021*, 5570224. [CrossRef]
28. Laaroussi, H.; Ferreira-Santos, P.; Genisheva, Z.; Bakour, M.; Ousaaid, D.; Teixeira, J.A.; Lyoussi, B. Unraveling the chemical composition, antioxidant, α-amylase and α-glucosidase inhibition of Moroccan propolis. *Food Biosci.* **2021**, *42*, 101160. [CrossRef]
29. Aboulghazi, A.; Bakour, M.; Fadil, M.; Lyoussi, B. Simultaneous optimization of extraction yield, phenolic compounds and antioxidant activity of Moroccan propolis extracts: Improvement of ultrasound-assisted technique using response surface methodology. *Processes* **2022**, *10*, 297. [CrossRef]
30. Matuszewska, E.; Klupczynska, A.; Maciołek, K.; Kokot, Z.J.; Matysiak, J. Multielemental analysis of bee pollen, propolis, and royal jelly collected in west-central poland. *Molecules* **2021**, *26*, 2415. [CrossRef] [PubMed]
31. González-Martín, M.I.; Revilla, I.; Betances-Salcedo, E.V.; Vivar-Quintana, A.M. Pesticide residues and heavy metals in commercially processed propolis. *Microchem. J.* **2018**, *143*, 423–429. [CrossRef]
32. Inmaculada González-Martín, M.; Escuredo, O.; Revilla, I.; Vivar-Quintana, A.M.; Carmen Coello, M.; Riocerezo, C.P.; Moncada, G.W. Determination of the mineral composition and toxic element contents of propolis by near infrared spectroscopy. *Sensors* **2015**, *15*, 27854–27868. [CrossRef]
33. Miguel, M.d.G.; Doughmi, O.; Aazza, S.; Antunes, D.; Lyoussi, B. Antioxidant, anti-inflammatory and acetylcholinesterase inhibitory activities of propolis from different regions of Morocco. *Food Sci. Biotechnol.* **2014**, *23*, 313–322. [CrossRef]
34. Touzani, S.; Al-Waili, N.; El Menyiy, N.; Filipic, B.; Pereyra, A.; El Arabi, I.; Al-Waili, W.; Lyoussi, B. Chemical analysis and antioxidant content of various propolis samples collected from different regions and their impact on antimicrobial activities. *Asian Pac. J. Trop. Med.* **2018**, *11*, 436–442. [CrossRef]
35. El-Haskoury, R.; Al-Waili, N.; Kamoun, Z.; Makni, M.; Al-Waili, A.; Lyoussi, B. Antioxidant activity and protective effect of propolis against carbon tetrachloride-induced liver and kidney injury by modulation of oxidative parameters. *Vet. World* **2021**, *14*, 3076–3083. [CrossRef]
36. Almuhayawi, M.S. Propolis as a novel antibacterial agent. *Saudi J. Biol. Sci.* **2020**, *27*, 3079–3086. [CrossRef]
37. Bouchelaghem, S. Propolis characterization and antimicrobial activities against *Staphylococcus aureus* and *Candida albicans:* A review. *Saudi J. Biol. Sci.* **2022**, *29*, 1936–1946. [CrossRef]
38. Santos, F.A.; Bastos, E.M.A.; Uzeda, M.; Carvalho, M.A.R.; Farias, L.M.; Moreira, E.S.A.; Braga, F.C. Antibacterial activity of Brazilian propolis and fractions against oral anaerobic bacteria. *J. Ethnopharmacol.* **2002**, *80*, 1–7. [CrossRef] [PubMed]
39. Abdullah, N.A.; Zullkiflee, N.; Zaini, S.N.Z.; Taha, H.; Hashim, F.; Usman, A. Phytochemicals, mineral contents, antioxidants, and antimicrobial activities of propolis produced by Brunei stingless bees *Geniotrigona thoracica*, *Heterotrigona itama*, and *Tetrigona binghami*. *Saudi J. Biol. Sci.* **2020**, *27*, 2902–2911. [CrossRef]
40. Kartal, M.; Yildiz, S.; Kaya, S.; Kurucu, S.; Topçu, G. Antimicrobial activity of propolis samples from two different regions of Anatolia. *J. Ethnopharmacol.* **2003**, *86*, 69–73. [CrossRef] [PubMed]
41. Przybyłek, I.; Karpiński, T.M. Antibacterial properties of propolis. *Molecules* **2019**, *24*, 2047. [CrossRef] [PubMed]
42. Tamfu, A.N.; Ceylan, O.; Cârâc, G.; Talla, E.; Dinica, R.M. Antibiofilm and anti-quorum sensing potential of cycloartane-type triterpene acids from Cameroonian grassland propolis: Phenolic profile and antioxidant activity of crude extract. *Molecules* **2022**, *27*, 4872. [CrossRef] [PubMed]
43. Mukherjee, S.; Bassler, B.L. Bassler Bacterial quorum sensing in complex and dynamically changing environments. *Nat. Rev. Microbiol.* **2019**, *17*, 371–382. [CrossRef] [PubMed]

44. Queiroga, M.C.; Laranjo, M.; Andrade, N.; Marques, M.; Costa, A.R.; Antunes, C.M. Antimicrobial, antibiofilm and toxicological assessment of propolis. *Antibiotics* **2023**, *12*, 347. [CrossRef]
45. Santos, L.M.; Rodrigues, D.M.; Kalil, M.A.; Azevedo, V.; Meyer, R.; Umsza-Guez, M.A.; Machado, B.A.; Seyffert, N.; Portela, R.W. Activity of ethanolic and supercritical propolis extracts in *Corynebacterium pseudotuberculosis* and its associated biofilm. *Front. Vet. Sci.* **2021**, *8*, 700030. [CrossRef]
46. Tennick, K.; Wafa, S.; Fearnley, H.; Gomez Escalada, M. Effects of propolis on the quorum sensing of selected biofilm producing bacterial species. *J. Apitherapy Nat.* **2018**, *1*, 8–19.
47. Louveaux, J.; Maurizio, A.; Vorwohl, G. Methods of melissopalynology. *Bee World* **1978**, *5*, 139–153. [CrossRef]
48. Council of Europe; European Pharmacopoeia Commission; European Directorate for the Quality of medicines and Healthcare. *European Pharmacopoeia*, 7th ed.; Council Of Europe: Strasbourg, France, 2010.
49. Boutoub, O.; El-Guendouz, S.; Manhita, A.; Dias, C.B.; Estevinho, L.M.; Paula, V.B.; Carlier, J.; Costa, M.C.; Rodrigues, B.; Raposo, S.; et al. Comparative study of the antioxidant and enzyme inhibitory activities of two types of Moroccan *Euphorbia* entire honey and their phenolic extracts. *Foods* **2021**, *10*, 1909. [CrossRef]
50. Boutoub, O.; Aazza, S.; El-Guendouz, S.; El Ghadraoui, L.; Miguel, M. Response surface methodology (RSM) for optimization of *Euphorbia resinifera* and *Euphorbia officinarum* extracts with antioxidant and anti-diabetic activities. *Pharmacogn. Mag.* **2022**, *18*, 940–952.
51. Apolónio, J.; Faleiro, M.L.; Miguel, M.G.; Neto, L. No induction of antimicrobial resistance in *Staphylococcus aureus* and *Listeria monocytogenes* during continuous exposure to eugenol and citral. *FEMS Microbiol. Lett.* **2014**, *354*, 92–101. [CrossRef]
52. Walker, J.N.; Horswill, A.R. A coverslip-based technique for evaluating *Staphylococcus aureus* biofilm formation on human plasma. *Front. Cell. Infect. Microbiol.* **2012**, *2*, 39. [CrossRef] [PubMed]
53. Sakimura, T.; Kajiyama, S.; Adachi, S.; Chiba, K.; Yonekura, A.; Tomita, M.; Koseki, H.; Miyamoto, T.; Tsurumoto, T.; Osaki, M. Biofilm-forming *Staphylococcus epidermidis* expressing vancomycin resistance early after adhesion to a metal surface. *BioMed Res. Int.* **2015**, *2015*, 9–11. [CrossRef] [PubMed]

antibiotics

MDPI

Article

Candida krusei M4CK Produces a Bioemulsifier That Acts on Melaleuca Essential Oil and Aids in Its Antibacterial and Antibiofilm Activity

Jéssica Mayra Mendes Araujo [1,2,†], Joveliane Melo Monteiro [1,2,†], Douglas Henrique dos Santos Silva [2], Amanda Karoline Veira [2], Maria Raimunda Chagas Silva [3], Fernanda Avelino Ferraz [2], Fábio H. Ramos Braga [3], Ezequias Pessoa de Siqueira [4] and Andrea de Souza Monteiro [1,2,*]

1 Rede de Biodiversidade e Biotecnologia da Amazônia Legal, BIONORTE, Saint Louis 65055-310, Brazil; jessicamendesaraujo1@hotmail.com (J.M.M.A.); joveliane-12@hotmail.com (J.M.M.)
2 Laboratório de Microbiologia Aplicada, Universidade CEUMA, Saint Louis 65075-120, Brazil; douglasbmdh@gmail.com (D.H.d.S.S.); amandavieiraalencar@gmail.com (A.K.V.); nanda_ferras@hotmail.com (F.A.F.)
3 Laboratório de Ciências do Ambiente, Universidade CEUMA, Saint Louis 65075-120, Brazil; marirah@gmail.com (M.R.C.S.); fabiobraga.ma@gmail.com (F.H.R.B.)
4 Laboratório de Química de Produtos Naturais, Centro de Pesquisas René Rachou Fundação Oswaldo Cruz, Belo Horizonte 30190-002, Brazil; ezequias.siqueira@gmail.com
* Correspondence: andrea003991@ceuma.com.br
† These authors contributed equally to this work.

Abstract: Surface-active compounds (SACs) of microbial origin are an active group of biomolecules with potential use in the formulation of emulsions. In this sense, the present study aimed to isolate and select yeasts from fruits that could produce SACs for essential oil emulsions. The *Candida krusei* M4CK was isolated from the *Byrsonima crassifolia* fruit to make SACs. This emulsification activity (E_{24}) was equal to or greater 50% in all carbon sources, such as olive oil, sunflower oil, kerosene, hexane, and hexadecane. E_{24} followed exponential growth according to the growth phase. The stability of emulsions was maintained over a wide range of temperatures, pH, and salinity. The OMBE4CK (melaleuca essential oil emulsion) had better and more significant inhibitory potential for biofilm reduction formation. In addition, bioemulsifier BE4CK alone on *Escherichia coli* and *Pseudomonas aeruginosa* biofilm showed few effective results, while there was a significant eradication for *Staphylococcus aureus* biofilms. The biofilms formed by *S. aureus* were eradicated in all concentrations of OMBE4CK. At the same time, the preformed biofilm by *E. coli* and *P. aeruginosa* were removed entirely at concentrations of 25 mg/mL, 12.5 mg/mL, and 6.25 mg/mL. The results show that the bioemulsifier BE4CK may represent a new potential for antibiofilm application.

Keywords: *Candida krusei*; essential oil; bioemulsifier; antimicrobial activity

Citation: Araujo, J.M.M.; Monteiro, J.M.; Silva, D.H.d.S.; Veira, A.K.; Silva, M.R.C.; Ferraz, F.A.; Braga, F.H.R.; Siqueira, E.P.d.; Monteiro, A.d.S. *Candida krusei* M4CK Produces a Bioemulsifier That Acts on Melaleuca Essential Oil and Aids in Its Antibacterial and Antibiofilm Activity. *Antibiotics* **2023**, *12*, 1686. https://doi.org/10.3390/antibiotics12121686

Academic Editors: Anna Duda-Madej, Joanna Kozłowska and Katarzyna Wińska

Received: 1 July 2023
Revised: 21 September 2023
Accepted: 25 September 2023
Published: 30 November 2023

1. Introduction

In recent years, the production of microbial Surface-Active Compounds (SACs) has attracted attention due to its diverse applicability, such as the ability to play an essential role in the solubility of insoluble compounds in water, in the binding of heavy metals, in the desorption of contaminants, in bacterial pathogenesis, and in adhesion and cell aggregation [1–4]. In addition to these characteristics, SACs have several advantages over synthetic surfactants; for example, low toxicity, lower critical micellar concentration (CMC), excellent biodegradability, and ecological acceptability [5]. The main microbial SACs are biosurfactants and bioemulsifiers; these have different chemical compositions and considerable biological activities [3,4,6].

SACs can exhibit antibacterial, antifungal, antiviral, and antitumor activities [1,7,8]. The properties of anti-adhesive and antibiofilm activities are also essential, as they can inhibit adhesion and colonization by pathogenic microorganisms and remove pre-formed biofilms on silicone rubber and other biomedical instruments [9].

Another relevant factor about SACs is their production by various microorganisms, including bacteria, yeasts, and filamentous fungi [10,11]. Yeasts have been constantly used as leavening agents for SACs, as they are generally considered to have a safe (GRAS) status. Organisms characterized by GRAS status do not pose risks of toxicity or pathogenicity, which allows for their available applications in the pharmaceutical and food industries [11,12].

In the current scenario, yeasts have been considered the future molecules, as they have a GRAS *status* and synthesize SACs that act both in bioremediation and the biological field [6]. Several studies have shown that surfactant compounds produced by these microorganisms have potential biomedical applications since they exhibit antibiofilm, antibacterial, anticancer, larvicidal, and immunomodulatory activities [13].

Oil-in-water emulsions are efficient delivery systems for oily compounds, such as essential oils, dispersing the lipid phase as a colloidal dispersion [12,14]. Emulsifying agents prevent droplet coalescence by decreasing interfacial tension. SACs have attracted significant interest in elaborating emulsions and nanoemulsions, considering several physical and chemical characteristics [14]. They can be used to reduce volatility, increase stability, increase bioactivity, and extend the shelf life of essential oils for different applications, such as food, cosmetics, and the preparation of pharmaceutical formulations [14].

Previously, it was demonstrated that yeasts isolated from environments can produce emulsifiers that present great stability of emulsions [6]. Considering their applicability in biotechnology, studying the diversity of yeasts associated with fruits and the study of their molecules with emulsifying potential becomes enjoyable. They think that emulsions for essential oils are formulations of interest in the pharmaceutical industry and the preparation of cosmetics, among other applications. This study aims to isolate yeast biopolymers to prepare essential oil emulsions and quantify the synergistic effect against bacteria and microbial biofilms.

2. Results

2.1. Isolation of Yeasts Producing Surface-Active Compounds

In the present study, yeasts (n = 13) were isolated from fruits *Anacardium occidentale*, *Byrsonima crassifolia*, *Citrus reticulata*, and *Platonia insignis* (Table 1). MALDI-QTOF-MS identified the yeast strains as *Candida metapsilosis* and *Candida krusei.*

Table 1. Yeasts isolated from fruits and substrate as the sole carbon source for growth.

Fruit	Popular Name	Identification	Substrate
Anacardium occidentale	Cashew	*Candida krusei* C1CK	Glycerol
Anacardium occidentale	Cashew	*Candida krusei* C2CK	Glycerol
Byrsonima crassifolia	Murici	*Candida krusei* M1CK	Glycerol
Byrsonima crassifolia	Murici	*Candida krusei* M2CK	Glycerol
Byrsonima crassifolia	Murici	*Candida krusei* M3CK	Glycerol
Byrsonima crassifolia	Murici	*Candida krusei* M4CK	Glycerol
Byrsonima crassifolia	Murici	*Candida metapsilosis* M1CM	Sunflower oil
Byrsonima crassifolia	Murici	*Candida metapsilosis* M2CM	Sunflower oil
Citrus reticulata	Tangerine	*Candida metapsilosis* T2CM	Sunflower oil
Platonia insignis	Bacuri	*Candida metapsilosis* B1CM	Glycerol
Platonia insignis	Bacuri	*Candida metapsilosis* B2CM	Glycerol
Platonia insignis	Bacuri	*Candida metapsilosis* B3CM	Glycerol
Platonia insignis	Bacuri	*Candida metapsilosis* B4CM	Sunflower oil

The isolated yeasts were inoculated separately in minimal minerals containing sunflower oil and glycerol at concentrations of 2%. Table 1 demonstrates the best source of carbon and energy for growth.

2.2. *Screening and Identification of SAC-Producing Microorganisms*

2.2.1. Determination of Emulsifying Activity (E_{24})

This study aimed to investigate the emulsification properties of the surface-active compound, so the E_{24} test was mainly used to determine the presence of SACs in the samples.

The most efficient yeast strain in producing surface-active compounds with emulsifying activity for several carbon sources was isolated from the fruit *Byrsonima crassifolia,* popularly known as murici and identified as *Candida krusei* M4CK. E_{24} was determined, and the values were indicated (Supplementary Table S1). The SACs produced by *C. krusei* M4CK were the only ones that showed emulsifying activity in all carbon sources.

The emulsification activity of SACs contained in the supernatant of MM medium with glycerol as a carbon source was appreciable. The SACs, called *Candida krusei* M4CK, showed E_{24} values of 55%, 56.75%, 50%, 68.5%, 63.6%, and 63.95% for olive oil, sunflower oil, frying oil (used), kerosene, hexane, and hexadecane, respectively (Figure 1). The values of the final surface tension (mN/m) in the medium after cell growth were determined using a tensiometer and ranged from 44 to 62 mN/m (Supplementary Table S2).

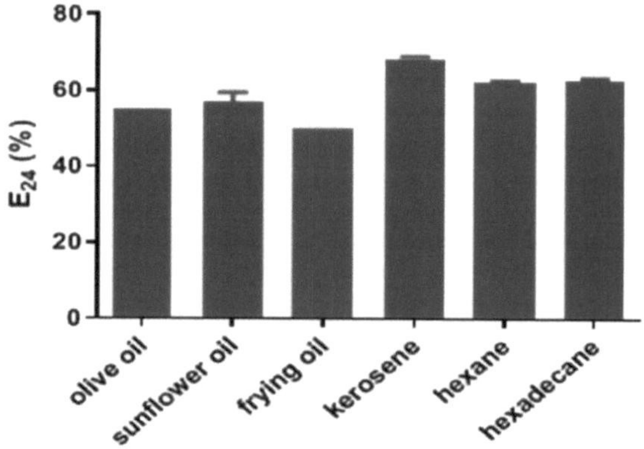

Figure 1. Emulsifying activity (E_{24}) presented by SACs produced by *Candida krusei* M4CK in different hydrophobic compounds.

2.2.2. Determination of Surface Tension

The production of biosurfactants and bioemulsifiers by yeast isolates can be identified indirectly through surface activity and emulsifying activity. In this study, it was observed that the compound BE4CK showed a reduction in surface tension from 72 mN/m to 44 mN/m. This result demonstrated no reduction in surface tension to values below 30 mN/m. For this reason, it was characterized as a bioemulsifier (BE4CK).

2.3. *Determination of Chemical Composition*

Analyzes of the chemical composition of SAC (BE4CK) revealed the presence of carbohydrates (46%), protein (38%), and a lower amount of lipid parts (3%). GC characterized the BE4CK that indicated in its structure; the major fatty acid was 9-octadecenoic acid, (E)-(58.43%), followed by methyl stearate (16.72%) and hexadecanoic acid (13.44%). Moreover, the monosaccharides were mannose (84.26% and glucose (15.74%).

2.4. *Growth Kinetics*

C. krusei M4CK growth and emulsifying activity in the mineral medium supplemented with 2% glycerol are shown in Figure 2. For the strain, bioemulsifier and E_{24} production follow an exponential growth phase growth. Bioemulsifier production continues even when microbial growth ceases. About 52.3% of E_{24} activity in kerosene was obtained in the exponential growth phase in the first 40 h; after this period, E_{24} activity increased, reaching a percentage $\geq$ 60% in 60 h–144 h. These results indicate that the emulsification activity is partially associated with growth, as the peaks of bioemulsifier production did not coincide. The highest stability and emulsification activity was achieved at 144 h (Figure 2).

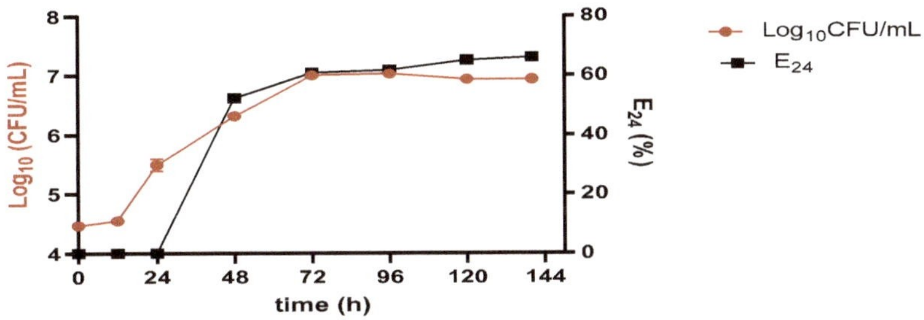

Figure 2. Kinetics of cell growth by *Candida krusei* M4CK and emulsifying activity by BE4CK. BE4CK: Bioemulsifier; CFU/mL: colony-forming unit per milliliters; E_{24}: emulsifier activity.

2.5. Evaluation of Bioemulsifier Stability

2.5.1. Effect of Temperature on the Emulsifying Activity

After being subjected to the autoclaving process at 120 °C for 15 min, BE4CK showed stability and maintenance of emulsifying properties since it emulsified all the hydrophobic sources (Figure 3). The E_{24} of the bioemulsifier was ≥50% in olive and sunflower oil, while for kerosene, hexane, and hexadecane hydrocarbons, it was greater than 60%. For frying oil (E_{24} of 40%), there was a 10% reduction in the primary emulsifying activity (Figure 3).

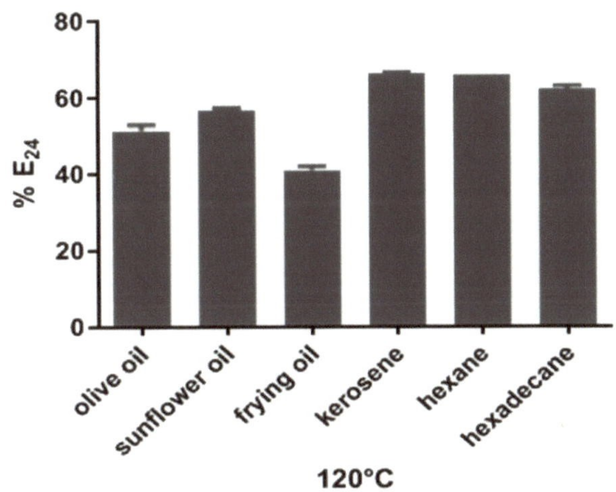

Figure 3. Emulsifying activity (E_{24}) of BE4CK after heating to 120 °C. Values indicate the mean of duplicate values and error bars represent the standard deviation (SD).

2.5.2. Effect of pH on the Emulsifying Activity

The SAC produced by *C. krusei* showed emulsifying activity for different sources at acidic and basic pH (Figure 4). At acidic pHs (3–6), emulsification activity was ≥44% for all substrates (sunflower, olive, and kerosene), while at basic pHs, emulsification was characterized only for kerosene hydrocarbons. In this source, there was no marked variation in E_{24} in the pH ranges of 3 to 9. From pH eight onwards, the bioemulsifier produced by the *C. krusei* M4CK strain could emulsify only kerosene (Figure 4). In the range of pH 10, there was no emulsification stability for any of the sources.

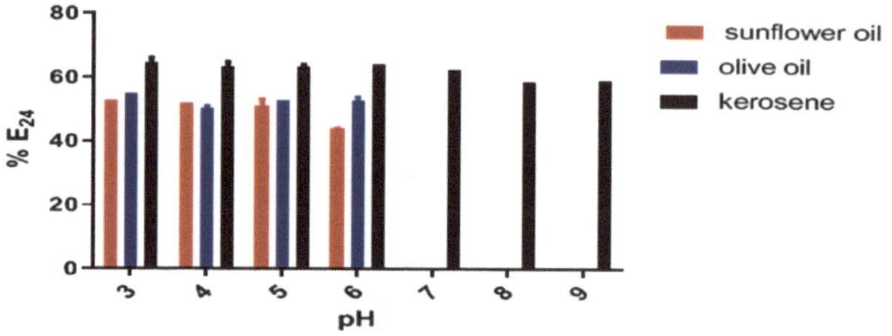

Figure 4. Emulsifying activity (E_{24}) of BE4CK in different pH ranges. Values indicate the mean of duplicate values, while error bars represent standard deviation in pH: hydrogen potential.

2.5.3. Effect of Salinity on Emulsifying Activity

It was observed that the E_{24} of the bioemulsifier in concentrations (10%, 30%, and 50%) of magnesium chloride ($MgCl_2$), sodium chloride (NaCl), and calcium chloride ($CaCl_2$) had contents ranging from 40 to 55% in vegetable oils (sunflower and olive) and for kerosene, a percentage $\geq$ 56%. The highest E_{24} value was obtained with kerosene, while the lowest was registered with olive oil (Figure 5).

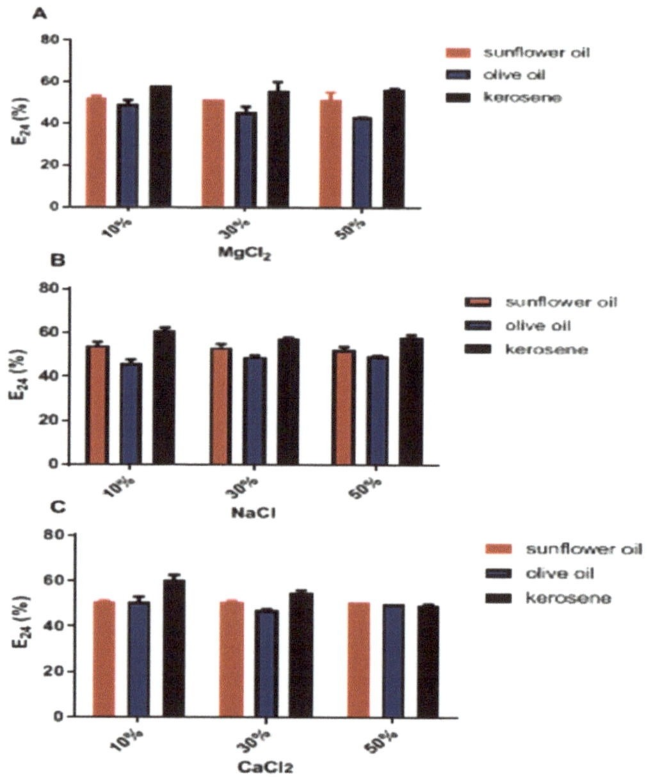

Figure 5. Stability of emulsifying activity in different salt concentrations. (**A**) $MgCl_2$: magnesium chloride; (**B**) NaCl: sodium chloride; (**C**) $CaCl_2$: calcium chloride. Values are three independent experiments' mean $\pm$ standard deviation (SD).

The stability of the emulsifying activity (E_{24}) was maintained at concentrations of 50% of the salts. It was also verified that as the salinity increased, there was a slight reduction in the emulsification rate. However, changes in salt concentrations do not influence E_{24} values (Figure 5).

2.6. Assessment of Bioemulsifier Cytotoxicity

By the toxicity test, BE4CK showed no toxicity in VERO cells at concentrations of 1024 μg/mL–2 μg/mL, and the percentage of cell viability was ≥88.89% (Figure 6).

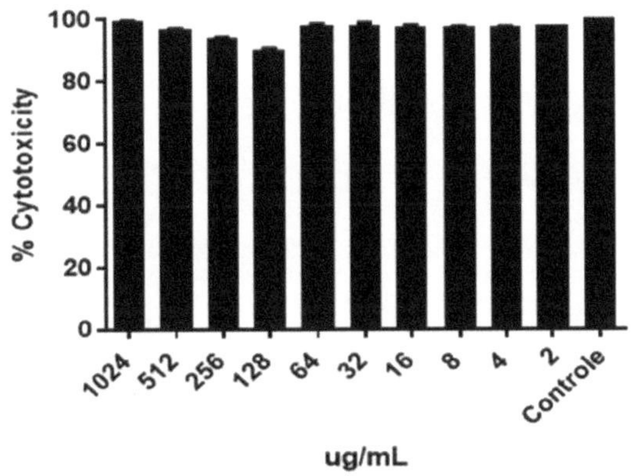

Figure 6. Evaluation of the cytotoxicity of the bioemulsifier BE4CK (1024 μg/mL–2 μg/mL) in VERO cells.

2.7. Determination of Antimicrobial Activity

When evaluating antimicrobial activity through the agar diffusion assay (Supplementary Figure S1), it was found that isolates of *E. coli* ATCC 25922, *E. faecalis* ATCC 35218, *P. aeruginosa* ATCC 27853, and *S. aureus* ATCC 25923 were significantly sensitive to the essential oil of melaleuca considering $p < 0.05$ (Table 2). In the case of BE4CK, the growth of microorganisms was not affected. The emulsion OMBE4CK showed an increase in the growth inhibition zone, corresponding to 34.5 mm for *E. coli* ATCC 25922. In comparison, melaleuca oil alone formed an inhibition zone of 28.5 mm for this microorganism. Thus, the compound combination showed antimicrobial activity, making *E. coli* ATCC 25922 more sensitive (Table 2 and Supplementary Figure S1).

Table 2. Antimicrobial activity of the melaleuca essential oil, bioemulsifier alone, and the combination of BE4CK and OMBE4CK (emulsion) against microorganisms.

Compound	Antimicrobial Activity (mm)			
	E. coli ATCC 25922	*E. faecalis* ATCC 35218	*P. aeruginosa* ATCC 27853	*S. aureus* ATCC 25923
BE4CK (25 mg/mL)	N	N	N	N
Melaleuca essential oil	28.5 ± 2.1	26 ± 1.4	17 ± 2.8	33.5 ± 0.7
OMBE4CK	34.5 ± 6.3	14.5 ± 2.1	-	24 ± 1.4

N—no antimicrobial activity. Values are three independent experiments' mean ± standard deviation (SD). ($p < 0.05$)—a significant difference in relation to concentrations.

Unlike BE4CK, the OMBE4CK emulsion and melaleuca essential oil showed significant inhibitory activity against *E. coli* ATCC 25922, *E. faecalis* ATCC 35218, and *S. aureus* ATCC 25923. The emulsion could not inhibit the growth of *P. aeruginosa* ATCC 27853.

2.8. Determination of Antibiofilm Activity

2.8.1. Inhibition of Biofilm Formation on Polystyrene

The antibiofilm potential of the bioemulsifier (BE4CK) produced by *C. krusei* M4CK was determined by its ability to impair and inhibit biofilm formation by *S. aureus* ATCC 25923, *E. coli* ATCC 25922, and *P. aeruginosa* ATCC 27853. The compound showed a dose-dependent effect, with the most significant activity at 25 mg/mL concentrations ($p < 0.001$). Inhibition percentages for these three isolates ranged from 51 to 75% at a concentration of 25 mg/mL of the bioemulsifier (Figure 7A–C). Applied at this concentration, BE4CK showed significant ($p < 0.001$) inhibitory potential against *S. aureus* ATCC 25923 (71%), *E. coli* ATCC 25922 (75%), and *P. aeruginosa ATCC* 27853 (51%). Even at lower concentrations (1.56 mg/L–3.12 mg/L) it was effective against *S. aureus* ATCC 25923 (16–40%), *E. coli* ATCC 25922 (18–34%), and *P. aeruginosa* ATCC 27853 (16–26%), respectively.

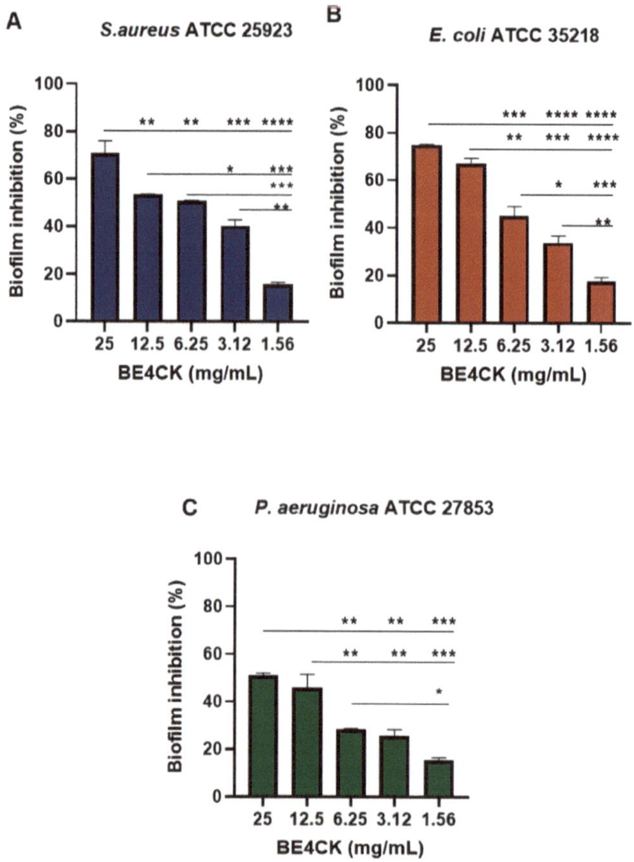

Figure 7. BE4CK inhibitory activity on biofilm formation by microorganisms. (**A**) *S. aureus* ATCC 25923; (**B**) *E. coli* ATCC 25922; (**C**) *P. aeruginosa* ATCC 27853. Values are three independent experiments' mean ± standard deviation (SD). * ($p < 0.05$), ** ($p < 0.001$), *** ($p < 0.001$), **** ($p < 0.0001$)—a significant difference in relation to concentrations.

The ability of the emulsion (OMBE4CK) resulting from the combination of BE4CK bioemulsifier and melaleuca essential oil to reduce biofilm formation by *S. aureus* ATCC 25923, *E. coli* ATCC 25922, and *P. aeruginosa* ATCC 27853 was evaluated using microtiter plates of 96 wells (Figure 8A–C).

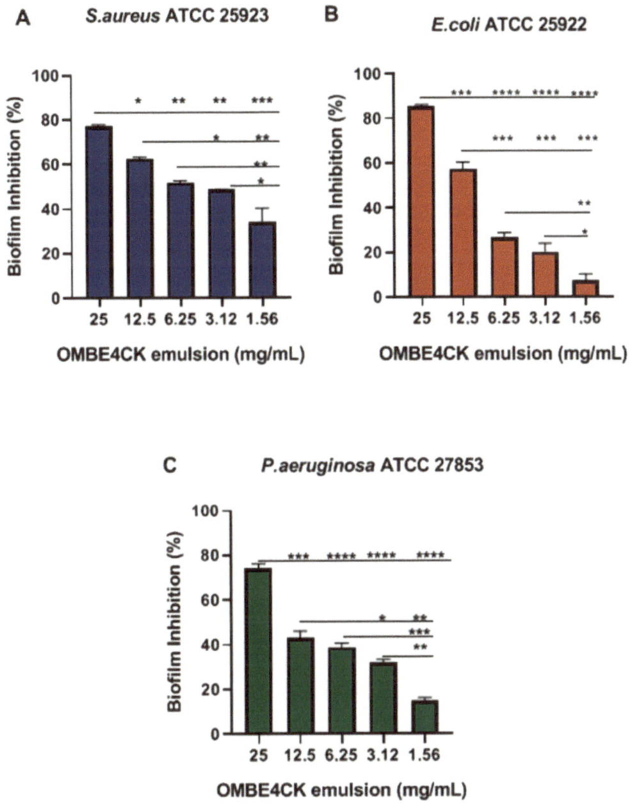

Figure 8. OMBE4CK inhibitory activity on biofilm formation by microorganisms. (**A**) *S. aureus* ATCC 25923; (**B**) *E. coli* ATCC 25922; (**C**) *P. aeruginosa* ATCC 27853. Values are the mean ± standard deviation (SD) of three independent experiments. * ($p < 0.05$), ** ($p < 0.001$), *** ($p < 0.001$), **** ($p < 0.0001$)—a significant difference in relation to concentrations.

For both strains, the use of emulsification at all concentrations resulted in significant inhibition of biofilm formation ($p < 0.05$). A biofilm inhibition of 77% for *S. aureus* ATCC 25923 (Figure 8A), 85% for *E. coli* ATCC 25922 (Figure 8B), and 74% for *P. aeruginosa* ATCC 27853 (Figure 8C) was obtained using the concentration of 25 mg/mL ($p < 0.001$). The antibiofilm activity of the emulsion (1.56 mg/mL–25 mg/mL) gradually decreased at low concentrations (1.56 mg/mL and 3.12 mg/mL), resulting in an inhibitory effect of 34% and 49% for *S. aureus* ATCC 25923 (Figure 8A); 7% and 20% for *E. coli* ATCC 25922 (Figure 8B); and 15% and 31% for *P. aeruginosa* ATCC 27853 (Figure 8C), respectively. Based on these results, the emulsion produced a significant antibiofilm effect against both pathogens.

2.8.2. Inhibition of Biofilm Formation in an Enteral Feeding Catheter

Quantitative Analysis

The antibiofilm potential of the emulsion was determined by its ability to inhibit biofilm formation by *S. aureus* ATCC 25923 in an enteral feeding catheter, which corresponds to a polyurethane medical device (Figure 9). In the 25 mg/mL concentration, the emulsion showed a significant ($p < 0.001$) inhibitory potential of 43% for *S. aureus* ATCC 25923. While at lower concentrations (3.12 mg/mL–12.5 mg/mL), the antibiofilm activity of the emulsion gradually decreased, resulting in an inhibitory effect varying significantly ($p < 0.05$) between 14 and 27%.

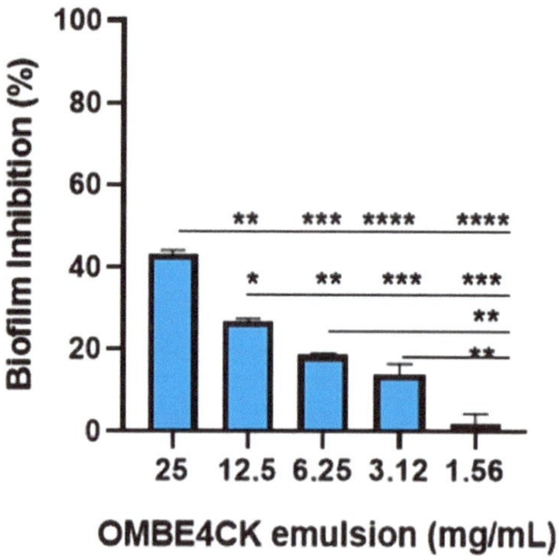

Figure 9. Inhibitory emulsion (OMBE4CK) activity on biofilm formation by *S. aureus* ATCC 25923 on the enteral feeding catheter surface. Values are mean ± standard deviation (SD) of three independent experiments. * ($p < 0.05$), **($p < 0.001$), *** ($p < 0.001$), **** ($p < 0.0001$)—a significant difference in relation to concentrations.

Semi-Quantitative Analysis

Through the semi-quantitative method of rolling catheters in a Petri dish containing TSA, catheter fragments exposed to the maximum concentration (25 mg/mL) of the OMBE4CK emulsion obtained a percentage of 4% contamination (with growth < 15 CFU/plate) by *S. aureus* ATCC 25923, whereas at the lowest concentrations (1.56 mg/mL–12.5 mg/mL), there was a range from 23% to 94% of colonization (with growth ≥ 15 CFU/plate) of *S. aureus* ATCC 25923 to polyurethane catheter surfaces (Table 3). At the concentration of 1.56 mg/mL of the emulsion, there was a 94% colonization of *S. aureus* ATCC 25923. Thus, biofilm inhibition depends on the emulsion concentration.

Table 3. Quantitative CFU/mL of *S. aureus* ATCC 25923 on the catheter surface.

Emulsion Concentration (mg/mL)	CFU/Plate		Classification
	N°	%	
25	4	14	Contamination
12.5	71	23%	Colonization
6.25	83	26%	Colonization
3.12	213	68%	Colonization
1.56	295	94%	Colonization
Control	314	100%	Colonization

CFU/mL: colony-forming unit per milliliter.

2.8.3. Biofilm Eradication

Biofilms formed by *S. aureus* ATCC 25923, *E. coli* ATCC 25922, and *P. aeruginosa* ATCC 27853 within 48 h and treated with the BE4CK were effectively disrupted at the highest concentration (25 mg/mL). At this concentration, the dispersion of biofilms formed by these microorganisms ranged from 10% to 55%, while at the lowest concentration (1.56 mg/mL), it ranged from 2% to 36% (Figure 10).

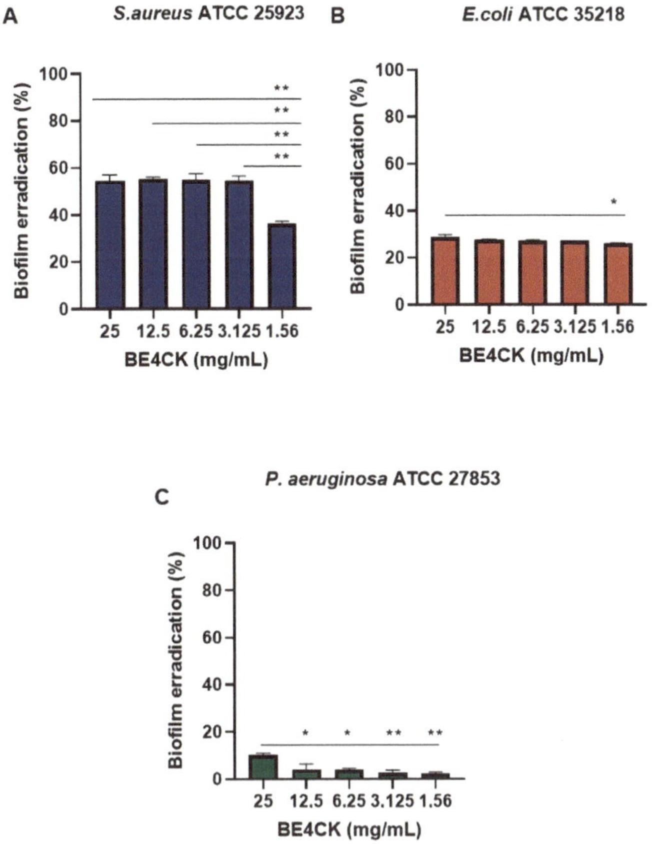

Figure 10. Eradication of the biofilm by the bioemulsifier BE4CK against microorganisms. Values are three independent experiments' mean ± standard deviation (SD). * ($p < 0.05$), ** ($p < 0.001$)—a significant difference in relation to concentrations.

The biofilm formed by *S. aureus* ATCC 25923 was susceptible to the action of the BE4CK at all concentrations used (Figure 10A). At 25 mg/mL–3.12 mg/mL concentrations, the compound promoted eradication rates ranging from 54% to 55%. While at the lowest concentration (1.56 mg/mL), it removed 36% of the preformed biofilm. The increase in the BE4CK concentration did not promote significant differences in removing the *S. aureus* ATCC 25923 biofilm. Thus, eradicating the biofilm did not depend on the concentration of the BE4CK.

The action of BE4CK on the biofilm of *E. coli* ATCC 25922 and *P. aeruginosa* ATCC 27853 showed little effective results (Figure 10B,C). Preformed biofilms were significantly ($p < 0.05$) eradicated at concentrations of 1.56 mg/mL to 25 mg/mL of BE4CK, with biofilm reduction ranges from 29 to 26% *E. coli* ATCC 25922 (Figure 10B) and 2 to 10% for *P. aeruginosa* ATCC 27853 (Figure 10C).

The emulsion OMBE4CK promoted significant dispersion ($p < 0.0001$) of the biofilm in all microorganisms tested (Figure 11A,B). The biofilms formed by *S. aureus* ATCC 25923 were utterly (100%) eradicated at all concentrations (1.56 mg/mL to 25 mg/mL) of the emulsion with no apparent dose–effect relationship between concentrations. The preformed biofilm by *E. coli* ATCC 25922 and *P. aeruginosa* 27853 was removed entirely ($p < 0.0001$) only at concentrations of 6.25 mg/mL–25 mg/mL, whereas at concentrations of 1.56 mg/mL and 3.12%, biofilm removal ranged from 39% to 77% for *E. coli* ATCC 25922 (Figure 11A) and 39% to 52% for *P. aeruginosa* 27853 (Figure 11B).

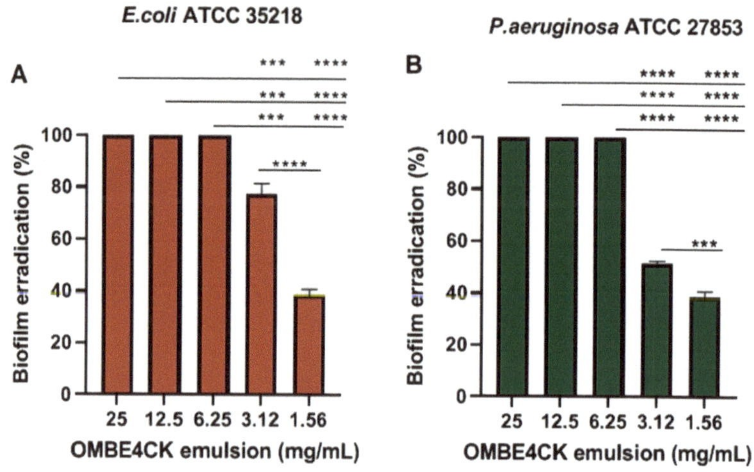

Figure 11. Biofilm eradication by OMBE4CK (emulsion) against microorganisms. (**A**) *E. coli* ATCC 25922, (**B**) *P. aeruginosa* ATCC 27853. Values are three independent experiments' mean ± standard deviation (SD). *** ($p < 0.0001$), **** ($p < 0.0001$)—a significant difference in concentrations.

3. Discussion

3.1. Production and Characterization of the Bioemulsifier Produced by Candida krusei M4CK

In the present study, the yeasts were identified as belonging to *Candida metapsilosis* and *Candida krusei* species. Yeast is widely present in the environment, and many colonize soils, foods, leaf surfaces, and flowers [15].

Candida krusei is an ascomycetic yeast and is widely distributed in nature. In addition, it is frequently identified in cocoa fermentation in different places [16]. Despite being the same yeast, publications related to industrial applications often use the species names *Pichia kudriavzevii*, *Candida glycerinogenes*, and *Issatchenkia orientalis* in preference to *C. krusei* due to the negative safety connotations of using a pathogen in a biotechnological or food context [17].

In this study, the yeasts were isolated from the fruit *B. crassifolia*, popularly known as murici; this fruit originates from Central and South America and has been used for food production and folk medicine [18]. The isolation of yeasts with the potential to produce surface-active compounds indicates that the producers of SACs are widely distributed on the surfaces, internally found in the pulp of various fruits and plants, and have the potential for use in emulsifying systems and different biotechnological processes [19,20].

The *C. krusei* M4CK strain had better growth and performance in producing SACs with glycerol as a carbon and energy source. This discrepancy may be related to the type of carbon sources since these compounds can act in different ways, such as altering the structure of the biosurfactants and bioemulsifiers produced and, consequently, their emulsifying properties [21].

As the type of carbon source also plays a critical role in the production of surfactant compounds by microorganisms [22], it appears that the absorption of glycerol can help to optimize the production processes of biosurfactant glycolipids, such as the production of sophorolipids (SL) by *Starmerella bombicola* [23]. Like the present study, glycerol was reported as one of the best carbon sources for glycolipid production by *Pseudomonas* sp. [24], *a substrate suitable for* higher production and maximum emulsifying activity of the bioemulsifier produced by *Bacillus* sp. [25].

The bioemulsifier of *C. krusei* M4CK was called BE4CK. It showed an emulsifying activity equal to or greater than 50% in olive oil, sunflower, after frying, and hydrocarbons such as kerosene, hexane, and hexadecane. The best emulsifying activities of SACs were demonstrated in the presence of hydrocarbons such as kerosene, hexane, and hexadecane, in which an $E_{24} \geq 63.6\%$ was reached. According to Willumsen and Karlson [26], E_{24} values equal to or greater than 50% are considered significant, and for Bosch et al. [27], an emulsion with this value is considered stable.

BE4CK can be classified as an amphiphilic biopolymer that contains monosaccharides (mannose and glucose) and long-chain aliphatic groups, such as 9-octadecenoic acid and hexadecanoic acid, in its polymer structure and is effective in emulsifying kerosene. The emulsification for BE4CK can be achieved through the interaction of hydrophobic (fatty acids) with hydrocarbons and the interaction of hydrophilics (monosaccharides) with water molecules. Furthermore, as it is a polymer, there is a greater probability of interaction due to the stearic phenomenon, which can lead to a more significant layer of emulsification of petroleum-derived hydrocarbons than vegetable oils. It is believed that there is a more outstanding balance of forces that stabilize kerosene emulsions, providing a higher emulsified layer [28]. Similar to the presented results, sophorolipids synthesized by yeasts of the genus *Meyerozyma* sp. were able to emulsify in hydrocarbons (kerosene, hexane, and hexadecane) and olive oil, and crude oil, the emulsifying activity for these compounds was more significant than 60% [28].

The bioemulsifier BE4CK showed a surface tension of the water of 44 mN/m. Other SACs, such as the glycolipid type produced by *Rhodotorula mucilaginosa* LBP5, reduced the surface tension of water to 40.7 mN/m [29]. However, it is observed that the surface tension of SACs produced by yeasts can be as low as 40 mN/m, followed by the glycolipid-type biosurfactant produced by *Rhodotorula* sp.CC01, whose critical micellar concentration tests indicated a value of 33.39 mN/m [30].

A surface tension reduction value similar to that obtained in the assay using BE4CK was observed for *Candida tropicalis* cultivated in n-hexadecane; this yeast produced an extracellular emulsifier that was able to reduce the surface tension of water from 70 mN/m to 49.5 mN/m. The same compound emulsified several hydrocarbons, including aromatic hydrocarbons [31]. Other SAC-producing yeasts have been identified, such as *Meyerozyma caribbica, Yarrowia lipolytica*, and *Trichosporon mycotoxinivorans* which, despite producing potent bioemulsifiers with high emulsification values in hydrocarbons, did not reduce the surface tension of water when tested [22,32,33].

E_{24} follows exponential growth for *C. krusei* M4CK from the growth phase up to 144 h, obtaining a higher emulsifying activity between 60 h and 100 h in the stationary phase. These data showed a linear correlation between possible cell concentration and emulsifying activity. This behavior can be explained by the fact that the emulsifier is being produced to improve the assimilation of the carbon source. In the production associated with growth, there are parallel relationships between growth, substrate utilization, and the production of surface-active compounds [22,34,35]. Bioemulsifier production kinetics partially associated with cell growth have also been reported for yeast species *Geotrichum* sp. and *Trichosporoon montevideense* [6].

Bioemulsifier effectiveness depends on emulsifying activity over various temperatures, pH, and salinity [36]. The bioemulsifier BE4CK maintained stable emulsions over a wide temperature range (120 °C) for sources of olive and sunflower oil ($E_{24} \geq 50\%$) and hydrocarbons ($E_{24} \geq 63\%$) such as kerosene, hexane, and hexadecane. However, for the frying oil (E_{24} of 40%), there was a 10% reduction in the emulsifying activity. Such results demonstrate that BE4CK is thermostable.

The stability of the bioemulsifier produced by *C. lipolytica* at high temperatures has also been reported at temperatures up to 100 °C [36]. Likewise, biosurfactant sophorolipids produced by *Candida keroseneae* showed high thermal stability with a maximum emulsification activity of 60% in kerosene and preserved 97.25% of its original activity at 120 °C [37]. The reduction in emulsifying activity may be related to the denaturation of the protein component of the bioemulsifier during heat treatment. This property depends on the molecule's chemical structure, as many emulsifiers are stable in significant temperature variations or show minor deviations in higher temperatures [22].

The stability of emulsification by BE4CK was maintained for sunflower and olive oil sources only at acidic pH (3–6). At the same time, BE4CK maintained emulsifier stability in kerosene in all pH ranges (3–9). The inhibition of oil activity and the pronounced reduction in kerosene were recorded at pH values above 6.0. The decrease in emulsifying activity is likely related to structural changes in bioemulsifiers caused by extreme pH conditions [33,38]. In addition to maintaining emulsifying activity at high temperatures and pH ranges, the bioemulsifier BE4CK carried the E_{24} in different concentrations of $MgCl_2$, NaCl, and $CaCl_2$. Due to its nature, the amphiphilic biopolymer BE4CK appears to be more stable and less susceptible to the action of ions for coalesce activity. Similar results to the present study were demonstrated for the bioemulsifier produced by *Pseudomonas stutzeri* NJtech 11-1. The crude bioemulsifier can tolerate 50% NaCl, $MgCl_2$, and $CaCl_2$ with only a 29.8% loss of E_{24}. *Meyerozyma caribbica* yeasts also produced stable bioemulsifiers at high salinity levels [39]. In contrast, a bioemulsifier produced by *Candida lipolytica* using burned motor oil as a hydrophobic substrate showed that its emulsifying stability at 10% NaCl concentration was inactive [36].

By the toxicity test, the bioemulsifier showed no toxicity at the concentrations. Cells remained viable and did not undergo cell death; cell viability was $\geq$88.89%. Toxicity tests with SACs of biosurfactants produced from species of the genus *Candida* also indicated the absence of toxicity of these biomolecules against marine bioindicators [40]. Our results were in accordance with previous work by [41], who reported no risk of toxicity bioemulsifier produced by *S. cerevisiae* against normal epithelial kidney cells at concentrations up to 250 µg/mL. However, further, in vivo tests are required to confirm the safety of bioemulsifier BE4CK.

3.2. Antibacterial and Antibiofilm Activity of the C. krusei M4CK Bioemulsifier with Melaleuca Essential Oil

When evaluating the antimicrobial activity through the diffusion assay, it was verified that the isolates of *E. coli* ATCC 25922, *Enterococcus faecalis* ATCC 35218, *P. aeruginosa* ATCC 27853, and *S. aureus* ATCC 25923 did not show sensitivity to the bioemulsifier BE4CK. However, when combined with melaleuca essential oil emulsion (OMBE4CK), they exhibited antimicrobial activity for *E. coli* ATCC 25922 isolates, *E. faecalis* ATCC 35218, and *S. aureus* ATCC 25923. The antimicrobial activity of the OMBE4CK emulsion was more efficient for *E. coli* ATCC 25922 and did not cause sensitivity in *P. aeruginosa* ATCC 27853. These data indicate that melaleuca essential oil can be a candidate for producing formulations with surface-active compounds synthesized by microorganisms.

Antimicrobial performance similar to the bioemulsifier BE4CK was demonstrated for sophorolipid produced by non-pathogenic yeasts of *Candida bombicola* (ATCC 22214); this biosurfactant could not inhibit the growth of *E. coli* alone [42]. But, in combination with antibiotics, it promoted a total inhibition of the development of the microorganism. This ability is not only related to the inhibitory effect increased by the additive action of two antimicrobial agents; it may also be associated with the performance of SL to facilitate the entry of molecules into bacterial cells [42]. So, the antimicrobial effect of biosurfactants is due to their potential to form pores within cell membranes; this feature can increase the effectiveness of antimicrobial agents [43].

Antimicrobial resistance to antibiotics is often related to the ability of microorganisms to form biofilms, which exhibit dozens of attributes within this structure, making them difficult to eliminate [44]. Given this fact and the antimicrobial capacity of the emulsion (BE4CK and OMBE4CK), the action of the bioemulsifier and the emulsion in the inhibition and eradication of biofilms formed by *E coli* ATCC 25922, *P. aeruginosa* ATCC 27853, and *S. aureus* ATCC 25923.

The compound showed a concentration-dependent effect, with the best biofilm formation inhibitory activity against bacterial pathogens. Higher BE4CK activity was found in. *S. aureus* ATCC 25923, while less activity was demonstrated for *P. aeruginosa* ATCC 27853. The BE4CK activity profile may be related to the cell wall characteristics of both bacteria.

The antibiofilm activity of the emulsion gradually decreased at lower concentrations. Melaleuca essential oil has already demonstrated antibiofilm activity against *S. aureus* and *E. coli* [45,46]. On the other hand, there is previous evidence of the synergism between SACs and essential oils; this fact reinforces the hypothesis of the ability of SACs to facilitate the entry of antimicrobial molecules into cells [47]. In addition, melaleuca essential oil emulsions prove to be an excellent microbial control strategy, as their improved tissue infiltration capacity and ability to stabilize the volatilization of terpenes, their main antimicrobial compounds, are considered. In this study, tea tree oil emulsions (OMBE4CK) were stable up to 250 (E_{250}) evaluation days, which enhances their use in formulations.

Biofilm formation on medical devices is one of the leading causes of nosocomial infections and represents a severe public health problem. As a possible alternative for solving this problem, the potential of the emulsion was also evaluated for inhibiting biofilm formation in polyurethane catheters at different concentrations. At the highest concentration (25 mg/mL), OMBE4CK showed a significant % inhibitory potential of 43% for *S. aureus* ATCC 25923. Other authors also reported this ability to interfere with biofilm formation. Mendes et al. [48] showed that a mixture of sophorolipids produced by *Starmerella bombicola* CBS 6009 with sodium dodecyl sulfate (SDS, commercial surfactant) demonstrated inhibitory activity of biofilm formation by *S. aureus* and *E. coli* [48]. Another study proposed by Pontes et al. [49] determined that the presence of sophorolipids on the surface of the catheter (silicone) decreased the hydrophobicity of the material and the formation of biofilm by *S. aureus* ATCC 25923 and *E. coli* ATCC 25922 [49].

The OMBE4CK emulsion showed better activity in removing biofilms compared to the isolated action of BE4CK. Biofilm eradication may be related to the bacteriostatic and bactericidal effects of the OMBE4CK emulsion.

Biofilms are microbial communities that are difficult to eliminate. The potential of the bioemulsifier BE4CK and the emulsion OMBE4CK in eradicating biofilms was evaluated to seek strategies to disperse this structure. The treatment with the bioemulsifier alone on the biofilm of *E. coli* ATCC 25922 and *P. aeruginosa* ATCC 27853 did not show effective results in all concentrations. In contrast, *S. aureus* ATCC 25923 biofilms were significantly eradicated.

The low dispersion of the biofilm of *E. coli* ATCC 25922 and *P. aeruginosa* ATCC 27853 can be attributed to the difficulty of the SACs in penetrating the biofilm matrix; in this sense, higher concentrations of SACs are necessary for the detachment of the preformed biofilm. A surface-active substance must penetrate the biofilm, and the substrate interface must detach from the biofilm [50]. After their penetration, they can change surface properties, leading to biofilms' separation from surfaces [50].

In the complete eradication of the *S. aureus* biofilm, the pit is probably related to the ability of the *bioemulsifier BE4CK* to increase the action of melaleuca essential oil, facilitating its entry through the biofilm matrix. The biofilm matrix presents a dynamic to be considered when applying a particular antibiotic compound. In this case, the action to dispose of the cells can be improved using SACs that alter the surface characteristics, such as hydrophobicity or hydrophilicity, in addition to the electrical charge characteristics (repulsion and attraction) [51]. Emulsified (containing essential oils) systems can be interesting in controlling microbial biofilms. The present study observed that the emulsion at higher concentrations was more effective than the bioemulsifier BE4CK in eradicating the *P. aeruginosa* biofilm.

4. Materials and Methods

4.1. Yeast Isolation

For the isolation and selection of yeasts of interest, fruits of the species *Anacardium occidentale*, *Byrsonima crassifolia*, *Citrus reticulata*, and *Platonia insignis* were collected from plants present in sites located in the Municipalities of São José de Ribamar and São Luís, Maranhão. The fruits selected for yeast isolation were in their intermediate stage of maturation. The harvested fruits were stored in thermal boxes with ice and transported to the Applied Microbiology Laboratory of CEUMA University for further processing.

The fruits were initially sanitized with 5% sodium hypochlorite to isolate SAC-producing yeasts and washed three times with sterile distilled water to remove sodium hypochlorite. Subsequently, approximately 20 g of fruit samples were inoculated into Erlenmeyer flasks with the addition of a minimal mineral medium (MM) containing (in g/L) 1.0 g K_2HPO_4, 1.0 g KH_2PO_4, 0.2 g $MgSO_4{\cdot}7H_2O$, 1.0 g NH_4NO_3, 0.002 g $CaCl_2{\cdot}2H_2O$, and 0.2% yeast extract with 2% sunflower oil or glycerol added as the only carbon source. The final pH of the medium was 7.2.

The flasks were incubated at 30 °C with agitation at 140 rpm for seven days. After incubation, 1 mL of the cultures were serially diluted in a sterile saline solution (0.85% NaCl). Aliquots of 50 μL of decimal dilutions were spread on the surface of the Sabouraud Dextrose Agar medium (SDA, Difco TM, Detroit, MI, USA) to isolate, quantify, and characterize colonies.

Samples of yeast colonies were inoculated in the SDA medium to obtain a pure culture; later, microcultures were performed for morphological characterization. The yeasts were kept in inclined tubes with the SDA medium and stored in cryotubes at –80 °C in Sabouraud broth (Difco) with 20% glycerol.

4.2. Screening and Identification of SAC-Producing Yeasts

4.2.1. Characterization of Emulsifying Activity (E_{24}) of Culture Supernatant

The yeast isolates were inoculated at an initial concentration of 0.1 OD_{600} nm ($\approx 1 \times 10^7$ CFU/mL) in Erlenmeyer flasks with 50 mL of a minimal mineral medium (MM) containing (in g/L) 1.0 g K_2HPO_4, 1.0 g KH_2PO_4, 0.2 g $MgSO_4{\cdot}7H_2O$, 1.0 g NH_4NO_3, 0.002 g $CaCl_2{\cdot}2H_2O$, and 0.2% yeast extract with the addition of 2% glycerol or 2% sunflower oil (as carbon sources for SAC production) as proposed by Monteiro et al. [6], with modifications. Each yeast isolate was incubated on the carbon sources to evaluate the production of SACs. The microbial cultures were incubated for up to 144 h at 30 °C under agitation at 140 rpm. After incubation, the cultures were centrifuged at 5000× *g* for 10 min, and the supernatants were used in the emulsifying activity (E_{24}) test. The cell-free supernatant's surface tension (TS) was determined using a Du Nouy ring-type tensiometer (Manual Tensiometer, model k6, Krüss GmbH, Hamburg, Germany). The TS measurement was performed at 20 °C with modifications). The microbial cultures were performed in duplicate.

The emulsifying activity was determined by the emulsification index (E_{24}), in which the supernatant was applied directly into test tubes in the proportion of 50% (2.5 mL) and 50% (2.5 mL) of different hydrophobic substrates (olive oil, sunflower oil, and fried foods, kerosene, hexane, and hexadecane). The mixture was subjected to vortex agitation for 2 min. After 24 h, the emulsification activity (E_{24}) was determined by dividing the height of the emulsified layer by the total size of the mixture, expressed as a percentage [52]. For tests, rhamnolipids (100 mg/mL) (Di-rhamnolipid, Sigma-Aldrich, St. Louis, MO, USA) were a positive control and the MM medium was a negative control. The emulsifying activity tests were performed in triplicate.

MALDI-TOF MS-based identification of yeast isolates was performed according to Bruker Daltonics (Bremen, Germany) using the protocol in Aslani et al. [53].

4.2.2. Determination of the Surface Tension of Water in the Presence of a Bioemulsifier

Surface tension (TS) was measured using a Du Nouy ring-type tensiometer (Manual Tensiometer, model k6, Krüss GmbH, Hamburg, Germany). The TS measurement was performed at 20 °C with modifications. For tests, sodium dodecyl sulfate (SDS, 100 mg/L) was the positive control and the MM medium was the negative control. The TS measurement was performed at 20 °C with modifications. The concentration of the bioemulsifier used varied from 0.01 mg/mL to 10 mg/mL. The round hanging from the scale's hook was immersed in the liquid. Then, the call was slowly pulled upwards, lowering the sample plate. Finally, the force applied to the ring during traction across the surface was recorded. The measurement was repeated three times; the platinum ring was washed three times with distilled water and three times with alcohol between measurements [6]. Measurement of the surface tension of the distilled water was used as a control.

4.3. Chemical Composition of SACs

Standard methods were used to determine the chemical composition of the bioemulsifier. The carbohydrate content was determined by the phenol–sulfuric acid method of Dubois et al. [54] using D-glucose as a standard. Protein content was measured as described by Bradford et al. [55] using bovine serum albumin as a standard. Lipid content was estimated using the K117—BIOCLIN monoreagent triglyceride kit. All experiments were performed in triplicate. The carbohydrate and lipid composition of the bioemulsifier (SAC) was determined by gas chromatography and mass spectrometry using the previously (GC-MS) developed protocol [6].

4.4. Growth Kinetics

The inoculum was prepared from *C. krusei* cells incubated in a solid Sabouraud medium for 24 h at 32 °C and washed three times with saline solution (NaCl 0.85%, *w*/*v*). After washing, the cells were resuspended in 50 mL plastic tubes containing 20 mL of the MM medium. Optical density readings at 600 nm were performed. The values obtained determined the volume to be transferred to the test flasks for a final concentration of 0.02 OD, corresponding to ($\approx 5 \times 10^6$ CFU/mL). The experiments were conducted in 500 mL Erlenmeyer flasks containing 200 mL of the MM medium added with 2% glycerol. The flasks were incubated at 32 °C under continuous agitation at 180 rpm for up to 144 h. The samples were removed at 12 h or 24 h to determine cell concentration and emulsifying activity. The tests were performed in three repetitions.

Cell growth was determined by removing the aliquots from the culture medium and obtaining decimal dilutions from which samples were removed, serially diluted, and plated in a solid Sabouraud medium to determine CFU/mL. The values obtained were converted into Log_{10} of CFU/mL [6]. The emulsification activity (E_{24}) described in Section 4.2.1 also evaluated the supernatant samples for emulsifying activity. The tests were performed in three repetitions.

4.5. Stability of the Bioemulsifier in Different NaCl, $MgCl_2$, and $CaCl_2$ Concentrations, Autoclavings, and pH Variations

The effects of different concentrations (10, 30, and 50%) of magnesium chloride ($MgCl_2$), sodium chloride (NaCl), and calcium chloride ($CaCl_2$), a temperature of 121 °C for 15 min (autoclaving), and different pH values (3–10) were analyzed in the activity of the bioemulsifier BE4CK. The emulsification activity (E_{24}) was evaluated after adjusting the pH minimum medium with sodium hydroxide (NaOH) or hydrochloric acid (HCl).

4.6. Cytotoxicity Test

The assay for evaluation of bioemulsifier cytotoxicity was performed using African green monkey kidney cells (VERO, ATCC), according to Sieuwerts et al. [56]. For the assay, the cells were washed with phosphate-buffered saline and resuspended in Dulbecco's Modified Eagle Medium (DMEM, Thermo Scientific, Erlangen, Germany) culture medium supplemented with 10% fetal bovine serum (FBS) with 100 IU/mL of penicillin and 100 µg/mL of streptomycin. Cell counting was performed to determine the cell number with an equal volume of trypan blue. Briefly, 100 µL of DMEM with 10% of FBS was added to the 96-well plate (TPP, Trasadingen, Switzerland). Cells were added to the wells (1.5×10^6 cells per well) and treated with 5000 µg/mL–2 µg/mL diluted in DMEM and PBS for the negative control. The plates were incubated for 24 h at 37 °C, 5% CO_2, and a 90% humidity incubator. After incubation, 20 μL of MTT (3-4,5-dimethylthiazol-2,5-diphenyltetrazolium bromide, Sigma-Aldrich) at 5 mg/mL was added into each well in the 96-well plate and incubated for four hours in 37 °C, 5% CO_2, and a 90% humidity incubator. After the incubation, 170 μL of the medium with MTT was removed from every well, and 100 μL DMSO (Sigma-Aldrich, USA) was added to each well to extract and solubilize the formazan crystal. Finally, the plate was read at a 570 nm wavelength using TP-Reader Plus (ThermoPlate, China). Each extract concentration (flavonoids) was assayed in triplicate in two independent experiments. The following formula calculated the percentage of proliferation: [OD sample − OD control] × 100% proliferation = OD control.

4.7. Well Diffusion Test in Agar

The tests were performed according to Johnson et al. [57] as modifications for the study of the antimicrobial activity of the SAC. For *Escherichia coli* ATCC 25922, *Enterococcus faecalis* ATCC 35218, *Pseudomonas aeruginosa* ATCC 27853, and *Staphylococcus aureus* ATCC 25923 were added in a Tryptic Soy Agar (TSA, Difco TM, Detroit, MI, USA) for 24 h at 30–37 °C. Subsequently, the isolates were suspended in 0.9% NaCl and adjusted to equal the standard turbidity to 0.5 on the McFarland scale (1.5 × 10^8 CFU/mL). The suspensions were used for seeding in the TSA medium stabilized at 37 °C (0.1 mL microorganism test suspension/10 mL medium). The TSA culture medium was poured into 90 mm Petri dishes (20 mL medium/dish) and allowed to solidify. Once the medium was solid, four 9 mm diameter wells were made with sterilized plastic straws and filled with one volume of SACs produced by *C. krusei*, melaleuca essential oil, and emulsion (a mixture of SACs and essential oil melaleuca). Plates were pre-incubated for one hour at room temperature to ensure adequate diffusion and finally incubated at 30–37 °C for 24 h. Inhibition halos were evaluated in mm. All tests were performed in triplicate.

4.8. Antibiofilm Activity

4.8.1. Inhibition of Biofilm Formation on Polystyrene

Suspensions of biofilm-forming isolates were prepared according to the methodology established by Merghini et al. [58], with modifications. For the tests, breaks of each isolate of *E. coli* ATCC 25922, *P. aeruginosa* ATCC 27853, and *S. aureus* ATCC 25923 were prepared in Mueller–Hinton broth (Difco TM, Detroit, MI, USA) plus 5% sucrose at a cell concentration of 0.5 on the McFarland turbidity scale (1.5 × 10^8 CFU/mL). The bioemulsifiers were diluted in Mueller–Hinton broth at 25 mg/mL, 12.5 mg/mL, 6.25 mg/mL, 3.12 mg/mL, and 1.56 mg/mL, with a final volume of 200 μL. Each well containing the bioemulsifier received 20 μL of the standardized bacterial suspension. The plate was then incubated for 24 h at 35 ± 2 °C.

The contents of each well were aspirated, the wells were gently washed three times with 200 μL of *phosphate-buffered saline* (PBS, pH 7.4), and the plates were kept at room temperature to dry. The adhered biomass was fixed with 200 μL of methanol (P.A 99% *v*/*v*) for 10 min. Then, 200 μL of a 0.5% crystal violet solution was added to each well. Subsequently, the crystal violet was removed, and the wells were washed three times with distilled water. Then, 200 μL of alcohol 95% (*v*/*v*) was added to each well, and the OD of the crystal violet extract was determined in a microplate reader at 570 nm. Using the formula below, the percentage of biofilm inhibition was calculated.

Percentage inhibition (%) = 100 − ((OD 570 nm of the treatment)/(OD 570 nm of the control) × 100).

This study also evaluated the activity of the emulsion resulting from the combination of the bioemulsifier and melaleuca essential oil (OMBE4CK) in inhibiting biofilm formation for each bacterium. For this, the same methodological procedure was followed. The used concentrations of OMBE4CK were 25 mg/mL to 1.56 mg/mL in emulsion.

To form the emulsion between the SAC and the melaleuca essential oil (commercially purchased), 50 mg of BE4CK (SAC) was used. This amount was dissolved in 10 mL of water for sterile injection; later, this solution was applied directly in test tubes in the proportion (2 mL) of the solution and (2 mL) the melaleuca essential oil. The mixture was subjected to vortexing for 2 min, and the emulsion formed was used in the tests at different concentrations (1.56 mg/mL–25 mg/mL).

4.8.2. Inhibition of Biofilm Formation in an Enteral Feeding Catheter

Quantitative Analysis

To evaluate the inhibition of biofilm formation in a polyurethane catheter (enteral feeding catheter, Embramed SP, Brazil), sterile segments of the device, measuring approximately 4 cm, were immersed in a TSB medium containing 0.5×10^5 CFU/mL of *S. aureus* 25923 suspensions and OMBE4CK (emulsion) concentrations (1.56 mg/mL–25 mg/mL). Catheter fragments were incubated for 24 h at 37 °C. After incubation, the catheters were washed with 200 μL of sterile saline three times and vortexed for 2 min. About 100 μL of the last wash was removed for serial dilutions in sterile saline. After the decimal dilutions, 30 μL of the sample were transferred to the surface of Petri plates containing the TSA medium, and the samples were homogenized with Drigalsky loops. After drying the samples, the plates were incubated for 24 h at 37 °C. After incubation, the colony-forming unit (CFU/mL) count was performed. All tests were performed in triplicate.

A semi-quantitative analysis of the inhibition of biofilm formation on the surface of a polyurethane catheter was performed by rolling the catheter segment in the Petri dish containing the TSA medium (Difco TM, Detroit, MI, USA). The semiquantitative technique distinguishes colonization (greater than or equal to 15 colonies) from contamination [59].

4.8.3. Biofilm Eradication

An evaluation of the influence of the bioemulsifier BE4CK on biofilms produced by *E. coli* ATCC 25922, *P. aeruginosa* ATCC 27853, and *S. aureus* ATCC 25923 were performed after the formation of microbial biofilms on the surface of the wells of 96-well polystyrene plates according to the established methodology by Morais et al. [60], with modifications. A total of 200 μL of Brain Heart Infusion (BHI, Difco TM, Detroit, MI, USA) with 2% sucrose were added to all wells; then, 10 μL of standardized bacterial suspension on the McFarland scale (~0.5×10^5 CFU/mL) was added, the positive control was performed with the addition of the BHI medium + sucrose, and 10 μL of bacterial suspension was added. For the negative control, only the BHI medium with sucrose was added. The plates were incubated for 48 h at 37 °C. After incubation, the wells were gently washed with 200 μL of sterile saline three times in each well; then, 200 μL of concentrations of 25 mg/mL were added; 12.5 mg/mL; 6.25 mg/mL; 3.12 mg/mL; and 1.56 mg/mL, which were incubated for 24 h. After incubation, washing was performed with 100 μL of a saline solution (NaCl, 0.85%). Subsequently, 100 μL of saline was added to the plate wells, and homogenization was carried out. A total of 50 μL was taken from the well, and decimal dilutions were performed for plating the samples. Viable cells were determined using microdroplet plating, for which approximately 10 μL were transferred to a Tryptic Soy Agar (TSA, Difco TM, Detroit, MI, USA) plate as a drop; after drying, the dish was incubated correctly for 24 h in the bacteriological oven. After incubation of the plates, the colony-forming unit (CFU/mL) count was performed.

This study also evaluated the activity of the emulsion resulting from the interaction between the bioemulsifier and melaleuca essential oil (OMBE4CK) in eradicating the biofilm for each isolate. For this, the same methodological procedure was followed. The used concentrations of OMBE4CK were 25 mg/mL, 12.5 mg/mL, 6.25 mg/mL, 3.125 mg/mL, and 1.56 mg/mL of the emulsion. All tests were performed in triplicate.

4.9. Statistical Analysis

Data were submitted for statistical analysis using the one-way ANOVA procedure in GraphPad Prism (version 8.0.2). All duplicate results are expressed as mean ± standard deviation. Differences were examined using the Tukey test, with a significance level of 95%. The significance levels were set at $p < 0.05$, $p < 0.001$, and $p < 0.0001$.

5. Conclusions

This study presented a bioemulsifier derived from *Candida krusei* M4CK isolated from fruit murici and showed an excellent ability to use glycerol as a substrate for growth and bioemulsifier production. This biomolecule showed promising properties in emulsifying different sources of vegetable oil and hydrocarbons, a lack of toxicity, and stability of emulsions in a wide range of temperatures, pH, and salinity. Furthermore, the bioemulsifier BE4CK and its combination with melaleuca essential oil (OMBE4CK) inhibited bacterial biofilm formation and eradicated bacterial biofilm on polystyrene surfaces and catheters. Such properties emphasize that the bioemulsifier can represent a new source for biomedical development in emulsion formulations containing essential oils and can be a strategy for reducing the colonization of medical devices and surface abiotics.

Supplementary Materials: The following supporting information can be downloaded at https://www.mdpi.com/article/10.3390/antibiotics12121686/s1, Table S1. E24 values from yeasts from *Anacardium occidentale*, *Byrsonima crassifolia*, *Citrus reticulata*, and *Platonia insignis* fruits. Table S2. Determination of final surface tension (mN/m) in medium after cell growth yeasts strains. Figure S1. Antimicrobial activity of BE4CK and emulsion (OE Melaleuca and BE4CK- OMBE4CK) by agar diffusion against *E. coli* ATCC 25922.

Author Contributions: Methodology, investigation, writing—original draft preparation, software, and visualization, J.M.M.A. and J.M.M.; writing—reviewing, D.H.d.S.S.; methodology and research, A.K.V. and M.R.C.S.; methodology and investigation, J.M.M.A., M.R.C.S. and F.A.F.; funding acquisition, F.H.R.B. and E.P.d.S.; project administration and writing—original draft preparation A.d.S.M. All authors have read and agreed to the published version of the manuscript.

Funding: This study was financed in part by Fundação de Amparo à Pesquisa e o Desenvolvimento Científico e Tecnológico do Maranhão (FAPEMA, BD-01413/19); and Conselho Nacional de Desenvolvimento Científico e Tecnológico (CNPQ, 434149/2018-7).

Institutional Review Board Statement: Not applicable.

Informed Consent Statement: Not applicable.

Data Availability Statement: Not applicable.

Acknowledgments: The authors thank Hélio Euclides S. dos Santos and Marinaldo for technical assistance.

Conflicts of Interest: The authors declare no conflict of interest.

References

1. Ceresa, C.; Tessarolo, F.; Maniglio, D.; Tambone, E.; Carmagnola, I.; Fedeli, E.; Caola, I.; Nollo, G.; Chiono, V.; Allegrone, G.; et al. Medical-Grade Silicone Coated with Rhamnolipid R89 Is Effective against *Staphylococcus* spp. Biofilms. *Molecules* **2019**, *24*, 3843. [CrossRef]
2. Badrul Hisham, M.D.; Ibrahim, N.H.; RamLi, M.F.; Abd-Aziz, N. Production of Biosurfactant Produced from Used Cooking Oil by *Bacillus* sp. HIP3 for Heavy Metals Removal. *Molecules* **2019**, *24*, 2617. [CrossRef] [PubMed]
3. Durval, I.J.B.; Mendonça, A.H.R.; Rocha, I.V.; Luna, J.M.; Rufino, R.D.; Converti, A.; Sarubbo, L.A. Production, Characterization, Evaluation and Toxicity Assessment of a *Bacillus cereus* UCP 1615 Biosurfactant for Marine Oil Spills Bioremediation. *Mar. Pollut. Bull.* **2020**, *157*, 111357. [CrossRef]
4. Yamasaki, R.; Kawano, A.; Yoshioka, Y.; Ariyoshi, W. Rhamnolipids and Surfactin Inhibit the Growth or Formation of Oral Bacterial Biofilm. *BMC Microbiol.* **2020**, *20*, 358. [CrossRef]
5. Sharma, D.; Misba, L.; Khan, A.U. Antibiotics versus Biofilm: An Emerging Battleground in Microbial Communities. *Antimicrob. Resist. Infect. Control* **2019**, *8*, 76. [CrossRef] [PubMed]
6. Monteiro, A.S.; Bonfim, M.R.Q.; Domingues, V.S.; Corrêa, A., Jr.; Siqueira, E.P.; Zani, C.L.; Santos, V.L. Identification and Characterization of Bioemulsifier-Producing Yeasts Isolated from Effluents of a Dairy Industry. *Bioresour. Technol.* **2010**, *101*, 5186–5193. [CrossRef]
7. Gaur, V.K.; Regar, R.K.; Dhiman, N.; Gautam, K.; Srivastava, J.K.; Patnaik, S.; Kamthan, M.; Manickam, N. Biosynthesis and Characterization of Sophorolipid Biosurfactant by *Candida* Spp.: Application as Food Emulsifier and Antibacterial Agent. *Bioresour. Technol.* **2019**, *285*, 121314. [CrossRef]
8. Kim, H.-Y.; Jung, H.; Kim, H.-M.; Jeong, H.-J. Surfactin Exerts an Anti-Cancer Effect through Inducing Allergic Reactions in Melanoma Skin Cancer. *Int. Immunopharmacol.* **2021**, *99*, 107934. [CrossRef]

9. Ceresa, C.; Fracchia, L.; Williams, M.; Banat, I.M.; Díaz De Rienzo, M.A. The Effect of Sophorolipids against Microbial Biofilms on Medical-Grade Silicone. *J. Biotechnol.* **2020**, *309*, 34–43. [CrossRef]
10. Kobayashi, Y.; Li, Q.; Ushimaru, K.; Hirota, M.; Morita, T.; Fukuoka, T. Isolation and Characterization of Novel Naturally Occurring Sophorolipid Glycerides. *Bioresour. Technol. Rep.* **2023**, *22*, 101399. [CrossRef]
11. Yalçın, H.T.; Ergin-Tepebaşı, G.; Uyar, E. Isolation and Molecular Characterization of Biosurfactant Producing Yeasts from the Soil Samples Contaminated with Petroleum Derivatives. *J. Basic Microbiol.* **2018**, *58*, 782–792. [CrossRef] [PubMed]
12. Santos, V.A.; Sarubbo, L.A. Formulation of a commercial biosurfactant for application as a dispersant of petroleum and by-products spilled in oceans. *Front. Microbiol.* **2016**, *7*, 1646. [CrossRef]
13. Atakpa, E.O.; Zhou, H.; Jiang, L.; Ma, Y.; Liang, Y.; Li, Y.; Zhang, D.; Zhang, C. Improved Degradation of Petroleum Hydrocarbons by Co-Culture of Fungi and Biosurfactant-Producing Bacteria. *Chemosphere* **2022**, *290*, 133337. [CrossRef] [PubMed]
14. Huang, K.; Liu, R.; Zhang, Y.; Guan, X. Characteristics of Two Cedarwood Essential Oil Emulsions and Their Antioxidant and Antibacterial Activities. *Food Chem.* **2021**, *346*, 128970. [CrossRef]
15. Alvarez, S.C.V.; Alaniz, M.J.L.; Furlani, M.V.M.; Vazquez, F.; Agresti, P.M.; Nally, M.C.; Maturano, Y.P. Bioprospecting of the probiotic potential of yeasts isolated from a wine environment. *Fungal Genet. Biol.* **2023**, *164*, 103767.
16. Chagas Junior, G.C.A.; Ferreira, N.R.; Andrade, E.H.d.A.; Nascimento, L.D.d.; Siqueira, F.C.d.; Lopes, A.S. Profile of Volatile Compounds of On-Farm Fermented and Dried Cocoa Beans Inoculated with *Saccharomyces* cerevisiae KY794742 and Pichia kudriavzevii. *Molecules* **2021**, *26*, 344. [CrossRef] [PubMed]
17. Douglass, A.P.; Offei, B.; Braun-Galleani, S.; Coughlan, A.Y.; Martos, A.A.R.; Ortiz-Merino, R.A.; Byrne, K.P.; Wolfe, K.H. Population Genomics Shows No Distinction between Pathogenic *Candida krusei* and Environmental *Pichia kudriavzevii*: One Species, Four Names. *PLoS Pathog.* **2018**, *14*, e1007138. [CrossRef]
18. Perez-Gutierrez, R.M.; Muñiz-Ramirez, A.; Gomez, Y.G.; Ramírez, E.B. Antihyperglycemic, Antihyperlipidemic and Antiglycation Effects of *Byrsonima Crassifolia* Fruit and Seed in Normal and Streptozotocin-Induced Diabetic Rats. *Plant Foods Hum. Nutr.* **2010**, *65*, 350–357. [CrossRef]
19. Konishi, M.; Maruoka, N.; Furuta, Y.; Morita, T.; Fukuoka, T.; Imura, T.; Kitamoto, D. Biosurfactant-Producing Yeasts Widely Inhabit Various Vegetables and Fruits. *Biosci. Biotechnol. Biochem.* **2014**, *78*, 516–523. [CrossRef] [PubMed]
20. Silva, G.F.D.; Gautam, A.; Duarte, I.C.S.; Delforno, T.P.; Oliveira, V.M.; Huson, D.H. Interactive Analysis of Biosurfactants in Fruit-Waste Fermentation Samples Using BioSurfDB and MEGAN. *Sci. Rep.* **2022**, *12*, 7769. [CrossRef]
21. Luna, J.M.; Rufino, R.D.; Sarubbo, L.A.; Campos-Takaki, G.M. Characterisation, Surface Properties and Biological Activity of a Biosurfactant Produced from Industrial Waste by *Candida sphaerica* UCP0995 for Application in the Petroleum Industry. *Colloids Surf. B Biointerfaces* **2013**, *102*, 202–209. [CrossRef]
22. Bhaumik, M.; Dhanarajan, G.; Chopra, J.; Kumar, R.; Hazra, C.; Sen, R. Production, Partial Purification and Characterization of a Proteoglycan Bioemulsifier from an Oleaginous Yeast. *Bioprocess Biosyst. Eng.* **2020**, *43*, 1747–1759. [CrossRef]
23. Claus, S.; Jenkins Sánchez, L.; Van Bogaert, I.N.A. The Role of Transport Proteins in the Production of Microbial Glycolipid Biosurfactants. *Appl. Microbiol. Biotechnol.* **2021**, *105*, 1779–1793. [CrossRef]
24. Phulpoto, I.A.; Yu, Z.; Hu, B.; Wang, Y.; Ndayisenga, F.; Li, J.; Liang, H.; Qazi, M.A. Production and Characterization of Surfactin-like Biosurfactant Produced by Novel Strain *Bacillus nealsonii* S2MT and It's Potential for Oil Contaminated Soil Remediation. *Microb. Cell Fact.* **2020**, *19*, 145. [CrossRef]
25. Rulli, M.M.; Alvarez, A.; Fuentes, M.S.; Colin, V.L. Production of a Microbial Emulsifier with Biotechnological Potential for Environmental Applications. *Colloids Surf. B Biointerfaces* **2019**, *174*, 459–466. [CrossRef]
26. Willumsen, P.A.; Karlson, U. Screening of bacteria, isolated from PAH-contaminated soils, for production of biosurfactants and bioemulsifiers. *Biodegradation* **1996**, *7*, 415–423. [CrossRef]
27. Bosh, M.T.; Robert, M.; Mercadé, M.; Espuny, M.J.; Parra, J.L.; Guinea, J. Surface active compounds on microbial cultures. *Tensides Surfactants Deterg.* **1988**, *25*, 208–211. [CrossRef]
28. Rehman, R.; Ali, M.I.; Ali, N.; Badshah, M.; Iqbal, M.; Jamal, A.; Huang, Z. Crude Oil Biodegradation Potential of Biosurfactant-Producing *Pseudomonas aeruginosa* and *Meyerozyma* sp. *J. Hazard. Mater.* **2021**, *418*, 126276. [CrossRef]
29. Vieira Neta, R.A.; Azevedo, M.; Camargo, M.A.; Delforno, F.P.; Bona, T.P.; Vicente, E.; Duarte, J.G.P. Potential of acid-hydrolysated brewer's spent grain as substrate for biosurfactant production by *Rhodotorula mucilaginosa* LBP5 oleaginous yeast. *Biomass Convers. Biorefinery* **2023**. [CrossRef]
30. Lin, X.; Zhou, H.; Zeng, F.; Jiang, L.; Atakpa, E.O.; Chen, G.; Xie, Q. A biosurfactant-producing yeast *Rhodotorula* sp. CC01 utilizing landfill leachate as nitrogen source and its broad degradation spectra of petroleum hydrocarbons. *World J. Microbiol. Biotechnol.* **2022**, *38*, 68. [CrossRef] [PubMed]
31. Singh, M.; Desai, J.D. Hydrocarbon Emulsification by *Candida tropicalis* and *Debaryomyces polymorphus*. *Indian J. Exp. Biol.* **1989**, *27*, 224–226.
32. Souza, F.A.S.D.; Salgueiro, A.A.; Albuquerque, C.D.C. Production of bioemulsifiers by *Yarrowia lipolytica* in sea water using diesel oil as the carbon source. *Braz. J. Chem. Eng.* **2012**, *29*, 61–67. [CrossRef]
33. Monteiro, A.S.; Domingues, V.Z.; Souza, M.V.D.; Lula, I.; Golcalves, D.B.; Siqueira, E.P.; Santos, V.L. Bioconversion of biodiesel refinery waste in the bioemulsifier by *Trichosporon mycotoxinivorans* CLA2. *Biotechnol. Biofuels* **2012**, *5*, 29.
34. Desai, J.D.; Banat, I.M. Microbial Production of Surfactants and Their Commercial Potential. *Microbiol. Mol. Biol. Rev.* **1997**, *61*, 47–64. [CrossRef] [PubMed]

35. Liepins, J.; Balina, K.; Soloha, R.; Berzina, I.; Lukasa, L.K.; Dace, E. Glycolipid Biosurfactant Production from Waste Cooking Oils by Yeast: Review of Substrates, Producers and Products. *Fermentation* **2021**, *7*, 136. [CrossRef]
36. Souza, A.F.; Rodriguez, D.M.; Ribeaux, D.R.; Luna, M.A.C.; Lima E Silva, T.A.; Andrade, R.F.S.; Gusmão, N.B.; Campos-Takaki, G.M. Waste Soybean Oil and Corn Steep Liquor as Economic Substrates for Bioemulsifier and Biodiesel Production by *Candida lipolytica* UCP 0998. *Int. J. Mol. Sci.* **2016**, *17*, 1608. [CrossRef]
37. Ganji, Z.; Beheshti-Maal, K.; Massah, A.; Emami-Karvani, Z. A Novel Sophorolipid-Producing *Candida keroseneae* GBME-IAUF-2 as a Potential Agent in Microbial Enhanced Oil Recovery (MEOR). *FEMS Microbiol. Lett.* **2020**, *367*, fnaa144. [CrossRef]
38. Daverey, A.; Pakshirajan, K. Production, Characterization, and Properties of Sophorolipids from the Yeast *Candida bombicola* Using a Low-Cost Fermentative Medium. *Appl. Biochem. Biotechnol.* **2009**, *158*, 663–674. [CrossRef]
39. Fan, Y.; Tao, W.; Huang, H.; Li, S. Characterization of a Novel Bioemulsifier from *Pseudomonas stutzeri*. *World J. Microbiol. Biotechnol.* **2017**, *33*, 161. [CrossRef] [PubMed]
40. Luna, J.M.; Santos Filho, A.; Rufino, R.D.; Sarubbo, L.A. Production of biosurfactant from *Candida bombicola* URM 3718 for environmental applications. *Chem. Eng. Trans.* **2016**, *49*, 583–588.
41. Elsaygh, Y.A.; Gouda, M.K.; Elbahloul, Y.; Hakim, M.A.; El Halfawy, N.M. Production and structural characterization of eco-friendly bioemulsifier SC04 from *Saccharomyces cerevisiae* strain MYN04 with potential applications. *Microb. Cell Fact.* **2023**, *22*, 176. [CrossRef]
42. Joshi-Navare, K.; Prabhune, A. A Biosurfactant-Sophorolipid Acts in Synergy with Antibiotics to Enhance Their Efficiency. *BioMed Res. Int.* **2013**, *2013*, 512495. [CrossRef] [PubMed]
43. Gudiña, E.J.; Rocha, V.; Teixeira, J.A.; Rodrigues, L.R. Antimicrobial and Antiadhesive Properties of a Biosurfactant Isolated from *LactoBacillus paracasei* Ssp. Paracasei A20. *Lett. Appl. Microbiol.* **2010**, *50*, 419–424. [CrossRef] [PubMed]
44. Rather, M.A.; Gupta, K.; Mandal, M. Microbial Biofilm: Formation, Architecture, Antibiotic Resistance, and Control Strategies. *Braz. J. Microbiol.* **2021**, *52*, 1701–1718. [CrossRef] [PubMed]
45. Iseppi, R.; Mariani, M.; Benvenuti, S.; Truzzi, E.; Messi, P. Effects of *Melaleuca alternifolia* Chell (Tea Tree) and *Eucalyptus globulus* Labill. Essential Oils on Antibiotic-Resistant Bacterial Biofilms. *Molecules* **2023**, *28*, 1671. [CrossRef]
46. Budzyńska, A.; Wieckowska-Szakiel, M.; Sadowska, B.; Kalemba, D.; Rózalska, B. Antibiofilm activity of selected plant essential oils and their major components. *Pol. J. Microbiol.* **2011**, *60*, 35–41. [CrossRef] [PubMed]
47. Mukherji, R.; Prabhune, A. Novel Glycolipids Synthesized Using Plant Essential Oils and Their Application in Quorum Sensing Inhibition and as Antibiofilm Agents. *Sci. World J.* **2014**, *2014*, 890709. [CrossRef]
48. Mendes, R.M.; Francisco, A.P.; Carvalho, F.A.; Dardouri, M.; Costa, B.; Bettencourt, A.F.; Ribeiro, I.A. Fighting *S. aureus* catheter-related infections with sophorolipids: Electing an antiadhesive strategy or a release one. *Colloids Surf. B Biointerfaces* **2021**, *208*, 112057. [CrossRef] [PubMed]
49. Pontes, C.; Alves, M.; Santos, C.; Ribeiro, M.H.; Gonçalves, L.; Bettencourt, A.F.; Ribeiro, I.A.C. Can Sophorolipids Prevent Biofilm Formation on Silicone Catheter Tubes? *Int. J. Pharm.* **2016**, *513*, 697–708. [CrossRef]
50. Abdollahi, S.; Tofighi, Z.; Babaee, T.; Shamsi, M.; Rahimzadeh, G.; Rezvanifar, H.; Saeidi, E.; Mohajeri Amiri, M.; Saffari Ashtiani, Y.; Samadi, N. Evaluation of Anti-Oxidant and Anti-Biofilm Activities of Biogenic Surfactants Derived from *Bacillus* Amyloliquefaciens and *Pseudomonas aeruginosa*. *Iran. J. Pharm. Res.* **2020**, *19*, 115–126. [CrossRef] [PubMed]
51. Silva, S.S.E.; Carvalho, J.W.P.; Aires, C.P.; Nitschke, M. Disruption of *Staphylococcus aureus* Biofilms Using Rhamnolipid Biosurfactants. *J. Dairy Sci.* **2017**, *100*, 7864–7873. [CrossRef] [PubMed]
52. Cameron, D.R.; Cooper, D.G.; Neufeld, R.J. The Mannoprotein of *Saccharomyces Cerevisiae* Is an Effective Bioemulsifier. *Appl. Environ. Microbiol.* **1988**, *54*, 1420–1425. [CrossRef] [PubMed]
53. Aslani, N.; Janbabaei, G.; Abastabar, M.; Meis, J.F.; Babaeian, M.; Khodavaisy, S.; Boekhout, T.; Badali, H. Identification of Uncommon Oral Yeasts from Cancer Patients by MALDI-TOF Mass Spectrometry. *BMC Infect. Dis.* **2018**, *18*, 24. [CrossRef]
54. Dubois, A.B.; Botelho, S.Y.; Bedell, G.N.; Marshall, R.; Comroe, J.H., Jr. A Rapid Plethysmographic Method for Measuring Thoracic Gas Volume: A Comparison with a Nitrogen Washout Method for Measuring Functional Residual Capacity in Normal Subjects. *J. Clin. Investig.* **1956**, *35*, 322–326. [CrossRef] [PubMed]
55. Bradford, M.M. A rapid and sensitive method for the quantitation of microgram quantities of protein utilizing the principle of protein-dye binding. *Anal. Biochem.* **1976**, *72*, 248–254. [CrossRef] [PubMed]
56. Sieuwerts, A.M.; Klijn, J.G.M.; Peters, H.A.; Foekens, J.A. The MTT Tetrazolium Salt Assay Scrutinized: How to Use This Assay Reliably to Measure Metabolie Activity of Cell Cultures in Vitro for the Assessment of Growth Characteristics, IC50-Values and Cell Survival. *Clin. Chem. Lab. Med.* **1995**, *33*, 813–824. [CrossRef]
57. Johnson, E.C.; Eseyin, O.A.; Udobre, A.E.; Ike, P. Antibacterial Effect of Methanolic Extract of the Root of *Aspilia africana*. *Niger. J. Pharm. Appl. Sci. Res.* **2011**, *1*, 44–50.
58. Merghni, A.; Dallel, I.; Noumi, E.; Kadmi, Y.; Hentati, H.; Tobji, S.; Ben Amor, A.; Mastouri, M. Antioxidant and Antiproliferative Potential of Biosurfactants Isolated from Lacto *Bacillus casei* and Their Anti-Biofilm Effectin Oral *Staphylococcus aureus* Strains. *Microb. Pathog.* **2017**, *104*, 84–89. [CrossRef]

59. Maki, D.G.; Jarrett, F.; Sarafin, H.W. A Semiquantitative Culture Method for Identification of Catheter-Related Infection in the Burn Patient. *J. Surg. Res.* **1977**, *22*, 513–520. [CrossRef]
60. Morais, I.M.C.; Cordeiro, A.L.; Teixeira, G.S.; Domingues, V.S.; Nardi, R.M.D.; Monteiro, A.S. Santos Biological and physicochemical properties of biosurfactants produced by *LactoBacillus jensenii* P 6A and *LactoBacillus gasseri* P 65. *Microb. Cell Factories* **2017**, *16*, 155. [CrossRef]

MDPI AG
Grosspeteranlage 5
4052 Basel
Switzerland
Tel.: +41 61 683 77 34

Antibiotics Editorial Office
E-mail: antibiotics@mdpi.com
www.mdpi.com/journal/antibiotics

www.ingramcontent.com/pod-product-compliance
Lightning Source LLC
Chambersburg PA
CBHW041409210926
51850CB00013B/3226
9783725836338